FORM AND STRATEGY IN SCIENCE

JOSEPH HENRY WOODGER

FORM AND STRATEGY
IN SCIENCE

Studies dedicated to Joseph Henry Woodger on the

Occasion of his Seventieth Birthday

Edited by

JOHN R. GREGG AND F. T. C. HARRIS

Springer-Science+Business Media, B.V.

ISBN 978-94-010-3605-4   ISBN 978-94-010-3603-0 (eBook)
DOI 10.1007/978-94-010-3603-0

Softcover reprint of the hardcover 1st edition 1964

CONTENTS

CONTENTS

CONTENTS

W. F. FLOYD AND F. T. C. HARRIS

JOSEPH HENRY WOODGER, CURRICULUM VITAE

Joseph Henry Woodger is a dedicated scholar, a teacher, a research worker in biology, the philosophy of biology, its theory and metatheory and a very delightful companion. His research covers four inter-dependent categories. His scientific work, as a descriptive embryologist at the laboratory bench, led him to an examination of the philosophical pre-suppositions not only of embryology but also of the greater part of the range of biological theory. Having achieved this he embarked upon a search for, found and successfully applied rigorous methods of deriving non-numerical statements such as those relating cells to their parts. He has developed, and continues to develop and apply these methods in the construction of minutely analysed and rigorously deduced biological theory. In order to search out the axioms implicit in biological theory, and by extension, in scientific theory, he developed the technique of methodological analysis that led him to his characterisation of the structure of scientific theory.

Woodger was born on the second of May, 1894. He was educated at Felsted School in Essex, at which he soon showed an interest in biology. His interest in living things originated with his early life in Norfolk. His family had been associated with the fishing industry in Great Yarmouth to an extent sufficient to convince his friends and associates that it was a Woodger who had invented the kipper. His small bearded figure, when seen vibrating down Regent Street to the London Library topped by a tan sou' -wester in the pouring rain, gives substance to the idea. Certainly his pleasure in the Norfolk broads has never diminished.

He went to University College, London, in the October of 1911 and enrolled under J. P. Hill in order to read Zoology. In each of his undergraduate years his ability was rewarded with silver medals or class prizes and culminated in 1914 with the award of an honours degree, the College prize in Zoology and the Derby Research Scholarship.

Of course the research on which he was starting was interrupted by the first of this century's great wars. His voluntary enlistment in the

1

April of 1915 is an example of the uncompromising moral earnestness that has characterised his work and life. He was commissioned in the Norfolk regiment, the regiment of his family's county. Fortunately for the early development of the axiomatic method in biology he did not serve long in the european war and was soon sent to Mesopotamia where his energy and determination earned him a mention in dispatches. The map reading and compass marching technique that he learned there has stood him in good stead subsequently for even now he steers a compass course homewards when the downs that border his home are enveloped in thick fog. But then, in the Mesopotamian desert, the lives of his infantry platoon and himself depended on his precision. In the November of 1917 he became protozoologist to the central laboratory in Amara. Despite the arduous conditions and extremely limited facilities, he made time from his heavy routine duties to carry out research. It was in these unlikely circumstances that his interest in the philosophy of science was first stimulated through his contact and friendship with Dr. Ian Suttie who was one of his comrades at the battle of Shumran Bend and the subsequent advance to Baghdad.

After his demobilisation in the February of 1919 he resumed his Scholarship and carried out research until he left University College in 1922. During this period he was appointed Assistant in Zoology and Comparative Anatomy, in 1919, and Senior Assistant in 1921. In the July of the following year the Department of Biology at the Middlesex Hospital Medical School was re-organised and Woodger was appointed to the newly created readership in the department. He was at once faced with a very heavy teaching time-table. He delivered and lectured alone not only the whole biology course but was also responsible for both the lectures and practical work in histology for those students attending the classes for the second medical examination and separately for the teaching relevant to the primary fellowship course. Despite this he was able to carry out work on the germ-line in birds and found time to write a much needed text-book for his biology students in which the illustrations, which are models of line illustration, came almost entirely from his own hand. He carried out the preliminary work for this book so thoroughly that some of the statements in it, for example those arising from his own observations on the anatomy of Periplaneta, are original and corrected inaccuracies in the publications to which he referred.

2

During these early years at the Middlesex Hospital Medical School he became fluent in German, somehow finding the time to teach himself the language from a linguaphone course. That he was able to carry out so many projects at the same time was to a large extent due to the encouragement and assistance of his wife, Doris, whom he had married in 1921. His only technical assistance at his department was that provided by his technician, C. W. Falconer, who came from University College with him, and remained with him throughout his entire career at the Middlesex Hospital, a career that saw the passage of 38 years.

In the spring of 1926 he was given a term's leave to study under Przibram in Vienna. Fortunately for the future development of his work his research plans received a severe set-back. When he arrived in Vienna his experimental material had already been collected for him. It had been decided that he should work on transplantation in annelids, but the species of worm gathered proved inoperable and the frozen ground could not be induced to yield a further harvest. Thus he was able to devote more time, than he could otherwise, to the intra-departmental discussions of Przibram and his assistants and so refreshed his interest in the questions that had been aroused by his discussions with Suttie. He at once realised that there were fundamental unanalysed assumptions in the theories then in circulation amongst biologists and that his training had not equipped him, nor was it likely to have equipped anyone else, to examine or identify these assumptions.

On his return to England he threw himself into a study of philosophy, or those aspects of philosophy that were the necessary pre-requisite for an analysis of biological theory. Within two years he had completed the necessary spadework and had gone on to analyse the assumptions implicit in the biological antitheses between vitalism and mechanism, structure and function, preformation and epigenesis, for example, and the theory of explanation in biology. This period of his work culminated in the publication of *Biological Principles* in 1929 for which he was awarded the degree of D.Sc. by his University. Even the most superficial reading of this book makes it clear that he had already appreciated that for his purposes it was necessary to develop more rigorous methods of examining biological statements and had embarked upon a study of logic. With remarkable intuition and insight for a man without formal mathematical training he saw that the mathematical logic of Whitehead and Russell's

Principia Mathematica could be adapted to his purpose and he forthwith set to work to master the three volumes of this formidable work and to apply their concepts in an attempt to make a rigorous analysis of the theory of a number of biological disciplines. Despite his very heavy teaching commitments and his pre-occupation with the very considerable labour involved in his theoretical research at that time he was able to co-operate with G. R. de Beer in an examination of the development of the rabbit cranium. He also found time to translate, and in co-operation with the author to reformulate, Von Bertalanffy's *Modern Theories of Development* which was published in 1933.

Amongst the philosophers with whom Woodger now came into contact was Professor K. R. Popper who introduced him to A. Tarski. In his analysis of the relation 'part of', a pre-requisite for the study of theories involving statements about structure, Woodger had independently developed a system that was similar to Leśniewski's *Mereology*. Tarski's excitement at the first development of an application of such a system was a considerable stimulus to Woodger. In 1935 he went to Poland in order to meet with the Polish School of Logicians and to discuss mutual ideas, especially with Łukasiewicz and Tarski with whom he had been in correspondence.

This phase of Woodger's work was completed with the publication of his *Axiomatic Method in Biology* which was directed to the development of rigorous theories and definitions dealing with aspects of cytology, embryology, the division and fusion relations which have subsequently been expanded into an axiom system by Lindenmayer, genetics on which Woodger, Gregg, Przełęcki and others continue to work, Taxonomy subsequently expanded by Gregg, Sklar, Van Valen and others and the concept of environment which Woodger and Gregg are expanding in their continuing work on the formalisation of theories of genetics. The *Axiomatic Method in Biology* was published in 1937, ten years after Woodger's visit to Przibram's department, and, his completely new approach to theoretical biology is now beginning to stimulate a new generation of theoretical biologists, particularly in America.

Woodger was awarded a Rockefeller Foundation Scholarship to enable him to visit the United States in 1938 where he went to work with Clarke Hull whose theory of behaviour was considerably influenced by Woodger's development of the axiomatic method, although Woodger and Clark Hull were finally unable to agree on the axioms to be chosen.

With the outbreak of the second world war Woodger's time was taken up with the difficulties of teaching under the frustrating conditions involved in the evacuation of his Department to Leeds. But since the war Woodger has returned on several occasions to take part in conferences on the application of mathematical logic, a measure of the considerable interest taken in his work in America, and after his retirement was invited to the tenure of annual visiting Professorships by three American Universities, appointments which he has most unfortunately found himself unable to accept. His university has acknowledged his work, by conferring on him the titles of Professor, in 1949, and Emeritus Professor when he retired in 1959.

In 1949 he was invited to give the Tarner Lectures by Trinity College Cambridge. In these he expanded more fully his view that a nominalistic attitude was the correct basis for the language of science.

His views have not only been promulgated through the media of scientific, learned journals and books but also through his encouragement of a private group, the 'Theoretical Biology Club' which met occasionally at week-ends in Oxford and Cambridge and at a cottage on the Norfolk Broads before the 1939–1945 war and subsequently in London. It included among its participants Joseph and Dorothy Needham, J. D. Bernal, C. H. Waddington, P. B. Medawar, W. F. Floyd, Dorothy Wrinch and L. L. Whyte.

Woodger has continually sought to introduce into the Medical Profession a greater awareness of the place of psychology in medicine. His small book, *Physics, Psychology and Medicine*, modestly called by him an essay, is devoted to an explanation of the relationship between sense data, minds and persons and was directed primarily to Medical Students and others whose concern is with whole people.

There are certain similarities between Woodger and another great innovator in the field of Biology, Darwin. Like Darwin Woodger has laboured long, insufficiently recognised amongs this contemporaries, in support of his ideas and held to them unswervingly and with courage. And like Darwin he lacks the ability to promulgate his ideas effectively amongst an audience unfamiliar with the tools of his trade and his basic ideas. To a large extent his courageous maintenance of his viewpoint

5

against the misunderstanding of his contemporaries has rested on his wife's support and encouragement.

Now that he has retired Woodger is continuing his work on the axiomatisation of theories of genetics without the distraction of administering a department and with the encouragement of the growing numbers of workers who are joining him in weaving together the threads that he first spun.

EDITORS' NOTE

All of the essays in this volume are new. With one or two exceptions, they have been written especially for the occasion. Primarily, they are intended to honour Joseph Henry Woodger – who, for the past four decades, has written with diligence and acumen about the logical forms and strategies of the natural sciences, especially biology. Beyond this, they are offered to the reader as a valuable collection of authoritative modern enquiries into the sorts of methodological issues that confront nearly all branches of scientific investigation.

The volume was initiated by inviting a few prospective contributors to write for it. Their enthusiastic endorsement at once gave impetus to the whole project. The list of willing contributors grew rapidly to the point where further recruiting had to be curtailed, thus excluding the services of many who would have wished their work to be included in a volume written to honour Joseph Henry Woodger. To all of these, we offer our sincere apologies.

Credit for the volume belongs entirely to those who have written it. They were left free to write about topics of their own choosing, our efforts having been confined to arranging their contributions in a natural order. The first nine essays are contributions to the general philosophy of science, two being concerned with metaphysics, three with various problems of simplification and four with semantics. The next three essays exemplify diverse techniques of logical analysis as well as contributing significantly to the topics of which they treat. A further six essays investigate a variety of biological and other scientific models. Finally, there are eight analyses of miscellaneous biological systems. Together, they provide clear insights into some of the problems of scientific philosophy. Furthermore, they give some indication of the range of minds to which the work of Joseph Henry Woodger has seemed interesting and significant. Our heartfelt thanks and gratitude are extended to their authors.

The frontispiece was supplied by Mrs D. Woodger and Christopher Woodger, the foreword by Sir Cyril Hinshelwood, and the bibliography of the publications of Professor Woodger by Michael Woodger. Our indebtedness to all of these persons is extreme.

JOHN R. GREGG
F. T. C. HARRIS

FOREWORD

Men of science are sometimes mistrustful of or at least impatient with philosophy. One of them, himself no stranger to hard thought, was one day heard to comment on his colleagues in another faculty and on their propensity to indulge in what he called "all this nonsense about thinking". Against this may perhaps be set a meeting of philosophers who decided to discuss the Second Law of Thermodynamics. When asked sardonically by a scientist whether they had disproved it, one of the philosophers replied: "No, we have concluded that it is not so much false as meaningless".

This curious appearance of cross purposes reflects something more than mere captiousness or misunderstanding. As to the "nonsense about thinking", it is perfectly true that an excessive formalisation of arguments does not usually assist clear thinking very much. Plenty of people would be nonplussed by a formal logical exercise of the type: all A is B, C is B: is C therefore A? But equate A to Frenchman, C to Germans and B to Europeans, and they would never run the slightest risk of going astray.

There is also a little more than meets the eye in the brush over the Second Law of Thermodynamics. The definitions of energy and of temperature may easily be given in such a way as to make the laws of thermodynamics seem to depend upon circular arguments. Indeed Eddington himself has an interesting passage in *The Nature of the Physical World* in which he argues that from one point of view the whole of Newtonian mechanics might be represented as one vast tautology. Yet this can hardly be all that there is to say about the matter; or the laws of thermodynamics and of mechanics could hardly yield so splendid a crop of practical results.

It is convenient to draw a distinction between two separate ways in which philosophy comes into relation with science, and this distinction is illustrated by what has just been said. On the one hand there is the external philosophical critique of what the man of science does. This is a subject in its own right, but the man of science is no more obliged to interest himself in it – though there is no reason why he should not – than the artist is obliged to study the forms of art criticism. Painters paint, and the critics usually cannot. Good scientists all think straight, and the best ones use their imagination to good effect. But none of them need to be, and most of them are not conscious of the logical analysis of what they do. Many people observably handle arguments unerringly as long as there is a concrete content in which they are interested, but reject the idea of any concern with the abstract general principles. (People to whom algebra is a permanently closed book seem to perform wonders with permutations when it comes to Football Pools.)

On the other hand, there are many ways in which philosophical considerations enter deeply into the heart of the scientific questions themselves. The problem, for example, of how far a scientific law is a tautology in the light of the basic definitions, and, if it is, how the law can yield fruitful results is one which cannot be evaded without loss of understanding.

The distinction, then, is one between philosophy *about* science and philosophy *in* science. The man of science is at liberty to abstain from the one, but he neglects the other at his cost.

Methodological matters are very properly the concern of the philosopher and logician, and in so far as the scientist interests himself in them he is joining their ranks. His own most important function is discovery, and the key operations leading to discovery usually take place in the subconscious, whence intuitions spring, where subtle analogies are somehow perceived – perhaps by the comparison of stored patterns – and where combinations leading to the solutions of problems are endlessly explored, as has been vividly described by Poincaré in an account of his own experience of mathematical discovery. None of this is much helped by a conscious critique of scientific method, though nobody would underestimate, at the stage of verification, such matters as sound statistical methods.

We soon come, however, to a region where scientist and philosopher meet on more even terms, namely in the analysis of the language of scientific theories.

And closely connected with this is the problem of what in fact constitutes a scientific explanation. Here we encounter an extraordinary

10

range of idioms and concepts. At the one extreme is highly abstract mathematical symbolism, used for example in parts of nuclear physics, which derives with absolute precision the relations between one thing and another without relating either in any simple way to the facts of common experience.

Are we or are we not satisfied with such systems (assuming that they work)? This is a question which every scientist has to answer for himself, and the answer he gives is the declaration of a philosophical attitude. At the other extreme are theories couched in almost poetical or metaphorical language. These, while exposed to grave criticism, can also be fruitful, or at least can be the vehicle of fruitful intuitions. Perhaps the writings of Freud provide one of the most striking instances of this. There can scarcely be any corpus of scientific writing so rich in content and yet, in some respects, so imprecise in its expression.

The man of science is certainly philosophising unconsciously when he decides whether or not to be satisfied with, for example, a formal mathematical system like that of the famous "wave mechanics". He may not analyse his attitude, but he sacrifices much that is profound and interesting if he does not. In another way he places himself at a disadvantage if he does not consciously realise how much of the structure of science rests not upon simple inductions and deductions but upon highly intricate arguments of coherence. A modern Socrates could greatly disconcert many a chemist or physicist in an interrogation about the relations of mass, force, motion and energy or about the classical atomic and molecular theory, yet his sport would be perverse and his scores illusory. The simple and conclusive answer is that these theories establish systems with a wonderful internal coherence by which vast numbers of diverse facts and phenomena are brought into relation. It is highly desirable that men of science should be aware of matters like this, if only because they are thereby rendered more capable of defending themselves against their critics.

There are, however, certain scientific questions into the very structure of which philosophical considerations are built. The philosophical aspects can be provisionally left on one side by a process of deliberate abstraction, and it is often convenient, or even necessary to do this. But they cannot be forgotten without the risk of grave error.

The most obvious meeting ground of science and philosophy is perhaps, in the fundamentals of physics, in the discussion of space, time, causality, simultaneity and other essential concepts. The principle of indeterminacy, Einstein's non-Euclidean geometry of space-time, the exclusion principle (a veritable corner stone in the structure of the world) can be and sometimes are represented as convenient formulations rather than as descriptions of any kind of ultimate reality. But this attitude evades a deep and important question. If, as in fact most physicists would maintain, these convenient descriptions are all that can ever be hoped for, what then after all must be their true status?

Einstein brought the observer as an indispensable participant into some of the basic laws of physics, but it is in biology that the deepest, most puzzling and at the present time most intractable problems of subject and object lie. Descartes may have stated the dualism of mind and matter in some respects too crudely, and has been, somewhat unjustly, caricatured by certain modern philosophers for doing so. Their own attempts to talk the problem out of existence by linguistic analysis are unconvincing. They seem to ignore the richness, coherence and power of the conventional scientific description of the outer world, the equal fulness and vividness of accounts of the conscious inner world of man, and the utter incapacity of the respective idioms to be transferred from the one purpose to the other.

A few behaviouristic biologists ignore the problem. They say that the admittedly highly suggestive machines which can now compute, "think" and learn are indistinguishable from "rational" beings by any operational test.

They say that ethical qualities can be "explained" by natural selection: and so on. It is quite true that the study of the machines may ultimately teach us a great deal about the working of the brain. It is also true (we may hope) that moral qualities have survival value and that those possessing them will be favoured by natural selection. But all this is irrelevant. Behaviouristic acts are accompanied by data of consciousness. These data are quite certain to any individual as far as he himself is concerned, and, by coherence arguments, at least as cogent as any in science, as far as other people are concerned as well. The relation of the data of the consciousness to the scientific description of the world is one of the greatest of problems in face of which science and philosophy at the moment stand nonplussed but in alliance.

PART I

PHILOSOPHY OF SCIENCE

MORTON BECKNER

METAPHYSICAL PRESUPPOSITIONS AND THE DESCRIPTION OF BIOLOGICAL SYSTEMS

Professor Woodger somewhere remarks that scientists who consider themselves above all metaphysical beliefs are merely up to their necks in them. This is not a popular opinion nowadays. The conception of the nature of metaphysical statements which prevails, at least among philosophers and scientists who are likely to be interested in the logic of science, entails the view that metaphysical statements, even though they have discernible logical relations among themselves, are completely disengaged from all systems of scientific statements. While science moves ahead, the metaphysical engine, which at one time was regarded as a source of motion, merely idles.

This picture of an endlessly circling metaphysics borne along on a platform of genuine empirical knowledge is, I think, totally erroneous – a segment of the larger contemporary view which denies that philosophy is at all concerned with substantive questions. The conception of metaphysics as redundant rests jointly on a simplified account of the nature of scientific theory, and on a misinterpretation of the way in which traditional metaphysics has accomplished its tasks. In this paper I do two things: first, I maintain that science does indeed have presuppositions whose functions in inquiry are exactly those that traditional metaphysical theories have exercised; accordingly, they may properly be termed "metaphysical presuppositions". Second, I examine the way that one such presupposition influences the conception of a biological system and how it is to be described.

I

The view that science has no metaphysical presuppositions is not supported solely by philosophical accounts of the nature of metaphysics. In addition, methodologists have analysed the conception of a "presupposition". I shall not distinguish all the various senses in which this term is used; but I shall assume that these analyses have established the following two important points. (1) There are no "presuppositions" of science in the sense of statements that are both premisses of substantive, empirical explanations, and are themselves unverifiable by the procedures – whatever they may be – of the empirical sciences. I agree, in other words, that every premiss of a scientific explanation is itself subject to the same kinds of verification as any other premiss or conclusion. For example, no empirical conclusions are entailed by metaphysical doctrines about space, time, or the uniformity of nature, or principles of limited variety. (2) Nor are there any "presuppositions" of science in the sense of general statements that must be true if (a) a particular theory, or type of theory, is true; or (b) if a particular methodological principal, e.g., one of Mill's Methods, a statistical rule of rejection or a procedure such as introspection, or phenomenological description, is justified. Of course, there are such general statements, viz., everything deducible from the theory or principle in question; but they are not "presuppositions" in the sense of antecedently accepted statements that influence the adoption of the theory or principle, and which indicate the presence in science itself of arbitrary, volitional, or otherwise unscientific components. Those statements which are so entailed are not the kind of presupposition that would offer aid and comfort to critics of the claim that scientific methods are the only reliable path to empirical knowledge.

But science does have metaphysical presuppositions in a sense which I shall now describe. Not science in general, but every particular scientist, holds some beliefs which are either not consciously formulated or else are given a very general formulation, and which do in fact exert influence on the form in which he attempts to cast his results and, more particularly, on the way in which he interprets his own aims and procedures. Our task is now to show that such beliefs exist, that they are not presuppositions in either of the above two senses, and that they are properly termed "metaphysical".

First, some examples: "Causal influences do not jump gaps in time"; "Goals are not agents of their own realization"; "All biological processes are physico-chemical in nature"; "Mere differences in spatial location cannot account for any other differences between two objects"; and "The behavior of any system is determined by factors within the spatio-temporal limits of the system". I shall subsequently examine this last statement in some detail; the others are of course formulations of doctrines that have stimulated considerable discussion among philosophers. I wish to call attention to four features these examples have in common and which are relevant to my argument.

1. The reader may well doubt whether he understands what any of them really mean. If so, these doubts are due not to any ignorance of science or the terminology of philosophy – there are no scientific treatises or philosophical dictionaries that would clear up their meaning in the way in which a treatise or dictionary can clear up the meaning of a phrase like the "main-line sequence". In short, they are stated in terms borrowed from everyday English or at best popular science; they do not employ a technical vocabulary. When someone says of such a statement, "I don't understand it", he does not mean, however, that he fails to understand English; he means "The usages of English do not dictate how it is to be reformulated so that it clearly applies to this or that case". Notice that it does not follow that the statement is unverifiable, even if no one understands the statement in this sense. One is employing a moot theory of verification if he holds either that linguistic rules must dictate – rather than do something weaker – the derivation of observable consequences of all verifiable statements, or that one does not understand a statement for which such dictatorial rules do not exist. One can of course propose that the terms "verification" and "understanding" be used in this way, but there is no compelling reason to do so, especially in view of the possibility that all sorts of important relations may exist between statements which on this usage "I don't understand" and statements which "I understand clearly".

2. These statements are indeterminate with respect to the kinds of evidence that could be cited for or against them. Their indeterminacy arises from two sources: first, they are stated with the help of terms, e.g., "goal", "causal", "material", "factor", which are vague in the sense that no rules exist – anywhere, in dictionaries, in "ordinary use", or Platonic heaven – which permit a non-arbitrary decision on borderline cases; and second, this vagueness is compounded by the fact that the statements as a whole are extraordinary contexts for the vague terms. Suppose that the term "distance" is precisely defined for expressions like "The distance between points P and Q"; this precise definition does not automatically provide a meaning for the expression "the distance between broomsticks A and B". In this case we need to add some specification of which points on the broomsticks are relevant, and then we can use the original

definition. This added specification is commonly provided by the circumstances of the measurement, e.g., whether we are interested in their mutual gravitational attraction, or the degree to which they illuminate each other, etc. Because we learn how to add these specifications as we learn the meaning of a term like "distance", a person who understands "the distance between points P and Q" will unhesitatingly say that he understands "the distance between objects A and B", although he actually has no method of determining the latter distance.

The difference between this case and a metaphysical statement is simply that in the latter case these further circumstances are not provided. This is the way the game of metaphysics is played; the metaphysical principle is formulated, discussed, and criticized, always with the circumstances of its application left indeterminate.

3. If these statements are to have any influence on the sciences, they must have some connections with statements of the degree of determinacy which we associate with straightforwardly empirical assertions. Here we must be careful to avoid the pit that, I believe, has so well guarded against the proper understanding of metaphysics in modern empiricist circles. It is true that in one sense, the meaning of a metaphysical statement must be specified in order to have bearing on the sciences. But if metaphysical statement M is reformulated in a way not required by its meaning, the reformulation S cannot, of course, be synonymous with M. Nor can we say, except by way of an opaque metaphor, that S is an analysis of the meaning of M, as if its meaning had parts that could be extracted for a closer look. Finally, we cannot say that S clarifies the meaning of M, in the way in which we can clarify our view through a microscope by changing its focus. In general, it seems to me quite obvious that it is logically impossible to clarify the meaning of any statement; we can clarify a *person's* meaning by coaxing from him accessory statements with different meaning. This means that what I shall call a specification S of metaphysical statement M is not a synonymous reformulation, an analysis, or a clarification. M and S are different statements with different meanings; to use an honorific term, the passage from M to a particular S in the course of an inquiry is a genuinely creative activity on the part of the metaphysician or scientist who employs M; that is, the specification of M in every case introduces meanings which are both logically and psychologically novel. There is no single relation, logical or psychological, that holds

between *M* and its specifications. We can get some grasp of the *kinds* of relations between *M* and *S* by considering analogous cases. *M* can be related to *S* as "All human actions are selfish" to "A man desires only his own pleasure"; as "Nature has no gaps" to "There are no smallest particles of matter"; as Kant's Categorical Imperative to an injunction against habitual self-indulgence; as Plato's rule that celestial motions are uniform and circular to Ptolemy's conception of eccentrics and equant-points; or as the Greek doctrine that irrationals are not numbers to Eudoxus' and Euclid's theory of proportions.

4. Metaphysical statements gain their usefulness because they are so formulated that while no particular specifications are *entailed*, they are susceptible of a great variety of temporary and provisional specifications. But no set of specifications is regarded as canonical: this is the essential point. In this respect they are rather like Biblical texts: believed by the faithful, subjected to different interpretations in a thousand sermons every Sunday by clergymen who are perfectly willing to retract their interpretations should anyone insist. Even when a metaphysician offers what we might take to be a canonical specification, it usually happens that the alleged specification is itself more metaphysics, or else even his followers will hold on to the principle and give up the specification.

In point of historical fact, metaphysical statements have always played a role in scientific as well as other kinds of endeavor. They are part of the intellectual machinery that endows a language with general applicability. If anyone admits that a natural language contains a part whose function is the regulation of the internal economy of larger parts of the language, he should be prepared to admit my thesis.

A large measure of those elusive entities, semantical rules for a natural language, are imbedded in metaphysical doctrines, as a metalinguistic rule is imbedded in a theorem. The mistake of those critics who attack metaphysics from the standpoint of the verification principle consists in demanding that semantical rules connect linguistic elements *directly* with experience, and overlooking the point that some linguistic elements, which do not differ in grammatical form from directly empirical statements, function as schemata directing the formulation of empirical statements into connected systems. Metaphysical statements are of course not immediately verifiable as they stand; and verifiable specifications of them are no longer metaphysical statements. If one emphasizes their unverifiability, the view that they are meaningless is likely to follow; other interpretations, such as the view that if true at all they could not be otherwise, is a result of emphasizing one *kind* of possible specification.

Although I believe this characterization of metaphysical statements is true as far as it goes, it is evidently incomplete. I have omitted for lack of space examination of the details of the various relations between M and S; and I have not considered with sufficient explicitness the fact that M's commonly occur in clusters – metaphysical systems, more or less articulated and comprehensive – which provide additional insight as to appropriate specifications of the various statements that make up the cluster. In other words, the system as a whole embodies suggestions for the specifications of its parts. I think, however, that I have shown that if there are indeed statements in the sciences with the logical status described above, we may as well frankly call them presuppositions and also admit that they are metaphysical, since statements which are traditionally termed "metaphysical" function in exactly the same way.

II

I wish now to illustrate the function of a metaphysical presupposition by examining one of particular importance to biological theory and showing how it is specified in practice. I hope my examination exhibits the following points: (1) The interest for a philosopher of a metaphysical presupposition lies in the influence it exercises in its indeterminate form, *prior* to specification; (2) its appropriate specifications form a class whose members are related by the fact that they all deal with a group of concepts that are themselves related internally by empirical principles; (3) since it suggests ways in which empirical statements are to be formulated, its *most direct* influence on our thinking comes out in belief about proper norms of scientific inquiry; and (4) its results can be beneficial in some contexts, deleterious in others. The presupposition which I shall examine is "The behavior of any system is determined by factors within the spatio-temporal limits of the system". From this point the letter M will be used to denote this statement.

M is one of a cluster of metaphysical propositions that, taken together, specify in the most general terms the subject matter of the natural and social sciences, in the sense in which fossils – as against the principles of evolution – are the subject matter of palaeontology. M operates against the background of the less determinate agreement that the term "system" denotes those things which scientific laws are about.

In the first place, M suggests what factors are relevant in developing criteria of identity for systems, i.e., rules that determine what is to count as *one* system, and as the *same* system over a period of time. Since, according to M, internality, in a spatio-temporal sense, of causal connection is part of the concept of a system, both spatio-temporal and causal considerations enter into the actual definitions of particular material systems. Exactly how they enter M does not specify. Methodologically, this means that provisional spatio-temporal limits for a system are to be set up, and adjusted on the basis of what is then learned about the causal connections within those limits. M does suggest that these adjustments be made in such a way that relevant causal factors turn out to be characterizations of material bodies within the adjusted spatio-temporal limits. In the physical sciences, where the concept of a system has gained its most determinate specification, the system is regarded as *defined* by a set of state-variables of a deterministic theory; thus every system is a system relative to the theory that defines it, e.g., a mechanical, quantum-mechanical, or thermodynamic system.

It might be objected that this conception of a system is too obvious to need spelling out, and that we hardly need invoke metaphysics to understand how it developed; that the conception is virtually dictated by the nature of scientific inquiry itself. But it should be noticed that the assumption that every science possesses a determinate subject matter does not automatically provide one with criteria of identity for the subject matter. Freud, for example, cast his theories in a form clearly intended to assimilate psychological systems to physical ones, but I think it is fair to say there are no clear criteria of identity for, e.g., the ego, the unconscious, etc. And we find in gestalt psychology, particularly in Lewin's conception of a life-space, the attempt to hang onto the metaphysical theory of which M is a part by generalizing the notion of "space". It is not obvious, to understate the case, the M is useful in this context. For the same reason some existential psychoanalysts and organismic biologists have attempted highly problematical reinterpretations of both "space" and "time".

The application of the criteria of identity derived from M in the region of biological theory leads to the following difficulty. The factors which determine what counts as a biological system – the things of professional interest to the biologists – and which yield the limits of the system in accordance with the spatio-temporal criterion, are drawn largely from outside biological theory proper. The whole organism is sharply distinguished in perception from its environment, and its activity as a whole engages our practical attention – we fear the tiger, trap mice, raise cotton, etc. When we dissect, we encounter naturally discriminable units, the bones, muscles, etc. With microscopic examination cellular and tissue structure immediately stands out. These structures – organisms, organs, cells – and others such as chromosomes and species, strike us forcibly, in Plato's phrase, as nature already carved at its joints, and the biologist naturally formulates his fundamental problems in terms of these "natural" systems: how does the organism as a whole develop, maintain itself, and reproduce?; what are the functions of the organs?; on what factors does organ pathology depend? etc. As a consequence – and this is the point I regard as highly important – the biologist is under extreme pressure *not* to adjust the spatio-temporal limits of his theoretical systems in accordance with the other criterion embodied in M, the criterion of causal determination. He is unwilling, in other words, to redefine his systems in the light of whatever empirical principles that unbiased investigation might reveal. The physical scientist, on the other hand, regards his own theories as indicating where to locate nature's joints.

The attempt to provide philosophical interpretations of biological theory that will maintain organisms, cells, etc., as the natural biological systems is a factor behind two significant developments in modern methodology, viz., the denial of final causes, and the doctrine that the behavior of a system as a whole is in all cases a resultant of the behavior of its spatio-temporal parts. The latter is clearly a specified form of M: the step from the view that all causation is internal to the view that the behavior of the whole is a resultant of the behavior of its parts is quite natural, especially if we have before us the model of Newtonian mechanics, in which precise meaning is given to the notion of a "resultant". If M is held jointly with a physical theory which permits us to regard the spatio-temporal parts of a system as themselves systems relative to the physical theory, some version of reductionism is likely to emerge.

The doctrine that there are (or are not) final causes is itself a highly indeterminate metaphysical presupposition. Accordingly, when I say that adoption of *M* leads to the denial of final causes, I mean that those contextual factors which make *M* subject to particular kinds of specification *also* lead us to regard the ordinary specifications of the doctrine of final causes as false. These latter specifications may or may not be influenced by the same contextual factors that provide the relevant specifications of *M*. For example, if a final cause is regarded as immaterial, but immanent in a process, somehow guiding it, there is no immediate conflict with the usual specifications of *M*; but there are other standard interpretations of final causes which do conflict with *M*. This is obviously the case if final causes are regarded both as agents and as future states, for they would then be causal determinants outside the temporal limits of the system. Or if final causes are given the interpretation of Aristotle, i.e., a final cause brings about motion – change – in the way that the beloved moves a lover, or food a hungry man, we encounter a more subtle conflict with *M*. An adherent of *M* need not deny that the presence of food influences a hungry man; but he is compelled to give an account of the influence by means of a theory in which the operating causal factor is not the food itself, but some counterpart of the food within the man. This point of view yields such formulae as "Not the goal, but our idea (image, anticipation) of it is the cause of our action"; "we react directly to changes in our own sense-organs, and only indirectly to the objects that brought about these changes". In psychology, the widespread uncertainty about the logical status of such concepts as "stimulus", "cue", "drive", "motive", and "incentive" is a reflection of metaphysical perplexity as to how the elimination of final causes should be implemented.

III

So far I have not seriously impugned the credentials of the metaphysical presupposition *M*. Even if we grant its role in prescribing criteria of identity for systems, and grant further that these criteria are difficult to apply in some regions of inquiry, we might still argue that the difficulties stem from the complexity of biological subject matter, and not from any weakness in our fundamental conception of biological systems and how they are to be described. Indeed, those theories, such as topological and phenomenological psychology, that have tried hardest to specify *M* after some other model than the physical sciences have not demonstrated their scientific respectability, at least in the opinion of many methodologists. But I hope I have succeeded in suggesting that M is certainly not an *a priori* truth, and that the form of a proposed theory ought not to be evaluated by reference to it alone, or to any other presuppositions.

I want now to examine one more specification of M, one which is tacitly accepted by virtually all scientists, and which is demonstrably in contradiction with their actual practice – a situation which is bound to result in the verdict that part of their practice is unscientific or otherwise disreputable. The specification of M is this: "All concepts which can enter into a theory of the behavior of a system are definable in terms of factors within the spatio-temporal limits of the system to which they apply". Call this the "rule of definitional isolation". Its connection with M is clear: since behavior of a system is caused by factors within it, there could be no reason to give an account of its behavior in terms of factors outside it. According to the rule of definitional isolation, all admissible theoretical concepts are, in traditional terminology, intrinsic rather than extrinsic denominations. I maintain that the theoretical use of extrinsic denominations in the description of biological systems requires abandonment of the rule of definitional isolation, and clearly exhibits the problematical status of M itself.

The distinction between extrinsic and intrinsic denominations is roughly this: suppose we consider a particular object, a fruit fly specimen, for example. There are two kinds of things we can say about the specimen. We may say that it is alive, possesses two wings, red eyes, three body segments, is one millimeter in length, weighs one grain; in short, as we say, we may describe the fly itself. These are intrinsic properties of the fly, or involve the application of intrinsic concepts. But we can also say that the fly was collected in Mexico; that its eggs were treated with X-rays; that its parents were dead at the time of capture; and that it is now kept in a milk bottle. These may all be true statements about the fly, and in this sense predicate properties of it; but they are only extrinsic properties. It would also be possible to draw up a list of borderline cases, but we may avoid these. I shall call a concept *extrinsic* whenever it is necessarily, in virtue of the definition of the concept, used to predicate an extrinsic property of a thing. Again speaking roughly, intrinsic concepts can be applied on the basis of an inspection or investigation of the single object or system that interests us. In order to apply extrinsic concepts we must investigate more than the system, or at least utilize our knowledge of the system, or at least utilize our knowledge of the system as evidence that certain states of affairs external to the system actually obtain. Examples of extrinsic concepts in this sense are "ancestor", "container", "tool", "medicine", "chair", "relic", "lawyer". To avoid a possible confusion, it should be noticed that intrinsic concepts are often used to predicate extrinsic properties of a system, as when we say of a particular house that it was destroyed by an earthquake. The concept of an earthquake is intrinsic, whereas being destroyed by one is an extrinsic property of the house. An extrinsic concept predicates an extrinsic property of a thing in virtue of its definition, not merely in virtue of the way it occurs in a proposition.

Two of the three types of extrinsic concepts that will concern us have been discussed by philosophers often enough to have received names. They have been termed historical and teleological or functional concepts. For the previously unnamed third type I adopt the term "polytypic concept", and I shall call the class of things to which the concept applies a "polytypic class".

I shall now briefly consider the logical features of these three types of concepts and point out the respects in which they are extrinsic. A concept C is said to be historical whenever there exists a definition of C, or at least a necessary condition for any definition of C, such that every application of C to a system is necessarily false unless a specifiable set of events have occurred prior to the time that C is applied to the system. In other words, an historical concept is defined in terms of historical antecedents of the system to which it applies. Examples from everyday language are "alumnus", "veteran", "convert", "offspring", etc. Such concepts are evidently well adapted for historical narration and for describing contemporary states of affairs in the light of past events.

The following definition of "functional concept" is intended in part as an explication of what some writers have called "teleological concepts", that is, concepts which in some sense involve an implicit reference both to a user and to a use or purpose or contribution made. I shall call a concept C "functional" whenever C is defined, or else there is a necessary condition which any definition of C must satisfy, such that every application of C to a system or process S is necessarily false unless there exists a process F in a distinct system S', and at least one state and environment of S' in which the occurrence or presence of S is necessary for the occurrence of F. This will be clearer if we consider an example. One might argue that the concept of the kidney is functional on the grounds that an object would not be termed a kidney unless there are organisms in which the object is capable of excreting urea. In this case, the kidney is the system S, the organism is S', and excretion of urea is the process or function F. "Kidney" then meets our definition because there is at least one environment and state of the organism in which the kidney is necessary for the excretion of urea. Other examples are "teacher", "pupil", "expert", "husband", etc.

A polytypic concept is one which is applied to a system not by virtue of its possession of a set of necessary and sufficient conditions of application, but by virtue of the resemblance of the system to a large subclass of systems that all bear, in Wittgenstein's phrase, a family resemblance to each other. Stated formally: I call a class C "polytypic with respect to a set G of properties" whenever every property in G is possessed by large numbers of the members of C; each member of C possesses a large number of the properties in G; and membership in C is determined by reference to possession of the properties in G. Notice that no property in G need be possessed by every member of C; it follows that no property in G, nor any subset of G, is either necessary or sufficient for membership in C. But each of the members of C will have many properties in common with many of the other members. Wittgenstein's working example is the class of "games"; and he maintains that many problematical philosophical concepts, e.g., "knowledge", "thinking", "understanding", "signification", also possess this logical feature.

The sense in which polytypic, historical, and functional concepts are extrinsic is clear; the application of any one of these types of concept to a system logically requires that certain propositions concerning states of affairs external to the system be true, namely, for polytypic concepts, that certain properties possessed by the system be widely distributed in a large number of other systems; for historical concepts, that certain events have happened in the past; and for functional concepts, that there are circumstances under which the system would contribute to the performance of a function, if these circumstances are ever realized. It is also clear that there is a problem involved in understanding how such extrinsic concepts could function in a systematic theory such as thermodynamics or relativity

26

mechanics. The concepts which are employed in theories such as these predicate no extrinsic properties of the system under consideration. In accordance with presupposition M, the physicist adjusts the boundaries of his systems in order to exclude the necessity for introducing extrinsic concepts. Notice that the usual analysis of the nature of a systematic theory is carried out in terms of a set of intrinsic variables, or parameters, and equations which describe how they vary with time. From this point of view extrinsic concepts are scientifically disreputable; polytypic concepts lack the precision that seems attainable for intrinsic concepts; historical concepts have the genetic fallacy already built in; and functional concepts seem to presuppose a discredited teleology. Accordingly, the actual uses of these concepts have been largely ignored in favor of arguments to the effect that lack of precision in scientific concepts is no more than sloppiness or a reflection of ignorance of relevant facts; and historical and functional formulations are in the end eliminable from all theoretical science.

I cannot examine in detail here the various theoretical uses of these types of concept; but to show that they do have such uses I shall cite some examples from the technical vocabulary of biology. "Mutant", "hybrid", "homology", "hypertely", "polyploidy", "displacement activity", "rudimentation", etc. are historical concepts; "lung", "gonad", etc., "male", "gamete", 'parasite", "predator", "cryptic coloration", "escape reaction" are functional concepts. The clearest cases of polytypic classes are afforded by the taxonomic groups: all groups such as "Ungulata" are segregated on the basis of family resemblances, and are hence polytypic with respect to the set of taxonomic diagnostic characters. Concepts such as "frog's egg" are historical, functional, and polytypic with respect to the intrinsic properties of eggs; many morphological concepts are functional and also polytypic with respect to the so-called morphological properties of organisms, e.g., shape, color, position, etc.

The foregoing arguments show that the biologist in fact introduces theoretical terms that possess definitional references outside the systems to which they apply; they do not in all cases readjust the boundaries of the systems to conform with the rule of definitional isolation. It is easy to see the reason for this: these references go in so many different directions that the resulting system would lack internal causal coherency.

IV

Let me now restate my argument and indicate what I take to be its upshot. I have maintained that scientists are influenced by metaphysical pre-suppositions, of which the doctrine that "the behavior of any system is determined by factors within the spatio-temporal limits of the system" represents a typical example. The influence of these presuppositions is exercised in two ways: on the philosopher's and scientist's account of the aim and strategy of science, and on the form in which the scientist tries to place his theories. They are indeterminate with respect to methods of verification; in particular they are neither analytic nor synthetic since they do not posses the specificity of meaning requisite for applying even the rough tests for analyticity in a natural language. They function as schemata that direct the formulation of connected systems of synthetic statements in which certain key concepts such as "cause", "end", "system", etc. are employed. The presuppositions themselves are stated with the help of these key concepts, and thus may be thought of as provid-ing some of their semantical rules – rules which determine the uses of these terms in contexts where their use would be otherwise undetermined. They do not *state* these rules – metaphysical assertions are not about statements, but are about concrete subject matters – but provide them through their specifications, as a theorem can provide rules of inference, or as a work of literature can provide knowledge of human nature without making statements about it. The presuppositions are subject to an indefinite number of specifications of meaning; these specifications are suggested and directed, but not entailed by related presuppositions and by the empirical subject matter under consideration.

I think that metaphysical doctrines are indispensable in the formula-tion of general theoretical systems; but that no presupposition, however plausible, ought to blind one to viable alternatives. I have tried to make this point by discussing an eminently plausible presupposition, showing how it operates in inquiry, e.g., by yielding criteria of identity, supporting methodological reductionism and the attack upon final causes, and laying down rules governing the logical properties of theoretical concepts. I indicated that however useful some specifications of the presupposition have proved in some contexts, their usefulness in all contexts is by no means clear, and that one such specification is contradicted by a useful procedure of concept formation.

I propose, therefore, not the elimination of metaphysics, but an awareness of the ubiquity of metaphysics, and of the fact that metaphysics can be good or bad.

The metaphysical presuppositions that constitute the working background of contemporary biological and psychological theory need serious examination and reconstruction. The symptoms of this need are the existence of methodological programs, supported by various groups, but which are not demonstrably useful; and general unclarity concerning the relations between various types of theory. For example, the relations between the languages employed in the description of motivation and perception on the one hand and the language of physiology on the other, have not been clearly worked out; nor have the connections between orthodox behavioral theories and e.g., stochastic models of learning. These are not simply questions of the logical analysis of scientific language; they are questions of the most appropriate metaphysics for exhibiting the connection between sets of concepts with distinct logical features.

The two difficulties in the prevailing metaphysical view of the nature of systems which I have discussed – the possible inappropriateness of its associated criteria of identity for biological and psychological systems, and the conflict of the rule of definitional isolation with actual practice in the description of biological systems – requires rather severe measures: either a reinterpretation of the notion of causal determination, or else the development of a metaphysical theory which shows clearly the admissibility of non-causal determination. Either alternative could provide alternative ways of understanding the concept of a system.

Pomona College, Claremont, California, U.S.A.

PETER ALEXANDER

SPECULATIONS AND THEORIES

Philosophers have been inclined, during the present century, to reject, or at least to regard with suspicion, the speculations of the metaphysician. This attitude sprang largely from a growing interest in scientific theories and the belief that these constituted the only appropriate and acceptable attempts to describe or explain the natural world. The desire for greater rigour within science, which led to the nineteenth century efforts to eliminate metaphysical elements from science [1], stimulated empiricist philosophers to distinguish the verifiable, factually meaningful statements of science from the unverifiable and allegedly factually meaningless utterances of metaphysicians and to reject these last as uninformative. In view of this close connection between science and empiricist philosophy, it is surprizing to find, in more recent years, a reawakening of interest in metaphysical speculations precisely among those philosophers who are interested in the philosophy and history of science. By such philosophers it is now frequently urged that we should take metaphysical speculations more seriously than has been customary of late since they have contributed, and may still contribute, in an important way, to the development of scientific theories.

My main purpose here is neither to add further arguments for the view that metaphysical statements are meaningless nor, on the other hand, to show that they are, after all, meaningful. Indeed, I shall give no more than a passing glance to such questions because I regard them as irrelevant to my aim, which is to make some tentative suggestions about the way in which metaphysical speculations influence scientific theorizing and about the features of both which make this possible. For this purpose, I shall accept Karl Popper's distinction between scientific and metaphysical statements on the grounds that scientific statements are, at least in principle, falsifiable and that metaphysical statements are unfalsifiable, even in principle.[2] I shall ignore the difficulties attached to this criterion because I do not know of a better one and because if we are to talk of the influence of metaphysics on science, as many people do, we need some means of distinguishing between them.

30

The alleged facts of the situation are fairly well documented; both historians and philosophers of science have recently devoted much time and space to showing how metaphysical speculations, and even myths, may develop into, or contribute towards the development of, scientific theories.[3] Karl Popper and, in greater detail, A. G. M. van Melsen, have traced the growth of the modern atomic theory from Parmenides *via* Democritus. Ernst Cassirer has given examples of other such influences in a number of his works. J. W. N. Watkins has recently gone further and argued that certain statements, such as "Every event has a cause", are, at the same time, metaphysical, in some sense factual and, moreover, useful to scientists. Stephan Körner has given examples of specific influences in modern physics and put forward a view to account for them. Paul Feyerabend has argued that our experience has been so constricted by the theories of classical physics that we could never have begun to think non-classically if we had had to rely on experience alone and could not have had recourse to abstract metaphysical theories or, in Einstein's words, "free creations of the mind". I shall also mention, later, the current emergence of a physical theory partly from the metaphysics of Leibniz and Hegel at the hands of David Bohm.[4]

Even a fairly superficial examination of these manifestations shows that we must make distinctions if we are to avoid an illegitimate blurring of the line between science and metaphysics. Mr. Watkins has said [5] "the logical empiricist is attempting the logically impossible when he tries to erect a *cordon sanitaire* between science and metaphysics" and I have heard these facts taken, in discussion, as showing that there is not really much difference between metaphysical and scientific theories after all.

Such statements, I believe, suffer from too great a generality owing to a failure to distinguish between different kinds of metaphysical statements. It is understandable that this failure should occur when we are emerging from a climate in which metaphysical statements were often regarded as meaningless, in a body, and therefore hardly worth talking about. But we can distinguish different kinds of metaphysical statements even if we do not wish to *be* metaphysicians or to accept metaphysical statements in the spirit in which they are usually put forward. When, for instance, Popper says "The Copernican system... was inspired by a neo-Platonic worship of the light of the sun who had to occupy the 'centre' because of his nobility" [6] and Watkins points out that metaphysical statements such as "Every event has a cause" are logically related to empirical statements they are talking, I think, of different kinds of metaphysical statements about which different things have to be said.

I hope I can take it as agreed that metaphysical speculations have been fruitful in stimulating scientific ideas and that the positivist's dismissal of them as meaningless, whether justified or not, has carried with it the disadvantage of helping to obscure the extent and importance of this stimulation. We have profited by the empiricists' attempts to bring out the difference between scientific and metaphysical statements even if they reached conclusions which many philosophers now consider extreme. The various attacks on inductivism, however, show signs of going, equally misleadingly, to the opposite extreme of denying not only that the more interesting scientific statements are derivable in a precise way from experience but also that they are even suggested by experience. [7] In our anxiety to do justice to the importance of metaphysics in this respect we must avoid, first, confusing different kinds cf metaphysical statements and, second, running together scientific and metaphysical theories. I shall discuss these two dangers separately.

If we mark off metaphysics from science by the falsifiability criterion we must resist the attendant temptation to treat the unfalsifiable class as homogeneous. It is all too easy to regard all metaphysical statements as of the same kind, the unfalsifiable, meaningless, nonsensical kind. I have said that I do not want to call them "meaningless" and I now want to argue that their unfalsifiability does not entail that it is unnecessary to distinguish different kinds of metaphysical statements.

I propose, for convenience of reference, to label two kinds of metaphysical statements "soft" and "hard". I call "soft" such statements as "Every event has a cause" and "Every mental event has a physical determinant"; I call "hard" such statements as "The sun must occupy the centre of the universe because of his nobility" and "The monads have no windows". Soft metaphysical statements involve only concepts which can appear in scientific statements, e.g. event, cause: hard metaphysical statements involve concepts which logically cannot appear in scientific statements, e.g. monad, soul. There may be borderline cases, that is, metaphysical statements to which we are not sure which label to apply, but even if we cannot put each metaphysical statement unequivocally into one of two sharply demarcated classes it is enough for my purpose if there are metaphysical statements which have the characteristics I am marking by the labels "soft" and "hard". I believe there are. On the colour spectrum, although we have a continuum, a clear distinction can be made between yellow and blue. Even if metaphysical and scientific statements lie on a continuum we can distinguish between hard and soft metaphysical statements and scientific statements which lie at the centres of certain stretches of the continuum.

There might be some disagreement about whether soft metaphysical statements are indeed metaphysical but if they are this appears to be closely connected with their generality, since they may be arrived at by generalizing scientifically respectable statements. Hard metaphysical statements, since the concepts they contain are not to be found in scientific statements, cannot be arrived at in this way. "Every cat eats fish" can be arrived at by generalizing from falsifiable statements of the form "This cat eats fish"; "Every event has a cause" can be arrived at by generalizing from statements of the form "This event has a cause" but this last statement is *not* falsifiable because failure to find a cause could never justify our rejecting it; "Monads have no windows" can be arrived at by generalizing from statements of the form "This monad has no windows" but this statement is also not falsifiable. Thus, the statements "This event has a cause" and "This monad has no windows" are alike in being unfalsifiable and so, by our criterion not scientific, but they differ in that the first might be verified by methods acceptable to science while the second could not. We can examine events in an attempt to find their causes but we cannot examine monads in an attempt to discover that they are windowless. Moreover, *some* statements about events and causes are falsifiable whereas no statements about monads are falsifiable. On the other hand, "This monad has no windows" differs from "This perfect gas behaves in such and such a way" since although we cannot point to perfect gases we could do so if they existed, and they *might* exist, but we could not point to monads even if they existed.

Most of the statements of the great system-builders, from the pre-socratics to Bradley, of the kind parodied by Ayer in "The Absolute is lazy", can be described as "hard" and I think it is even true to say that their systems, in so far as they can be regarded *as systems*, are entirely composed of such statements. They assert the existence of, or purport to describe, monads, substance, absolutes and other entities inaccessible to scientific investigation.

It must be remembered, however, that some hard metaphysical statements look like soft ones, or even like scientific statements, since the same word may be used for different concepts. For instance, statements about atoms defined as indivisible material particles are *metaphysical* because the scientific statement "Some atoms have been divided" does not falsify the statement "All atoms are indivisible" and *hard* because no statement about indivisible material atoms can be reached by generalizing from scientific statements about atoms. Thus progress in science is sometimes to be regarded as resulting from the removal of metaphysical statements rather than the falsification of empirical ones. The division of an indivisible atom is a logical impossibility so that when the atom was split this showed not that Democritus was wrong in thinking his atoms indivisible but that his concept of the atom had been replaced by a different one. Logical considerations alone could have shown that it was never a scientific concept, as Leibniz might have concluded.

In the rest of this essay when I refer to metaphysical systems or theories I shall be referring to systems or theories composed, or largely composed, of hard metaphysical statements.

Watkins has urged the respectability of metaphysical statements on the grounds that they are factual assertions about the world, that they unite heterogeneous areas of existence, that they often inspire new programmes for empirical investigation, that they sometimes foreshadow scientific theories and that they are capable of inconclusive confirmation, though not of proof or falsification.[8] He seeks to account for this by what he calls their "all and some" character ("Every mental event has *some* physical determinant") and two consequences of this. First, they are "weak entailments" of falsifiable empirical statements; for example, "Every mental event has a physical determinant" is weakly entailed by the falsifiable statement "Every mental event is determined by what the person ate for breakfast". Second, they are inconsistent with certain empirical statements; for example, "Every mental event has a physical determinant" is inconsistent with the falsifiable statement "No mental event has a physical determinant". It is because there are such logical relations between metaphysical and empirical statements that Watkins made the assertion I have already quoted about the impossibility of erecting a *cordon sanitaire* round metaphysics.

I do not wish, here, to criticize Watkins' view except to the extent of pointing out that he has been able to exhibit these logical relations only by concentrating on soft metaphysical statements and I fail to see how he could do the same for hard ones. I cannot see how "Monads have no windows" or "Only one substance can be granted and that is absolutely infinite" could be put in the "all and some" form, or shown to be even weak entailments of, or to be inconsistent with, any empirical statements. Yet metaphysical statements of this sort equally deserve our respect if only because some of them have inspired new programmes for investigation and foreshadowed scientific theories. Indeed, they seem more likely to do so just because they use concepts which are not used in scientific theories, and so are able to suggest *new* scientific concepts. It looks as if some of the metaphysical statements which influence science in the ways Watkins describes do not have the logical relations to empirical statements which he attributes to them.

People sometimes speak as if systems such as those of Leibniz and Spinoza entail statements such as "Every event has a cause", but this is clearly not so if the events and causes referred to are the kinds of things with which science is concerned since the necessary connections in these systems do not hold between such things as can be the objects of scientific in vestigation. The most these systems can do is to *suggest the adoption* of such a principle but this might also be suggested by empirical investigations. Once we have hit upon the device of causal explanation and found it help-ful we may decide to explain as much as possible and express our decision by saying "Every event has a cause" or "Always look for causal laws".

Part of the strength of Körner's view [9] that metaphysical statements function in science as "directives" or "regulative principles" lies in the close connection of soft metaphysical statements with experience and its detailed explanation. But I doubt if the function of hard metaphysical statements can be put just in this way. They may suggest directives but do not themselves function as directives. "Monads have no windows" would not be of great assistance to scientists in the form "Never look for windows in monads" and it cannot be put in the form "Some scientific objects do not have direct but only indirect causal relations between them, so don't waste time looking for direct relations when it is likely that you

35

are dealing with these objects". It cannot be put thus because monads are not scientific objects and the windows they are said to lack are not the causal connections of science. Nevertheless such a directive might be *suggested* by Leibniz's statement.

We may accept a directive on purely pragmatic grounds or we may go further and hold that it works because it says something about how the world is. Hard metaphysical statements do not function as directives but as suggestions about how the world is, in the light of which we may formulate directives for our future guidance. "Every event has a cause" is unlikely to suggest new structural features for a theory since it is formulated after causal explanation has been successful, but the idea of the sun's nobility or the windowlessness of the monads may suggest such features.

It seems, then, that the functions of the neo-Platonic view of the sun in connection with the Copernican theory and of Leibniz's system in connection with Bohm's theory are not to be understood by showing that metaphysical statements are "all and some" statements or that they function as directives. It is often said that metaphysical systems are "transformed into" physical theories and this is perhaps correct if we put sufficient weight on "transformed" and remember that everything is what it is and not another thing. I think the relation may be put more precisely by saying that metaphysical theories serve as models for scientific theories by suggesting new concepts or new kinds of structure, where "model" is used in the sense of *pattern* or artist's model. The scientist may, like the artist, be more or less representational, more or less abstract, but, in this context at least, not *tachiste*. The transformation which is his task is the construction of a falsifiable theory which embodies the suggested structure or concept.

Acceptance of all this must not lead us to blur the distinction between metaphysical and scientific theories. The effect of showing that soft metaphysical statements are weak entailments of empirical statements is, at least, to move the *cordon sanitaire* and, at most, to lead us to see that it is a mistake to attempt to draw firm lines, while allowing us to continue to hold that there are important distinctions to be made both within and between the groups of statements which we have failed to separate sharply. Hard metaphysical statements are, I claim, not related to empirical statements by any such logical relation. That is why I have been careful to say that metaphysical statements *suggest* new scientific theories or modifications of old ones.

For this reason, I may say in passing, I am unhappy when Popper, talking of the development of myths into scientific theories, says "It would hardly contribute to clarity if we were to say that these theories were nonsensical gibberish at one stage of their development (i.e. in the myth stage) and would then suddenly become good sense in another".[10] If there were some logical relation between the myth or the metaphysical theory and the scientific theory the development of one into the other would perhaps show that myths and metaphysical theories were not, after all, meaningless. But fruitful suggestions may come to the scientist from any source whatsoever, in more or less irrational and accidental ways or in ways which we would regard as highly rational, and perhaps the only common relation between what suggests and what is suggested is that of resemblance, which may be no more than a resemblance of structure.

The familiar story of Kekulé's snake, even if it is apocryphal, is illuminating. The percentage composition of benzene was known but its structural formula was not. All efforts to show that it was some form of chain of six carbon and six hydrogen atoms had failed. Kekulé is said to have dozed off while worrying about the problem and dreamt of a snake which, after going through many contortions, finally put its tail in its mouth. He awoke with the solution that the structural formula of benzene involved a ring. This is as good a way of arriving at a new idea as any and part of the strength of current attacks on inductivism lies in the fact that such things happen. But if the suggestion had been made by a collection of nonsense sounds instead of a visual image, would the fact that a falsifiable statement about the ring structure of benzene resulted from these sounds go any way towards showing that the sounds were meaningful after all? And would the hypothesis of the ring structure be any the worse if it didn't? We might admit the influence of myths and metaphysical theories without committing ourselves to any decision on their meaningfulness although, admittedly, the adopting of metaphysical theories for scientific purposes usually appears to have a more rational foundation than this. I mention this merely to emphasize that I am not packing a great deal into the idea of suggestion. All sorts of things may suggest new ideas for science and whether what suggests is meaningful or not must be decided by examining the details of each particular case.

At the opposite extreme from the story of Kekulé's snake we have examples like the growth of the modern Atomic Theory out of Democritus' Atomism, where a great deal of the content of the metaphysical theory as well as its structure suggest the content and structure of the scientific theory. But still the word "suggest" is more appropriate than some stronger word such as "entail".

If it is objected that I am making a psychological rather than a logical point I shall not be greatly moved, for if there is a connection, of whatever sort, between metaphysical and scientific theories, some knowledge of the nature of this connection may help us later to discover what characteristics of metaphysical theories make this possible. In the meantime, if the connection is psychological rather than logical, the distinction between them by means of the falsifiability criterion remains intact and so, consequently, does the entailed distinction on the grounds that metaphysical theories are not explanatory, at least in the way in which scientific theories are.

Let us now look a little more closely at an example which shows how the process of transforming a metaphysical into a scientific theory sometimes works, for the double purpose of reinforcing what I have said so far and of bringing out some further points. I shall choose for this purpose Bohm's theory [11] which has the advantage of being topical and, for me at least, the advantage that I have had the opportunity of discussing it with its author. I shall mention only as much of it, and its genesis, as is strictly relevant to my theme.

Bohm's use of Leibnizean metaphysics springs from his dissatisfaction with the present position in quantum theory according to which we are unable to give a causal specification for certain events within the atom. Leibniz's monad theory suggests to Bohm a world in which there are relatively independent systems (windowless monads), the development of each being determined by its own inner law, each generating for itself its own space and time and each being connected with, i.e. not independent of, all the others at a deeper level (pre-established harmony). Thus, things are seen as unconnected, and events as not determining one another, on one level, e.g. the level of quantum mechanics, while at a deeper level, e.g. that of "sub-quantum mechanics", those same things are seen as connected and those same events as determined, through events at this deeper level.

Night does not cause day but the movements of the planets cause the alternation of dark and light.

Leibniz's two clocks constitute an analogy for his relatively isolated systems within a wholly deterministic system. But anything with a periodic "beat" may function as a clock and this allows us to have a whole system of clocks within clocks. One clock is an organized system of sub-clocks and these in turn are organized systems of sub-sub-clocks, and so on, where on each level relative independence and determination may both be found and what is seen as undetermined on one level may be seen as determined by a lower level. To use the analogy with ordinary mechanical clocks, we see each clock ticking away according to its own inner law and independently of others. When we find that both were constructed according to the same mechanical principles we see them as connected by these principles. But the clocks function similarly because each is composed of materials which in turn obey their own inner laws; the steel in both functions in the same way, and differently from the copper which in turn functions in the same way in both clocks. Now the replacing of monads by clocks and the defining of clocks in terms not of springs and cogs and hands but of periodic beats allow a fairly accurate reproduction of the structure of Leibniz's system, but also mark the transition from a metaphysical to a scientific theory, since periodic beats are the sorts of things that can be observed and measured, allowing at least the theoretical possibility of the falsification of any statement in the theory. One of the chief features of the theory is that an inner structure is postulated even for the so-called "fundamental particles".

I am not here concerned with the correctness of the physical theory: even if it proves to be impossible to test in practice, tests are possible in principle. It seems worth noting, too, that physicists, dismayed though they may be by the talk of Leibniz, are coming to recognize that this physical theory is one to be reckoned with and cannot be dismissed as the idle speculation of a crank. I am concerned with the way in which it was arrived at and the attitude of its inventor to it. Bohm constructed his theory partly, as he says, by taking seriously certain aspects of Leibniz's theory, which he recognizes as being in principle unfalsifiable. He also recognizes that his task is to produce from it a theory which shall be in principle falsifiable. The metaphysical theory is clearly being used as a kind of model or pattern for a scientific theory and the suggestive power

39

of the theory has no tendency to make the two theories one, to make the two sorts of theory more like one another than we supposed or to change our views about the logical properties of either the original metaphysical theory or the finished scientific theory.

What features of metaphysical theories account for the kind of fruitfulness for science which I am considering? Bohm's use of Leibniz seems somehow more reasonable than Kekulé's use of his dream snake. I am inclined to think that this is partly due to the apparent fortuitousness of Kekulé's dream. We can choose Leibniz's theory rather than Spinoza's but we cannot choose to dream of a snake at the appropriate time. Apart from this, perhaps the main difference is in complexity, that is, in the extent and character of the suggestions made.

One important feature of both examples is that some formal structure is taken from the model and used in a different context. There is a general formal similarity between metaphysical and scientific theories in that both are deductive. The more advanced a science is the more highly formal it is, so that it seems more likely that a metaphysical theory will suggest useful theories for a highly developed than for an undeveloped science. Systematic metaphysicians build just the sorts of theories that are likely to be useful as models for the theoretical physicist. There is something comparable here to the influence of Leibniz on Russell in connection with *Principia Mathematica*. However, within the general deductive character of metaphysical theories there are vast differences between them in content and detailed structure and this means that one metaphysical theory may be more suggestive than another in a given scientific context. A metaphysical theory may, by virtue of its structure, assist in the development of the purely mathematical part of a scientific theory [12] but it may do more and make suggestions about its interpretation and here its content, as well as its structure, becomes important. The kinds of entity (e.g. monads) which are formally organized by the metaphysician may suggest analogous kinds of entity (e.g. clocks) which can be organized in a similar way by the scientist. The scientist is a system builder but in the present connection the metaphysician is the system builder *par excellence* since, unlike the scientist, he is free from the necessity of connecting his system with the details of actual experience and so can concentrate on internal consistency. At the same time his systems are superior to those of the mathematician for the stimulation of scientists because they purport to talk about constituents of the universe; in an odd way they combine some of the advantages of both mathematical and scientific theories. Thus the scientist can find likely systems ready-made, choose the appropriate one as his model and refashion it in such a way as to give consequences which are in principle falsifiable.

My second, and perhaps more controversial, suggestion to account for the fruitfulness of metaphysical theories is that they can be themselves regarded as being, to a limited but not negligible extent, controlled by experience. The ways in which metaphysical theories are built are complex and I do not wish to minimize this complexity but it seems to me that *one* of the factors which may contribute to them is ordinary everyday experience. Thus when Popper speaks of theories as "freely invented ideas" [12] and Einstein of "free mental creations" I think they are both right and wrong. Our invention of theories *is* free in the sense that one rather than another is not forced upon us by experience, but may be dependent in the sense that we may draw our materials from experience. It is clear that some parts of scientific theories are guided by what Cassirer calls "the pressure of experience", felt through ordinary observations and laboratory experiments, but that the upper reaches of theories are considerably less influenced by this pressure. It may well be that our experience of the world is so much coloured by our three-dimensional theories that experience alone could never have led us to think four-dimensionally. But to say that these upper reaches are influenced by metaphysical theories is not, I suggest, to say that they are influenced by free mental creations *rather than* by experience but to say that they are, at least in part, influenced by experience by another, longer and more devious, route. Experience, and the ways in which we can talk about it, are so complex that the same set of experiences may suggest many different theories, some metaphysical, some scientific, some more fruitful than others, some more acceptable than others for different people or at different times or in different situations.

In this new context, it is perhaps helpful to bring out the difference between scientific and metaphysical theories in another way, by reminding ourselves of the most general questions to which they are intended to provide answers. The scientist asks "How can we account for the way the world is, as observation reveals it to us?" and his answers are guided by his ideal of testability through observation. The metaphysician, combining an assertion with his question, asks "Since the world cannot be just as it appears to us in observation, how it is really?" and his answers are guided by his ideal of intelligibility. My contention is that at least *some* of his clues for his answers may come from certain aspects of what he observes.

No doubt the primary aim of the metaphysician in constructing his theories is the solution of certain *logical* problems which stand in the way of intelligibility. How are we to make change, or motion, or time, understandable? There are anomalies in our experience of the world but reality "cannot accept inconsistent predicates" so how are these anomalies to be removed? But, however the details of metaphysical theories have in fact been arrived at, the results are such that they could have been arrived at by concentrating on certain features of our ordinary experience of the world, taking these as basic and explaining the other features of our experience in terms of them. That is, they might have been constructed by abstracting from experience those features which fit the metaphysicians' ideal of intelligibility and extrapolating by saying that everything has some character which certain things in our experience have or appear to have, *or* by saying that something has certain of the characters which things in our experience appear to have but in greater abundance, perfection or completeness. The differences between different metaphysical theories would then depend, in part, upon the particular features of experience which were singled out for special emphasis. At least we can say that most metaphysical systems contain concepts which could have been based on experience, but only some of those concepts, and that the specific character of each system is determined by the particular selection of such concepts which appears in it. This is not intended to be a historical point about the way in which metaphysical systems were, in fact, consciously constructed: it is a point which can be made even if, as a matter of historical fact, there is no evidence that their inventors consciously went through any such process as I have outlined. All I wish to say is that most, and perhaps all, metaphysical systems depend upon concepts which have this character and which could have been reached by the route I have outlined. But having been suitably cautious about the history of philosophy, I shall risk being bolder and say that I am inclined to think that, whatever the conscious process actually involved in the constructing of existing metaphysical systems and whatever the arguments used in support of them, it may well be that the final products were in fact dependent, in some such loose way as I have suggested, upon experience. If this were not so, it would be difficult to see how such ideas could arise and how we can understand them even as well as we do.

All this ought to be supported by detailed examinations of particular metaphysical theories but this is impossible within the limits of this essay. Perhaps a few brief comments on various systems will show that this is a view worth discussing.

Our ordinary experience contains apparently static elements as well as obviously changing elements. The system of Parmenides could be regarded as that of a man more impressed by the static elements and the system of Heracleitos as that of a man more impressed by the changing elements. Parmenides could be interpreted as saying that everything really has the static character which many things appear to have while Heracleitos could be interpreted as saying that everything really has the changing character that many things appear to have. Leibniz's and Spinoza's systems could be regarded as being similarly related. It is as if Leibniz concentrated on the unlimited possibility of division of physical objects and the continuity and indivisibility of mental experience and as if Spinoza concentrated on the connectedness of observable things rather than on their separateness. Clearly, purely logical considerations were always in fact involved but Leibniz, at least, seems explicitly to support my view by constantly drawing attention to physical divisibility and the actual divisions in nature. He writes at times almost as if this were evidence supporting his system. We might also regard the more or less observable connections between mental and bodily events as having prompted Spinoza to attempt to avoid the Cartesian dualism. Hegel's system looks like an attempt to give due weight to the alternation between unity and diversity which we encounter as we continually extend our experience in breadth and depth. Bradley was clearly profoundly influenced by experience but attached greater importance to the fact that things can be seen as related than to the fact that they can be seen as independent *things*. The atomists in general can be regarded as stressing the fact that they can be seen as things.

If we concentrate on waking perceptions of the world we may attach importance to the claim of perceptions to be externally caused, and seek to remove discrepancies by asserting the existence of external objects which yet do not have the qualities they appear to have. But if we concentrate on the resemblances between dreams and waking perceptions we are more likely to attach importance to the claim of perceptions to be internally caused and seek to remove discrepancies by asserting nothing external to the mind. Ideas of Gods may be arrived at by extrapolation of certain human characteristics which, as Hume has it, are augmented without limit; and such a process is also seen at work in the development of myths and the inclusion of anthropomorphic elements in metaphysical theories.

And so on. The examples could be multiplied indefinitely. We may not now be very happy about the arguments put forward in support of metaphysical theories, we may regard the theories themselves as failing to do what their creators intended them to do and as being of no immediate help to us except in the way of satisfying some desire for unity; but most of them are of such a kind that they could have been built up by a judicious use of ordinary experience of the natural world and it may even be possible that such experience in fact went towards their formation in an important way. The relation between metaphysical theories and experience is vague, uncomfortable and difficult to characterize precisely, but perhaps we can get somewhere near it by saying that certain features or aspects of experience serve as models, or at least could serve as models, again in the sense of patterns, for metaphysical theories.

The importance of this is, it seems to me, that it is *just because* metaphysical theories have this character, *just because* they involve concepts having such connections with experience, that they may in turn serve as models for scientific theories. The points of contact with experience may be the growing points of theories which will account for experience in the scientific way. Because Leibniz's theory arose partly from difficulties about space and time, because minds were free from these encumbrances and could be conceived as creating them, because a close examination of nature revealed developing organizations which could be parelleled by the development of rudimentary minds, and so on, it turns out that his theory is fruitful for Bohm because he, too, is concerned with problems of space and time in physics and needs, to account for the phenomena, some theory involving the idea of relatively independent but ultimately related systems, though not, of course, mind-like systems. Leibniz concentrated on aspects of experience suggesting unrelatedness and so had to show, for the sake of intelligibility, how this could grow out of relatedness and, for the sake of saving the phenomena, how other aspects of experience suggesting relatedness could be explained. Thus he had grappled with a problem similar to that facing Bohm when nature appears to present features, such as the positions of a given orbital electron at times t and $t + 1$, which are unrelated where we expect, and hope to find, relations.

Metaphysical theories, then, may be regarded as springing from a rejection of certain features of experience and a retention of others. They are constructions from which certain features, regarded as repugnant, are absent. When the scientist faces features of experience which, on the theory he has, he finds repugnant, one of the possible moves is to look for a ready-made theory which avoids just such features and to see if it will give clues for the reconstruction of his scientific theory.

Thus, in this context, metaphysical theories do not suffer from the disability that there is no rational means of choosing between them. What rational grounds can there possibly be for choosing between two internally consistent theories if neither has empirical consequences? Here we have an answer, at least for the present context. The pressure of experience urges us towards one rather than another because it, in turn, grew out of, or, at least, is such that it could have grown out of, the same sort of pressure of experience. Experiments, and our present theory, lead us to say that light behaves sometimes in a wave-like way, sometimes in a particle-like way, and that this is ultimate, admitting of no further explanation. If we find this duality awkward we may look for a theory which overcomes such a duality, which contains enough points of similarity with the physical situation to fit it and with the present physical theory, if possible, to enable us to avoid scrapping it altogether. We should be likely to choose one which allows us to talk of different levels such that on one level the wave-particle duality is ultimate but on a deeper level the duality is resolved in a single concept. Leibniz's system is *prima facie* more likely to be helpful than, say, Spinoza's, since on no level does Spinoza allow a real plurality. On the other hand, in a different situation where we were faced with the kind of problem which a field theory would solve we might obtain more help from Spinoza than from Leibniz, since he was mainly concerned with showing that nothing is really distinct from anything else. The choice of the scientist is not arbitrary and not ultimately free from the pressure of experience since both he and the metaphysician have been subjected to the same pressure.

I do not want to suggest that the fact that scientific theories have been influenced by metaphysical speculation constitutes the only reason for taking an interest in such speculation but only that even those who are most unfavourably disposed towards metaphysics should admit that this constitutes one reason. Neither do I wish to deny that men are capable of free creation. All I have argued is that the creation of metaphysical and scientific theories may be less free and more dependent on comparing, contrasting and abstracting than is sometimes supposed and that it is because they share this character that there are connections between them.

University of Bristol, Bristol, England

REFERENCES

1. See, e.g., the works of Ernst Mach, Heinrich Hertz, Pierre Duhem and Henri Poincaré.
2. Karl Popper, 'The Demarcation between Science and Metaphysics', in *Conjectures and Refutations*, Routledge, 1963, pp. 253 ff.
3. See, e.g., Karl Popper, 'The Nature of Philosophical Problems and Their Roots in Science, *The British Journal for the Philosophy of Science* 3, 124; A. G. Van Melsen, *From Atomos to Atom*, Duquesne University Press, Pittsburg, 1952; Ernst Cassirer, *Determinism and Indeterminism in Modern Physics*, Yale University Press, New Haven, 1956; J. W. N. Watkins, 'Between Analytic and Empirical', *Philosophy* 32 (1957); J. W. N. Watkins, 'The Haunted Universe', *The Listener*, November 21 and 28, 1957; J. W. N. Watkins, 'Confirmable and Influential Metaphysics', *Mind* 67 (1958); S. Körner, *Colston Symposium on Observation and Interpretation*, Butterworth, 1957, p. 97; P. Feyerabend, 'An Attempt at a Realistic Interpretation of Experience', *Proc. Arist. Soc.* 1957–1958, and 'Complementarity', *Arist. Soc. Supp.* 32 (1958); and various works by A. Koyré, M. Jammer, H. Butterfield, G. Sarton, G. di Santillana, and E. A. Burtt.
4. D. Bohm, *Causality and Chance in Modern Physics*, Routledge and Kegan Paul, London, 1957; idem, *Colston Symposium on Observation and Interpretation*, Butterworth, 1957, p. 33.
5. Watkins, 'The Haunted Universe', *The Listener*, November 21, 1957, p. 886.
6. Popper, 'The Demarcation Between Science and Metaphysics', p. 257.
7. See, e.g., Feyerabend, 'An Attempt at a Realistic Interpretation of Experience', *Proc. Arist. Soc.*, 1957–1958, p. 153.
8. Watkins, 'The Haunted Universe', *The Listener*, November 21, 1957, p. 838.
9. Körner, *Conceptual Thinking*, Cambridge, 1955, Dover Publications, 1959.
10. Popper, 'The Demarcation between Science and Metaphysics', p. 257.
11. *Colston Symposium on Observation and Interpretation*, Butterworth, 1957, p. 33; Bohm, *Causality and Chance in Modern Physics*, Routledge and Kegan Paul, London, 1957. Dr. Bohm should not held to be responsible for the use I make of his theory.
12. See Popper, 'The Nature of Philosophical Problems and their Roots in Science', *The British Journal of Philosophy of Science* 3.

WILLARD VAN ORMAN QUINE

ON SIMPLE THEORIES OF A COMPLEX WORLD

It is not to be wondered that theory makers seek simplicity. When two theories are equally defensible on other counts, certainly the simpler of the two is to be preferred on the score of both beauty and convenience. But what is remarkable is that the simpler of two theories is generally regarded not only as the more desirable but also as the more probable. If two theories conform equally to past observations, the simpler of the two is seen as standing the better chance of confirmation in future observations. Such is the maxim of the simplicity of nature. It seems to be implicitly assumed in every extrapolation and interpolation, every drawing of a smooth curve through plotted points. And the maxim of the uniformity of nature is of a piece with it, uniformity being a species of simplicity.

Simplicity is not easy to define. But it may be expected, whatever it is, to be relative to the texture of a conceptual scheme. If the basic concepts of one conceptual schema are the derivative concepts of another, and *vice versa*, presumably one of two hypotheses could count as simpler for the one scheme and the other for the other. This being so, how can simplicity carry any peculiar presumption of objective truth? Such is the implausibility of the maxim of the simplicity of nature.

Corresponding remarks apply directly to the maxim of the uniformity of nature, according to which, vaguely speaking, things similar in some respects tend to prove similar in others. For again similarity, whatever it is, would seem to be relative to the structure of one's conceptual scheme or quality space. Any two things, after all, are shared as members by as many classes as any other two things; degrees of similarity depend on which of those classes we weight as the more basic or natural.

Belief in the simplicity of nature, and hence in the uniformity of nature, can be partially accounted for in obvious ways. One plausible factor is wishful thinking. Another and more compelling cause of the belief is to be found in our perceptual mechanism: there is a subjective selectivity that makes us tend to see the simple and miss the complex. Thus

47

consider streamers, as printers call them: vertical or diagonal white paths formed by a fortuitous lining up of the spaces between words. They are always straight or gently curved. The fastidious typesetter makes them vanish just by making them crooked.

This subjective selectivity is not limited to the perceptual level. It can figure even in the most deliberate devising of experimental criteria. Thus suppose we try to map out the degrees of mutual affinity of stimuli for a dog, by a series of experiments in the conditioning and extinction of his responses. Suppose further that the resulting map is challenged: suppose someone protests that what the map reflects is not some original spacing of qualities in the dog's pre-experimental psyche or original fund of dispositions, but only a history of readjustments induced successively by the very experiments of the series. Now how would we rise to this challenge? Obviously, by repeating the experiments in a different order on another dog. If we get much the same map for the second dog despite the permutation, we have evidence that the map reflects a genuinely pre-experimental pattern of dispositions. And we then have evidence also of something more: that this pattern or quality space is the same for both dogs. But now I come to the point of my example: we cannot, by this method, get evidence of pre-experimental quality spaces unlike for the two dogs. By the very nature of our criterion, in this example, we get evidence either of uniformity or of nothing. An analysis of experimental criteria in other sciences would not doubt reveal many further examples of the same sort of experimentally imposed bias in favor of uniformity, or in favor of simplicity of other sorts.

This selective bias affords not only a partial explanation of belief in the maxim of the simplicity of nature but also, in an odd way, a partial justification. For, if our way of framing criteria is such as to preclude, frequently, any confirmation of the more complex of two rival hypotheses, then we may indeed fairly say that the simpler hypothesis stands the better chance of confirmation; and such, precisely, was the maxim of the simplicity of nature. We have, insofar, justified the maxim while still avoiding the paradox that seemed to be involved in trying to reconcile the relativity of simplicity with the absoluteness of truth.

This solution, however, is too partial to rest with. The selective bias in favor of simplicity, in our perceptual mechanism and in our deliberate experimental criteria, is significant but not overwhelming. Complex

48

hypotheses do often stand as live options, just as susceptible to experimental confirmation as their simpler alternatives; and in such cases still the maxim of simplicity continues to be applied in scientific practice, with as much intuitive plausibility as in other cases. We fit the simplest possible curve to plotted points, thinking it the likeliest curve pending new points to the contrary; we encompass data with a hypothesis involving the fewest possible parameters, thinking this hypothesis the likeliest pending new data to the contrary; and we even record a measurement as the roundest near number, pending repeated measurements to the contrary.

Now this last case, the round number, throws further light on our problem. If a measured quantity is reported first as 5.21, say, and more accurately in the light of further measurement as 5.23, the new reading supersedes the old; but if it is reported first as 5.2 and later as 5.23, the new reading may well be looked upon as confirming the old one and merely supplying some further information regarding the detail of further decimal places. Thus the "simpler hypothesis", 5.2 as against 5.21, is quite genuinely ten times likelier to be confirmed, just because ten times as much deviation is tolerated under the head of confirmation.

True, we do not customarily say "simple hypothesis" in the round-number case. We invoke here no maxim of the simplicity of nature, but only a canon of eschewing insignificant digits. Yet the same underlying principle that operates here can be detected also in cases where one does talk of simplicity of hypotheses. If we encompass a set of data with a hypothesis involving the fewest possible parameters, and then are constrained by further experiment to add another parameter, we are likely to view the emendation not as a refutation of the first result but as a confirmation plus a refinement; but if we have an extra parameter in the first hypothesis and are constrained by further experiment to alter it, we view the emendation as a refutation and revision. Here again the simpler hypothesis, the one with fewer parameters, is initially the more probable simply because a wider range of possible subsequent findings is classified as favorable to it. The case of the simplest curve through plotted points is similar: an emendation prompted by subsequent findings is the likelier to be viewed as confirmation-cum-refinement, rather than as refutation and revision, the simpler the curve. *

We have noticed four causes for supposing that the simpler hypothesis stands the better chance of confirmation. There is wishful thinking. There is a perceptual bias that slants the data in favor of simple patterns. There is a bias in the experimental criteria of concepts, whereby the simpler of two hypotheses is sometimes opened to confirmation while its alternative is left inaccessible. And finally there is a preferential system of score-keeping, which tolerates wider deviations the simpler the hypothesis. These last two of the four causes operate far more widely, I suspect, than appears on the surface. Do they operate widely enough to account in full for the crucial role that simplicity plays in scientific method?

Harvard University, Cambridge, Massachusetts, U.S.A.

---

* I expect that Kemeny has had all this in mind. He remarks the kinship of the rule of significant digits to that of simplicity on page 399 of 'The Use of Simplicity in Induction', *The Philosophical Review* **62** (1953) 391–408.

N. RASHEVSKY

THE DEVIOUS ROADS OF SCIENCE

A mathematician of ancient Greece is credited with the saying "There are no Royal Roads in mathematics". This is usually understood as meaning that it is impossible to understand the proof of a complicated theorem without having first understood the proofs of all preliminary theorems and lemmas that precede it. To put it differently, a person unfamiliar with geometry, cannot open a book on geometry in the middle and read with understanding, whatever is printed there. He has to read *all* the preliminaries.

This is different from books on purely descriptive matters. There is nothing in the way of opening in the middle a book which contain descriptions of different styles of architecture and reading it intelligently from the middle, as long as functional relations between the different styles are not discussed, in other words as long as the book *is purely descriptive.*

The absence of royal roads seems to be characteristic of sciences in which in addition to pure description there is also the question of understanding the "why" and not only the "what".

The ancient saying about royal roads seems, however, to have a much more general meaning than the one discussed above. It seems to apply to the development of *any* science, be it inductive, deductive, or even purely descriptive. Between the inception of a new idea and its more or less final embodiment there lies a tortuous path of trials and errors, of partial successes and failures. Compare, for example, the original formulations of the principle of conservation of energy of Robert Meyer and H. Helmholtz with the formulation given in any modern textbook of physics and you will see what I mean. The road from the early formulation by Max Planck of his quantum hypothesis, through Bohr's theory of the atom, to the present day quantum mechanics is another example. Thus we may generalize the ancient saying by stating "There are no royal roads in the development of Science", or "The Ways of Science are Devious".

A good example of such devious ways is given by the development in the last four decades of a borderline science which may be properly designated as mathematical biology, in the broadest meaning of those words.

Occasional scattered attempts to apply mathematical methods to biology have been made as far back as the end of the last century. It is, however, only since the twenties that important attempts at a systematic development of mathematical biology began to be made. From the very beginning this approach ramified in two seemingly very different directions. On one hand the mid-twenties witnessed the appearance of Alfred J. Lotka's *Elements of Physical Biology.* [1] This was followed in the early thirties by Vito Volterra's *Leçons sur la Théorie mathématique de la Lutte pour la Vie* [2], which, while differing in many respects from Lotka's book, is very similar to it in some basic ideas.

On the other hand that decade witnessed the first publications of J. H. Woodger on the axiomatic method in biology, which perhaps are best represented by his later book of the same title. [3] The books of Lotka and Volterra on one hand, and the books of Woodger on the other, look as different as Goursat's *Cours d'Analyse Mathématique* and the *Principia Mathematica.* Both approaches, however, are equally basic and important.

The analytical approach of Lotka and Volterra had perhaps more appeal to biologists who were interested in possible immediate applications of the theory to their experimental needs than had the logico-mathematical approach of Woodger. Lotka, as a physical chemist by training, approached the biological world as a physico-chemical system. He and Volterra derived a number of *quantitative* relations many of which could be compared to available *measurements* made by experimental biologists. Woodger, on the other hand, emphasized the *logical* foundations of biological concepts and the *qualitative relations* inherent in many biological phenomena. In those days not only the non-mathematicians but also some applied mathematicians were inclined to consider mathematics, as far as its practical applications went, as essentially a *quantitative* science in spite of the fact that such "qualitative" branches as the theory of groups, theory of sets, topology, and the theory of relations, were in existence. To a large extent this attitude may have been due to the circumstance that at that time biologists, though by far not always appreciating the importance of mathematics, already fully appreciated the importance of physics to biology. The mathematical physicist was essentially dealing with analytical methods; to him mathematics was basically a *quantitative* science. The only notable exception were the use of theory of groups in crystallography and later in some aspects of quantum mechanics. No one hardly ever thought in those days of applying to physics symbolic logic or the general theory of relations.

To the "practical" applied mathematician treatises like *Principia Mathematica* had little appeal. An applied mathematician, whether theoretical physicist or engineer, had sometimes to cope with problems of very great analytical difficulty. In his search for solutions he would, however, never consult the *Principia*, because that book was of no value for his immediate *practical* purposes. It dealt with basic fundamental principles; and even if the applied mathematician was not completely uninterested in those basic principles, he had no time to dwell on them and they were of no immediate use to him. Even amongst some pure mathematicians we did find at best a lack of sufficient interest in the *Principia*. No less a person than Henri Poincaré at several occasions showed a thinly veiled skepticism toward the work of Russell, Peano, Burali-Forti and other mathematical logicians. [4]

No wonder therefore that Woodger's work, which is largely based on the application of the methods of the *Principia* to the field of biology, found a far greater response amongst logicians than among the more "practically" minded biologists. And yet Woodger's work was not without "practical" implications. Frequently, however, those were masked by rows of formulae of symbolic logic, whereas the work of the other school of mathematical biologists had a more direct, intuitive, appeal to the physical aspects which underlie biological phenomena.

The work of Lotka and Volterra deals primarily with the quantitative laws of interactions of species. Towards the end of the twenties the same method began to be systematically applied to the theory of individual organisms and the interactions of their parts, by N. Rashevsky. [5] Subsequently this *Mathematical Biophysics* was developed by a number of other investigators and covered a very wide range of biological phenomena.

Methods of classical mathematical analysis are used throughout all this work with one exception. In 1943, W. McCulloch and W. Pitts showed that a number of phenomena in the central nervous system were isomorphic to sets of statements in a two-valued Boolean algebra.

Unfortunately this important work has never been developed further by its authors. Some contributions have been subsequently made by other investigators. About the same time symbolic logic was introduced as a mathematical tool in a number of engineering problems, notably in the theory of systems of relays.

Hundreds of papers have been published in mathematical biophysics covering many different problems. All those studies may be characterized as constructing physical or physico-mathematical models of various biological phenomena. Many of those models have led to a good *quantitative* explanation of known numerical relations or have even predicted correctly the quantitative outcome of new experiments. Some aspects of cell division and cell respiration, cardiovascular phenomena, lung function, active transport, nerve excitation and conduction, reaction times, discrimination, color vision, mortality, and many other phenomena have been successfully treated theoretically. A journal, *The Bulletin of Mathematical Biophysics*, has been established in 1939 in this field and is steadily increasing its circle of contributors.

Yet in spite of all this unquestionable success, one conspicuous shortcoming of this approach began to be apparent several years ago. As stated above, all those developments are based on constructing *models*. A mathematical biology should aim at more than that. It must formulate some general mathematical principles, akin to those of physics. A model is good for a particular case. A general principle applies to all cases. A general principle does not in itself *explain* a given situation as does a model. But it provides a background for the construction of an explanation.

Two general principles have so far been proposed. One of them deals with the determination of organic form by organic function. It proved to be useful in many respects. It uses, however, the same classical analytic, quantitative approach, that has been used in the construction of models. The other principle emphasizes the qualitative relations in and between organisms, and makes a distinct deviations in the direction which has been indicated decades ago by Woodger, though being very different from the actual approach of Woodger.

As the physico-mathematical models of various biological phenomena were successfully developed, it became eventually apparent that in spite of all the success this approach lacked a very essential element. While it is important to know, actually or hypothetically, why cells divide, how the blood flows, etc., this knowledge does not tell the whole story. It does not tell us about the relations of those different phenomena within the same organism. It does not tell, for example, that an error in the mechanism of the central nervous system may produce a faulty muscular movement which results in our cutting a finger. This results in a flow of blood from the wound and subsequently in a healing of the wound, in which cell division plays a prominent part. This sequence of events is in itself not quantitative in nature. Its description is a statement of a series of qualitative relations between sets of different phenomena.

This qualitative character becomes even more apparent when we consider the basic similarities between different organisms, similarities which exist *in spite of* quantitative differences and even in spite of differences of physico-chemical mechanisms involved. Thus both a paramecium and man do, in response to appropriate stimuli, move toward food; they both ingest the food; digest it; excrete the indigestible particles, and assimilate the rest. Not only are all those phenomena *quantitatively* different in a paramecium and in man, but in the two cases their mechanisms are quite different. Yet the important *relational* similarity is manifest.

The act of locomotion in man is much more complex than in paramecium. But the more numerous elements of the act of locomotion in man relationally correspond to the fewer elements of the act of locomotion in paramecium. The same may be said about ingestion, digestion, and other biological functions.

Thus there is a correspondence between the biological properties of all organisms. Such correspondences are called in mathematics *mappings.* In particular when several elements of one object corresponds to fewer elements of the other, we speak of many-to-one mappings.

Several branches of mathematics deal with mappings. For the purpose of formulating mathematically the above discussed generally known fact of relational similarity N. Rashevsky [6] chose in 1954 topology, which may loosely be described as the geometry of relations. It was shown that well known facts could be stated in the form of the following general principle: To every organism there corresponds an abstract topological space, the points of which represent the different biological properties of the organism. The topological spaces of different organisms map in a particular manner on each other. The manner of mapping is one called in topology *continuous mapping*.

A number of empirically verifiable conclusions can be drawn with the help of this principle. Those conclusions are not quantitative in nature. They are of the nature of existential statements. Thus for example, it can be shown that emotional disturbances in some higher animals and man can lead to gastrointestinal disturbances. This is a well-known fact, especially in man. Its *explanation* in terms of an appropriate model does not present difficulties. But in order to become interested in making an appropriate model, the fact itself must be taken as known. Its existence, however, follows from the above-mentioned principle. Another conclusion is the existence of animals which possess glands, the secretions of which serve to catch their prey. The spider is a classical example.

This type of theoretical developments, while qualitative rather than quantitative in its nature, is nevertheless still so different from Woodger's relational biology, as to preclude any comparison. Yet one essential feature of Woodger's work, namely the relational character of the approach, is manifest.

Further development of the above-mentioned principle of mapping revealed that the use of topology was an unnecessary restriction. On one hand R. Rosen [7] showed that much more general results can be obtained by using the so-called theory of natural equivalences, or categories, a branch of mathematics relatively recently developed by S. Eilenberg and S. MacLane, which is much more general than topology. On the other hand it was found that instead of considering continuous mappings of topological spaces we may consider mappings of sets, such that some given relations are preserved by those mappings. The same conclusions are reached as by the use of topology. In this approach we consider the organism as a set of properties which make the organism recognizable as such. The sets which correspond to different organisms can be mapped onto each other so as to preserve certain basic relations. Thus from a very different point of view than that used by Woodger and for a very different purpose we arrive at the application of the general theory of relations to biology. Again we find a very great difference between Woodger's approach and the more recent one described above. But it is a noteworthy fact that the road taken thirty years ago by the quantitative branch of mathematical biology has led now in a rather circuitous manner to the use of the mathematical tools advocated by Woodger at the very beginning.

It must be strongly emphasized that the relational approach to biology outlined above in no way supplants the earlier quantitative or *metric* approach. The two are equally important in biology. Neglecting either one of them would be a great error. It must, however, be said that in some respects the relational approach is more basic. It does emphasize the unity of the organism as well as of the organic world as a whole, a fact which the quantitative approach, thus far at least, almost ignores. The metric approach is more likely to emphasize the *quantitative differences* between organisms, whereas the relational approach emphasizes the *qualitative similarities* between them.

We have already emphasized the difference which still exists between Woodger's approach and ours. Yet this difference seems now to be more in the subject-matter treated rather than in the basic methodological points of view, though even here the two approaches are quite different. The reason for this difference was perhaps best formulated by Professor Alfred Tarski in a conversation with the author a few years ago. He remarked: "The difference between Woodger's approach and yours is due to the fact that Woodger is interested in the logical, while you are interested in the biological aspects of the problems." This seems to be very true. Yet we must remember that sooner or later the logical and the biological aspects of the problem are bound to meet. Even now it is hardly possible to draw a sharp line everywhere between them.

Both in pure mathematics and in theoretical physics the development of the "mathematical" and of the "physical" aspects preceded by centuries the study of the logical aspects. What would have happened if the *Principia Mathematica* were published in the days of Euler or Gauss? Some may be inclined to say that this would have been impossible. We are inclined to take a safer point of view and say that this would have been highly paradoxical. Something of a similar nature of paradoxicality seems to have happened more than thirty years ago when Woodger published his work on the logical foundations of biology at the time when biology as a whole was much further behind in its development than was mathematics at the time of publication of the *Principia Mathematica*. We have here again an example of the fact that the development of science does not follow the straight "logical" or "didactic" road. Is this following of a tortuous road inherent in the nature of scientific development or has it hitherto been purely accidental? Will the development of science eventually abandon its devious roads and follow a straight shortest path? We do not think so. But the discussion of this fascinating problem falls outside of the scope of this article.

The University of Chicago, Chicago, Illinois, U.S.A.

REFERENCES

1. Alfred J. Lotka, *Elements of Physical Biology*, Williams and Wilkins Co., Baltimore, 1925; reprinted under the title *Elements of Mathematical Biology*, Dover Publications, Inc., New York, 1956.
2. Vito Volterra, *Leçons sur la Théorie mathématique de la Lutte pour la Vie*, Gauthier-Villars, Paris, 1933.
3. J. H. Woodger, *The Axiomatic Method in Biology*, Cambridge University Press, Cambridge, London, 1937. See also his *Biology and Language*, Cambridge University Press, London, 1952.
4. H. Poincaré, *Science et Méthode*, E. Flammarion, Paris, 1909.
5. For a comprehensive summary of that work see N. Rashevsky's *Mathematical Biophysics* Vol. I and II, third edition, Dover Publications, Inc., New York, 1960.
6. N. Rashevsky, 'Topology and Life. In Search of General Mathematical Principles in Biology and Sociology', *The Bulletin of Mathematical Biophysics* 16 (1954) 317–348; also a number of subsequent papers in the *Bulletin*. See also reference 5, Volume II.
7. R. Rosen, 'The Representation of Biological Systems from the Standpoint of the Theory of Categories', *The Bulletin of Mathematical Biophysics* 20 (1958) 317–341.

COMPLEXITY AND ORGANIZATION

Living systems are quite generally spoken of as "complex" and "organized", and it may be assumed that these terms fulfil descriptive needs, at least at an intuitive level. Yet, upon examination in terms of physical science, they may appear vague and even conflicting. Complexity seems to imply multiplicity, and organization to suggest orderliness; in which case both might be expected to find description in terms of entropy. For our purposes the entropy of a system may be defined as the logarithm of the number of microscopic properties that have to be measured at a given instant [1] in order to describe the system, and so, in the above-connotation complexity ought to be associated with high entropy, and organization with low entropy.

Let me illustrate with two examples from non-living systems. In a gaseous system the molecules may occupy many positions in the system, and many internal states within themselves; so they have many microscopic properties to be described, and therefore high entropy. A crystal, on the other hand, has low entropy, since the molecules are arranged in a close spatial pattern and therefore can be described in terms of relatively few properties. May we not, then, think of the gaseous system as complex, the crystal as orderly, and if so is there not a conflict in the terms complex and organized if they are applied to non-living systems in the same way they are used in describing living systems?

The situation seems even more clouded when we consider some other terms commonly used in thermodynamics and in *information theory*. The terms randomness and disorder are associated with entropy; in the above examples it is clear that greater randomness and greater disorder are both contained in the expression, greater entropy. But it is also customary to associate entropy and probability, greater randomness and greater entropy being the more probable condition; thus an orderly system such as a crystal might seem less probable that a gaseous one, although intuitively we may think of a crystal as probable enough. The crystal is only to be regarded as having low entropy in a restricted sense; in wider view it is part of a larger system including the crystal and its environment, and this inclusive system has higher entropy than the crystal itself. To illustrate, let us think of the crystal at the moment of its formation from a super-saturated solution; in the process of crystal formation heat energy has to be lost to the environment, even though the reaction may go on under thermostated conditions, and the over all result will be an increase in entropy within the *[crystal – environment system]*. The second law of thermodynamics says that all real systems tend to go toward greater entropy, greater randomness, greater disorder, and greater probability, all these terms becoming synonymous when defined under comparable conditions. The act of "going" implies a movement in time, although no measurement in the time parameter is implied. A historical nuance creeps in. In the words of Eddington this law is "Time's Arrow"; although it is in no sense a quantitative measure of time.

If we are to compare the entropies of two systems it is clear that they must be measured in terms of some common extensive dimension. For example, we may make the comparison in terms of single molecules in which case we use the Bolzmann constant, or we may compare moles of molecules using the gas constant. This is common procedure in physics and chemistry. Theoretically it should be possible to measure the entropy of a living cell, at a given instant, in the same dimensions. There would be many microscopic properties to measure, but we might expect to find that the entropy content per mole would be less than that of a gas, although greater than that of a crystal. This would differ from part to part of the cell, the desoxyribose nucleic acid (DNA) of the chromosomes no doubt approaching the low entropy of a crystal. Would we from such a measurement decide that the cell was less probable than a gas, but more probable than a crystal? Obviously such a measurement would not help us much in our quest for an adequate description of living systems.

Can information theory help us with our problem? Let us think in terms of a machine of the digital computer type, which operates by choosing between two messages presented to it and recording the answer. By selecting the right answer this machine accumulates one binary unit or *bit* of information each time it chooses between two messages. By asking the machine the appropriate questions we ought to be able to describe all the microscopic properties of the system in terms of bits of information, and in this way measure the entropy, which would be equal to the number of bits. In accumulating this information the machine itself would have stored "negentropy". We might say that it has, locally, increased order – in doing this it would, of course, have to degrade a certain amount of energy thus increasing the entropy of the *[machine – environment system]*.

But is the mere ability to decrease entropy in a local region at the expense of increasing the entropy of the surroundings a sufficient index to describe an organized system? A crystal in forming from a supersaturated solution does a good job of decreasing entropy locally, yet I doubt that we would want to call the *[super-saturated solution – crystal system]* an organized one. Surely we would not want to draw very close analogy between such a system and a living organism. We might find closer analogy between the living organism and the computing machine.

The fact that the machine could, with adequate help, measure the total information content of a cell seems of no particular significance to us in the present instance. What is suggestive is that the machine itself has been able to collect and store negentropy, and we may ask if the cell too is able to collect, as negentropy, information about itself. The cell certainly has to do something of this kind in the course of replicating itself in exact image by the process of mitosis. It is clear that somewhere in this process the cell has to collect the requisite information to describe the pattern for the replication of its parts. That is, the cell contains some *genetic pattern*, according to which it reconstructs two images of itself (daughter cells), passing the pattern on to each. Collection of the information required for describing this pattern should be measurable in terms of negentropy, and this is presumably what is meant when the term "genetic information" is used. It is important, then, to disinguish between the total information content of the cell, or total entropy, and the transferable or reproducible information content which we refer to as genetic information, and which it would seem proper to measure as negentropy.

But negentropy – whether accumulated by a machine or by a living cell – is a number, and cannot be compared directly with the entropy of non-living systems in terms of extensive dimension. That is, we cannot use the Boltzmann constant or the gas constant to relate the amount of negentropy stored by the machine to that stored by the cell nor to the entropy of a gas or a crystal. [2] It would seem that the amount of negentropy stored can only be related – in our present state of knowledge at least – to the system that has stored it, whether this be a mechanical device or a living organism. Thus, there is no means of direct comparison; and a distinction between machines, cells, and their non-living environment, cannot be made on this basis.

A trouble with thermodynamics as a means of making such distinctions is that it is essentially quantitative and neglects the qualitative differences between the things it describes. Just as we have to supply the right questions to the machine if we are to get a meaningful answer, so we can only use the second law of thermodynamics for any specific purpose by supplying information which is not contained in that law itself. For example, if we want to measure the free energy or the entropy change of a chemical reaction we must specify the chemical reactants, the second law only giving general quantitative relationships which we have to apply in the particular case – when we speak of the thermodynamic properties of a substance we mean the way in which that specific substance will behave with regard to the laws of thermodynamics. Again, since there is only one position for the molecules in a perfect crystal at zero degrees absolute, and hence only one microscopic property to describe, the entropy at that temperature is zero; presumably the crystal might be composed of one of many kinds of molecules, but thermodynamics would not distinguish between them. The second law is notorious for its inability to distinguish between mirror-image optical isomers, although living systems have no trouble in doing so – the law has no geometrical content. So for all its power in description and analysis, the second law of thermodynamics is limited in what it can describe, and it should not surprise us to find it inadequate for the complete description of living systems, or for that matter, computing machines. This, of course, does not mean that either one disobeys the second law, as has been sometimes improperly claimed for living systems.

Elsewhere I have suggested[3] that the complete description of living systems entails a minimum of three functional properties. The thermodynamic or energetic property is one of these, the other two being a kinetic property and one of spatial pattern. While these three properties may be distinguished for analytic purposes, they cannot really be separated without destroying the "living" character of the system. The direction of the reactions that take place in the living organism are determined by the thermodynamic properties of the reactants; but their rates are determined largely by enzyme catalysts. Thus rate as well as free energy may determine which of those reactions possible may take precedence. By coupling of reactions, which again involves rates, some of the processes may go in direction opposite to that which would be predicted thermodynamically – with, however, no offense to the second law of thermodynamics in an overall sense. The genetic information that is stored by the cell in the process of its reproduction, has clearly a geometrical, spatial component, as is implied when we speak, by analogy, of a template for replication. This information we now believe to be contained, largely at least, in configurations of DNA molecules, which have, of course, a geometrical component. So the genetic information would not seem to be describable exclusively in terms of negentropy – which in itself is only a number. That is to say, although by asking it the appropriate questions the machine might be made to describe all of the genetic information of a cell, including spatial aspects, this description would in the end contain something in addition to a number of bits of information. That something we would ourselves have added in the formulation of the questions the machine was asked, just as in determining the free energy of a chemical reaction we have had to supply information regarding the kind of reactant molecules that were to take part. The thing we will have contributed would have had its thermodynamic cost including the free energy supplied by our own metabolism – an inclusive thermodynamic balance sheet would have to be drawn up in terms of the *[Man – machine system]*.

But supposing we can only describe the living organism in terms of no less than three kinds of properties – thermodynamic, kinetic and spatial. Does this set living systems apart from non-living ones, although made up of the same parts and obeying the same general laws? I suppose that upon analysis it would be found that a computing machine also has kinetic and spatial properties that have to be included in its description as well as thermodynamic properties. In this case, the measured negentropy alone could give only an incomplete description of the machine, just as it does of the living cell; that is, we ourselves would have to supply information in asking the machine questions about itself, even though we might conceive of a machine which could ask itself sufficient questions for its own reproduction.

Where shall we search, then, for the difference we intuitively feel to exist between the computing machine and the living organism? I am driven to the conclusion that the difference is to be sought principally in the

63

evolutionary history of the two systems – that is, it may be only on such a basis that a general distinction can be made. The machine is the product of Man, and hence its evolution is closely associated with his evolution. A man chipped out the first stone tool – which might be thought of as the progenitor of the machine – perhaps half a million years ago or more. The exact point in time is vague, and must remain so; but the history of Man and the machine have been related ever since. Of course, all biological evolution is back of the cultural evolution of Man, so we have in a way to include that too in the evolution of the machine. This is not the place, I think, to discuss the similarities and differences of the intimate mechanisms involved in the evolution of Man and machine, but only to point out the essential dependence of the latter upon the former. It is not necessary that these mechanisms be completely understood, I think, to maintain that the differences between a living organism and a computing machine, great as they are, may have to be described in any general picture, largely in terms of their respective evolutions. [4]

I am afraid I have said a good deal about "organization" in relation to machines and living organisms without arriving at any satisfactory definition of this term. I think, on the other hand, that the term "complexity" if given a strictly thermodynamic, numerical definition – as seems appropriate – is far from synonymous with organization, and even contrary in meaning. But, after all, my real object in this brief essay has not been the exact definition of words, but an attempt to explore in rather abstract, basic fashion, some major aspects of living systems.

National Cancer Institute, National Institutes of Health;
Princeton University, Princeton, New Jersey, U.S.A.

REFERENCES

1. There is a degree of uncertainty in the measurement of the entropy at any instant, but for simplicity this may be neglected in the present discussion. It may be more customary to regard the entropy as the average logarithm of the microscopic properties of the system taken over a brief period of time; but this introduces an uncertainty because the system may have to be considered as changing toward greater entropy as time proceeds.
2. This was pointed out by the late Dr. John Von Neumann in his Vanuxem lectures at Princeton University in 1953.
3. H. F. Blum, 'On the origin of self-replicating systems', *Rythmic and Synthetic Pro-*

64

cesses in Growth (ed. by D. Rudnick), Princeton University Press, Princeton, 1957
and 'On the Origin and Evolution of Living Machines', *American Scientist* **49** (1961)
474–501.
4. I may again refer to Von Neumann who in the same lecture cited above, entertained
the possibility of a machine that could duplicate itself, but expressed his inability to
imagine a machine that could create itself. I would refer this difficulty to our inability
to describe the multi-faceted evolutionary process that lies back of any machine, real
or imaginary, that we may conceive. See also H. F. Blum, 'On the Origin and Evolution of Human Culture', *American Scientist* **51** (1963) 32–47.

EVERT W. BETH

THE RELATIONSHIP BETWEEN FORMALISED
LANGUAGES AND NATURAL LANGUAGE

I have selected the present topic as it seems to be the one on which contemporary British Analytic Philosophy most strikingly disagrees with related currents in scientific philosophy both on the European Continent and in the United States. This choice does not imply any wish on my part to attack British Analytic Philosophy. In fact, my information about this movement is too fragmentary to justify any such intention. But I do hope that the following observations may elicit a fruitful discussion on some problems of major importance, the more so as the subject in itself is still highly controversial both in Europe and in America.

By many representatives of scientific philosophy the construction and further investigation of formalised languages are believed to present considerable interest from a purely philosophical point of view. I assume that this opinion is mainly based on two considerations. First, the study of formalised languages is usually expected to throw more light on the problems of the foundations of mathematics, a problem traditionally discussed in a philosophical context. Second, it is often expected to clear the way for the establishment of a truly scientific logic and thus for the introduction of scientific methods in the field of philosophical enquiry.

Both considerations seem to be abandoned by British Analytic Philosophy. As to the first, recent developments in the study of foundations increasingly emphasise the more "technical" aspects of its problems at the expense of their philosophical aspects; at the same time, philosophy tends to turn away from scientific to more "humane" problems. And as to the second consideration, it is maintained that philosophical analysis should not be based on investigations concerning formalised languages but rather on a study of concepts as expressed by means of natural language.

These tendencies in British Analytic Philosophy are typically illustrated by G. E. Moore's argument against Russell's theory of descriptions [1] and by Max Black's [2] and P. F. Strawson's [3] objections to Tarski's semantic definition of truth. If I should wish to attack British Analytic Philosophy, I might either try to follow Russell's example [4] and show that the notion of *common usage*, which plays such an important rôle in these discussions, is to a large extent fictitious, or I might join Y. Bar-Hillel [5] and J. Cohen [6] and try to adapt this notion to my own purpose.

However, as I said before, I have no wish to attack British Analytic Philosophy. Nevertheless I should like to suggest that its representatives tend to underestimate the philosophical importance of the construction and investigation of formalised languages and that a more solid, more powerful, and altogether richer philosophy could be established if their concern with concepts as expressed by means of natural language were combined with an equally vivid interest in the results which can be obtained by a study of formalised languages.

In other words, I intend to re-examine the two above-mentioned considerations in the light of my personal experience in the investigation of formalised languages. In particular, I wish to discuss the following points [7]:

(i)   The unification of pure mathematics;
(ii)   The discovery of the paradoxes and Hilbert's formalism;
(iii)   Formal calculi and formalised languages;
(iv)   Syntax and semantics;
(v)   Formalised languages and natural language;
(vi)   Natural language as a means of expression for deductive theories;
(vii)   Metalogic and linguistics.

On the basis of this preliminary discussion, I hope to be able to submit a few remarks on some problems in philosophical analysis.

In its earlier stages of development, pure mathematics was concerned with objects of various kinds: with *figures* in geometry, with *numbers* in arithmetic, with *variable magnitudes* and with *functions* in analysis. But Fermat and Descartes already pointed out, in their analytic geometry, how figures could be reduced to (real) numbers. Further progress in this direction was made during the nineteenth century, in spite of the expansion of the domain of mathematics which took place at the same time. The last steps were made possible by Frege's *logicism* and by Cantor's *theory of sets*. Thus in 1903 Russell [8] could proclaim the unification of pure mathematics as an accomplished fact: "... all pure mathematics deals exclusively with concepts definable in terms of a very small number of fundamental . . . concepts, and . . . its propositions are deducible from a very small number of fundamental . . . principles . . . ."

In as much as pure mathematics applies a deductive method, there is only one manner in which this claim can be established. The following steps are required:

(i) enumeration of the fundamental concepts;

(ii) enumeration of the fundamental principles;

(for the sake of brevity it will be convenient to refer to these fundamental concepts and principles of pure mathematics as the *absolute concepts* and *principles*; actually, various systems of absolute concepts and principles are available at present;)

(iii) definition of the specific notions of the various specialised domains of pure mathematics in terms of the absolute concepts;

(iv) proof of the specific axioms of these various domains on the basis of the absolute principles and of the definitions of their respective specific notions.

An enumeration of absolute concepts and principles must be based on an extensive and profound analysis of extant pure mathematics; in this respect Russell could take advantage of previous work by Peirce, Dedekind, Frege, Cantor, and Peano. Once the necessary absolute concepts and principles have been made available, however, the acceptance of a unified treatment of pure mathematics depends entirely on the lucidity of the definitions under (iii) and on the rigour of the proofs under (iv). Now it proves impossible to attain the level of lucidity and rigour, which is needed here, if we use natural language as a means of expression. Therefore, as realised already by Frege, unless we are ready to abandon every hope of achieving a unified treatment of pure mathematics we have to construct a suitable formalised language and to use it as a means of expression.

At this point, two questions may be raised, namely:

(i) Are the objections to natural language as a means of expression in a unified treatment of pure mathematics not exaggerated?

(ii) Are the definitions and proofs, which can be formulated by means of a suitable formalised language, indeed more lucid and more rigorous than those which could be stated by means of natural language?

It is relatively easy to answer these questions, as both Frege [9] and

Russell [10] have published, besides the "official" statement of their arguments by means of a formalised language, an expository treatment of the same subject in which natural language is used. Now Frege, although somewhat dull, is certainly a lucid and penetrating stylist, while Russell is both brilliant and lucid. Nevertheless, in both cases the expository treatment is difficult to follow and unconvincing, unless it is used as an explanation to the formal argument. But if we study side by side the formal argument and the expository treatment, then no doubt will remain as to the superiority and the conclusive character of the first.

It is well known that the results of Frege, Cantor, and Russell were soon severely menaced by the discovery of a series of paradoxes. If Frege and Russell had been satisfied with using natural language, this discovery might have appeared less troublesome. In a formalised language, however, the emergence of a paradox must be taken much more seriously; this shows once more the far-reaching implications of the introduction of formalised languages.

In order to eliminate the paradoxes, a painstaking revision is required of the system of absolute principles and of the formalised language which is used as a means of expression. The results of such a revision, based on Russell's *theory of logical types* [11], were published in 1910–1913 by Whitehead and Russell. [12]

In spite of numerous attempts, nobody has so far been able to derive a paradox on the basis of *Principia Mathematica* and it seems rather unlikely that future attempts should prove more successful. However, Russell never actually *proved* that the safety measures implied by his theory of types provide complete protection. In fact, his approach to the problem could hardly be expected to lead to such a proof.

Hilbert's *formalism*, on the other hand, seems to be more promising in this respect. In order to understand Hilbert's approach, it will be helpful to return to the considerations by which Frege und Russell had been compelled to take the paradoxes so seriously.

A formal proof in a formalised language results from the application of certain *rules of inference*, the statement of which is one of the most essential steps in the construction of such a language. These rules of inference have a *formal* character, that is, they can be stated by means of purely "typographical" terms, without any reference to an interpretation

69

of the symbols involved. As an illustration, I mention the *modus ponens*, which can be stated as follows:

From two premisses U and U → V, draw the conclusion V.

Once the rules of inference are stated, we know exactly, which proofs are admissible within the formalised language under consideration and which are not. Therefore, if we are faced with a proof of a paradox (that is, of two contradictory sentences U and $\bar{U}$) within that formalised language, our only reaction will consist in checking whether each step in the proof has been made in accordance with the rules of inference. If one single step is discovered which is not based on some rule of inference, we are in a position to reject the proof. But if each step turns out to be an application of some rule of inference, no way of escape is left; the paradox must be taken to be proved, however strange the argument may appear with a view to the interpretation of the symbols involved.

The problem as to, whether or not, the rules of inference for a given formalised language permit us to prove a paradox, is strikingly similar to a chess problem. The given initial position of the pieces corresponds to the absolute principles, the rules for moving the pieces to the rules of inference, and the final position of the pieces, to the conclusion of the proof. The question as to whether or not a paradox can be proved corresponds to the question as to whether or not a certain position of the pieces can be achieved.

There can be no doubt that for Frege and Russell the *symbols* appearing in a formalised language are substitutes for *words* and thus are, in the same way as the words in a natural language, endowed with a certain definite *meaning*. In this sense Russell [13] writes, for instance: "The doctrines just mentioned are, in my opinion, quite indispensible to any even tolerably satisfactory philosophy of mathematics, as the following pages will show. . . . the fact that they allow mathematics to be true, which most current philosophies do not, is surely a powerful argument in their favour."

Therefore, such symbol systems as constructed by Frege and Russell can serve as a means of expression for *thought*, and thus may be referred to as *languages*. They differ from natural languages in two important respects: (1) their internal structure admits of a precise and exhaustive description, and (ii) their means of expression are strictly limited. This difference we take into account by referring to these symbol systems as *formalised* languages.

One might believe that by adopting Hilbert's formalistic approach we do not change the situation essentially. The fact that in the context of a proof of formal consistency we may forget about the meaning of the symbols does not imply that the symbols do not or cannot have meanings.

Nevertheless, if we look more closely at the investigations of Hilbert and his school, a shift of interest becomes apparent. In order to establish the formal consistency of a formalised language L, we may proceed as follows. We introduce a system L' having a more simple internal structure; we prove:

(i) if L is inconsistent, then L' is also inconsistent;

(ii) L' is consistent; it then follows that L must be consistent.

It will be clear that in this argument the meaning of the symbols which appear in L' may be entirely irrelevant. It is by no means necessary that any reasonable meaning can be associated with the symbols in L'. Thus by adopting Hilbert's formalistic approach we are compelled to make allowance for the construction of symbol systems L' which cannot be considered as formalised languages in the above sense.

Therefore, it is convenient to adopt the following terminology. A symbol system L whose internal structure admits of a precise and exhaustive description (on similar lines as the internal structure of a formalised language), but whose symbols either have a meaning which is left out of consideration or do not allow of any reasonable meaning, is called a *formal calculus*. If with the symbols of a formal calculus L a definite meaning has been associated, then L will be called a *formalised language*.

The above conceptions call for the development of two domains of science:

(i) In the first domain we discuss the internal structure of formal calculi as well as certain notions which are used in describing these structures. This domain, which has developed from Hilbert's metamathematics, has been systematised by Carnap under the name of *logical syntax*. [14]

(ii) In the second domain we discuss the meanings associated with the symbols in formalised languages, as well as the notions which are used in characterising these meanings and in describing the relations between the symbols and their meanings. It has been shown by A. Tarski that this subject can be treated in a precise and rigorous manner; he thus created a new domain which is known as *logical semantics*. [15]

If the conceptions of Frege and Russell are taken to the letter, then it seems that we ought altogether to abandon natural language as a means of expression for pure mathematics and to adopt instead a suitable formalised language. In fact, Frege and Russell have restated certain rather extensive parts of pure mathematics by means of formalised languages, but in later years, although numerous formalised languages have been constructed and studied, they are seldom actually used in a more or less systematic presentation of a mathematical theory. This development can be explained as follows:

(i) Proofs expressed by means of a formalised language are nearly always long and tedious; in addition, they are hard to follow and difficult to print.

(ii) Generally speaking, we are not anxious actually to see the proofs of certain theorems as expressed by means of a given formalised language L; we are satisfied if we have sufficient evidence to show that L provides a basis for such proofs.

(iii) Certain methods developed in Hilbert's metamathematics sometimes enable us, instead of proving theorems one by one, to show that all theorems of a certain type are provable in L.

(iv) Thanks to an experience of many years, logicians know that certain parts of the proof of certain theorems are responsible for most of the trouble encountered in an attempt to express pure mathematics by means of a formalised language.

Therefore, if we wish to show that a certain formalised language L provides an adequate means of expression for pure mathematics, we usually do not take the trouble of actually proving a great many theorems, starting from the absolute principles. We restrict ourselves to pointing out that the crucial parts of certain proofs, as meant under (iv), can be carried out in L. This task is sometimes simplified if we first establish a general result on provability, as suggested under (iii).

The fact that we have become more and more familiar with proofs as expressed by means of formalised languages has still another effect. In many cases proofs are stated by means of natural language, but in such a manner that their "translation" into a formalised language is only a matter of routine. In fact, natural language, as used in mathematical proofs, shows signs of being influenced by formalised language.

Although this trend contributes to the tendency to avoid or restrict actual use of formalised languages, it does not eliminate the above essential objections to natural language as a means of expression for pure mathematics. Therefore, further construction and study of formalised languages is not made superfluous.

Modern logic, as a study of formalised languages (and, to a certain extent, also of natural language), is exclusively concerned with language as a means of expression for pure mathematics (or, more generally, for deductive theories). As I have never seen any cogent argument to the contrary, I take it for granted that for this specific purpose formalised languages surpass natural language to an extent which justifies the modern logician's predominant concern with formalised languages. Nevertheless, there are also facts which call for a certain concern with natural language as well, namely:

(i) formalised languages always appear embedded, so to speak, in natural language;

(ii) natural language can, under certain conditions and to a certain extent, be used as a means of expression for deductive theories.

Nevertheless, I am convinced that in a logical analysis of natural language (as a means of expression for deductive theories) we ought to start from the insights obtained as a result of the construction and the study of formalised languages. It may be true that in this manner no completely satisfactory results will ever be obtained, but this is due to the fact that natural language cannot be fully adapted to the purpose of serving as a means of expression for deductive theories.

There remains of course a great variety of purposes for which natural language does provide an adequate tool. It is the specific task of *general linguistics* to find out how natural language manages to serve all its different purposes so well. Although in this context the purpose of serving as a means of expression for deductive theories is of very slight importance (in fact, most languages have never actually served this purpose), it is interesting to see that certain recent trends in general linguistics show a definite affinity with, and perhaps even an influence from, modern logic and, in particular, logical syntax; I mention *structural linguistics* (Zellig S. Harris) and *glossematics* (L. Hjelmslev) without, of course, passing a value judgement on these doctrines.

We may also anticipate that the current practice of using natural language as a substitute for formalised languages will have a certain influence on future developments in natural language itself.

I should like now to discuss a few concrete problems and, in the first place, the distinction between *analytic* and *synthetic* sentences (or propositions). In this connection, it may be interesting if I quote a fragment of a platonic dialogue which was recently discovered.

Socrates: Mother, are all bachelors unmarried?

Mother: Yes, my dear.

Socrates: Aren't they allowed to marry?

Mother: Yes, they are, but when they are married they are no longer bachelors.

Socrates: But why, Mother?

Mother: That is in the law, for otherwise they would have to pay those high income taxes. But now you must stop asking questions, for I have to go to a confinement.

I do not think that the way in which such matters are discussed on the level of everyday discourse has very much changed since Socrates was a little boy. Therefore, let us now turn to what many representatives of modern logic would consider a much more promising approach; let us set up a deductive theory *T* which involves the term under consideration.

The primitive notion of our theory *T* will be *married*, and its "universe of discourse" will consist of all grown-up *male persons*. For the moment, no axiom will be introduced, but we state:

Definition 1. A bachelor is an unmarried person.

On the basis of this definition, we can prove:

Theorem 1. All bachelors are unmarried.

In this manner, the statement in Theorem 1 receives a clear logical status: it is *provable* in the theory *T*. However, I do not believe that this fact provides us with an adequate solution of Socrates' problem, even though the reply which it suggests may be better than the answer given by his mother.

I think that we can describe the situation correctly if we characterise the statement under consideration as *analytic* in the sense that its logical status is similar to that of a theorem in a deductive theory. If Socrates' mother were right, then the statement should rather be called *synthetic*. (As the introduction of these notions is usually ascribed to Kant, I may observe that he used the terms in a quite different sense. Kant considered as synthetic sentences those which are based on a deductive procedure, and as analytic sentences those which are not.[16] However, I do not wish to dwell upon this matter as we are concerned with the notions as they appear in contemporary discussions.) I do not wish to go more deeply into the distinction between analytic and synthetic sentences, but I should like to make a remark which is also helpful in other connections.

In the usage of natural language we can make a distinction between three different levels, namely:

(i) A low level, where assertions are made with complete indifference as to their meaning, their truth, and their foundation.

(ii) An intermediate level, where speakers do care about these requirements, but are unable to give relevant explanations, if interrogated.

(iii) A higher level, where speakers are able to present reasonable (although not necessarily correct) explanations concerning the meaning, the truth, and the foundations of the assertions which they make.

Level (i) is characterised, not so much by deliberate abuse of language in lying or scolding, as rather by mere paroting[17]; on this level such sentences as: "all bachelors are unmarried" are used as a kind of proverb.[18] Level (ii) is typically represented by Socrates' mother and level (iii) by every competent schoolmaster.

On each level, our problems take a different shape and the situation is made still more involved by the existence of still higher levels by which the levels (i)-(iii) are influenced. On these higher levels, definitions are used to specify extant word-meanings, to introduce new word-meanings, and even to create new words. In scientific language, even though it is not necessarily formalised, definitions play very much the same rôle as in formalised languages where their application can be very accurately described. This is important, as often word-usage on the higher levels penetrates to lower levels.

I think that the notion of *truth* provides a case in point. It has been argued by P. F. Strawson[19] that the semantic theory of truth can be accepted only as a description of a certain technical usage of the word "true" (with reference to sentences in formalised languages) and that it is not valid with respect to the actual use of the word "true" and thus cannot be relevant in connection with the philosophical problem of truth.

Strawson thus makes a distinction between the technical and the non-technical use of the word "true". I believe that we should make a distinction between two kinds of non-technical use: the use which has been interpreted (rightly or wrongly) as semantic use and which I shall denote as *emphatic use*, and the use which is clearly non-semantic and which I shall denote as *casual use*.

Strawson's argument seems to rest on the view that casual use, exemplified by such phrases as "That's true" and "It's true that . . . but . . .", is the *normal* use of the word "true", and that its emphatic use constitutes, so to speak, an incidental shift from its normal use. The semantic theory then wrongly starts from the assumption that emphatic use is the normal case; therefore it only considers those sentences which present emphatic use of the word "true" and on the basis of an analysis of these sentences establishes rules for using this word by which its casual use is excluded or mis-interpreted.

Now there is a striking fact which makes this whole construction extremely unplausible. Tarski's paper in which the semantic theory of truth was first developed appeared originally in Polish (1933), then in German (1945), and recently in English (1956).[20] If the main word involved in this investigation had been taken in a somehow shifted meaning, one might expect considerable difficulties in its translation, manifesting themselves, for instance, in footnotes justifying the manner in which the word was rendered. However, it is obvious from the texts that in this respect the translators did not encounter any difficulty. Tarski's original *Pojęcie prawdy* appears as *Wahrheitsbegriff* and as *Concept of Truth*, without any further comment. So it is clear that the word "true", in its emphatic use, has exact equivalents both in Polish and in German (and in many other languages as well). In its casual use, on the other hand, the word "true" is often rather difficult to translate. [21]

In my opinion, this fact alone would already be sufficient to compel us to consider the emphatic use of the word "true" (and of its various equivalents in other languages) as normal, and to explain its casual use as derivative.

This view is further corroborated by historical facts. The words involved are of different origin, and they seem to have acquired their present meaning only recently. It seems rather likely that they obtained this meaning when they were used to render the Greek word ἀληθής and its derivates which frequently appear in the New Testament, where they must undoubtedly be interpreted in accordance with emphatic use. The casual use of these words constitutes a later development. [22]

Furthermore I wish to show that the semantic theory provides a correct description of the emphatic use of the word "true". It does not matter that Strawson's account of this theory is not very accurate. For, with a view to our present discussion, we may agree to denote as a semantic theory of truth every conception which, if applied to sentences in natural language, leads to the Liar Paradox.

In order to prove this contention, I have to show that such a conception is indeed accepted by those people who use and understand the word "true" in its emphatic sense. As a criterion for the acceptance of such a conception, we may take the awareness of these people that their way of using and understanding the word "true" leads to the Liar Paradox [23]; for it is easy to see that for a man who uses and understands the word differently the Liar Paradox will be completely pointless.

Now there is abundant historical material to show the intensity of this awareness among philosophers during a period which, roughly speaking, starts with Plato and has not yet reached its end [24]; and thus I feel that my contention has been conclusively established.

The last topic which I should like to discuss is the so-called *"theory of meaning"*. It has been observed by Whitehead [25] that "European philosophy is founded upon Plato's dialogues", a remark which is strikingly illustrated by another fragment recently discovered.

Socrates: Father, are all priests unmarried?

Father: Yes, my boy.

Socrates: Aren't they allowed to marry?

Father: Yes, they are, but when they are married they are no longer priests.

Socrates: Is that in the law, Father?

Father: Oh no! An unmarried clergyman is called a priest, and a married clergyman is called a minister. It is just a matter of changing names. If an army captain is transferred to the navy he will be called a commander.

Socrates' father seems to adhere to what G. Ryle [26] has described as the *'Fido'-Fido principle*. This principle expresses the view "that to mean is to denote, in the toughest sense, namely that all significant expressions are proper names, and what they are the names of are what the expressions signify", or "that to every significant grammatical subject there must correspond an appropriate *denotatum* in the way in which Fido answers to the name 'Fido'". Now, as Ryle observes, "the word 'dog', if assumed to denote in the way in which 'Fido' denotes Fido, must denote something which we do not hear barking, namely either the set or class of all actual and imaginable dogs, or the set of canine properties which they all share. Either would be a very out-of-the-way sort of entity".

I believe that here we meet with a situation where our attitude with respect to natural language may differ from that with respect to formalised languages. Let us first consider a formalised language T constructed in accordance with Russell's theory of logical types; let MT be the corresponding metalanguage.

In T we can define a certain class N to be the set of all natural numbers. Then we can make in MT the following statement:

The symbol 'N' of T denotes the set of all natural numbers.

Now certainly the set of all natural numbers is at least as much of an out-of-the-way sort of entity as the set of all actual and imaginable dogs. However, if we agree to discuss such entities by means of the object-language T, why should we refuse to discuss them in the metalanguage MT? And, on the other hand, if we succeed in giving a nominalistic interpretation of the theories expressed by means of the object-language T, we may expect this interpretation to apply as well to theories expressed by means of the metalanguage MT. [27]

So it seems rather obvious that, guided by the 'Fido'-Fido principle, we can establish a relatively simple and lucid theory of meaning for the language T and also for other formalised languages.

Let us now turn to natural language. There is a long experience to show that if we wish to discuss the above out-of-the-way sort of entity by means of natural language, we must be prepared for all kinds of trouble. Hence, if the 'Fido'-Fido principle does not provide a suitable basis for a theory of meaning for natural language, this may be due not so much to this principle as rather to the semantic structure of natural language itself. In other words, it remains to be seen whether natural language admits at all of a reasonable theory of meaning.

Although the above remarks were concerned with concrete problems, they seem to suggest a general conclusion. The analysis of notions and beliefs as expressed in everyday life by means of natural language can be of considerable importance to philosophy if it is combined with investigations on scientific ideas and theories as expressed by means of scientific language, formalised or not, but it will inevitably tend to become philosophically sterile and even seriously misleading if it is detached from or even opposed to such investigations and, in particular, if it is conducted in a spirit which is foreign or even hostile to scientific thought.

Therefore, I may express the hope that Analytic Philosophy may find ways and means to establish a more cordial understanding with other currents in the continually expanding domain of scientific philosophy.

University of Amsterdam, Amsterdam, The Netherlands.

BIBLIOGRAPHY

Bar-Hillel, Y. (1), 'Analysis of "Correct Language" ', *Mind* **55** (1946).

Beth, E. W. (1), 'Critical Epochs in the Development of the Theory of Science', *The British Journal for the Philosophy of Science* **1** (1950).

Beth, E. W. (2), *La crise de la raison et la logique*, Paris – Louvain, 1957.

Beth, E. W. (3), 'Über Lockes "Algemeines Dreieck"', *Kant-Studien* **48** (1957).

Beth, E. W. (4), *The Foundations of Mathematics*, Amsterdam, 1959.

Black, M. (1), 'The Semantic Definition of Truth', *Analysis* **8** (1948).

Carnap, R. (1), *The Logical Syntax of Language*, New York – London, 1937.

Carnap, R. (2), *Meaning and Necessity*, Chicago, Ill., 1947.

Carnap, R. (3), 'Empiricism, Semantics, and Ontology', *Revue Internationale de Philosophie* **4** (1950).

Cohen, J. (1), 'Mr. Strawson's Analysis of Truth', *Analysis* **10** (1950).

Dugas, L. (1), *Le psittacisme et la pensée symbolique – Psychologie du nominalisme,* Paris, 1896.

Frege, G. (1), *Die Grundlagen der Arithmetik*, Breslau, 1884.

Greenwood, D. (1), *Truth and Meaning*, New York, 1957.

Leibniz, G. W. (1), 'Nouveaux essais sur l'entendement humain' in R. E. Raspe (ed.), *Oeuvres philosophiques de Leibnitz*, Amsterdam – Leipzig, 1765.

Mannoury, G. (1), *Les deux pôles de l'esprit*, Paris, 1933.

Mannoury, G. (1), *Les fondements psycho-linguistiques des mathématiques*, Bussum – Neuchâtel, 1947.

Moore, G. E. (1), 'Russell's "Theory of Descriptions"' in P. A. Schilpp (ed.), *The Philosophy of Bertrand Russell*, Evanston – Chicago, 1944.

Russell, B. (1), *The Principles of Mathematics*, 2nd. ed. London – New York, 1938.

Russell, B. (2), *Introduction to Mathematical Philosophy*, London, 1919.

Russell, B. (3), 'The Cult of "Common Usage"', *The British Journal of Philosophy* 3 (1952).

Rüstow, A. (1), *Der Lügner*, Erlangen, 1908; Leipzig, 1910.

Ryle, G. (1), 'The Theory of Meaning' in C. A. Mace (ed.), *British Philosophy in the Mid-Century*, London, 1957.

Stegmüller, W. (1), *Das Wahrheitsproblem und die Idee der Semantik*, Wien, 1957.

Strawson, P. F. (1), 'Truth', *Analysis* 9 (1949).

Tarski, A. (1a), *Pojęcie prawdy w językach nauk dedukcyjnych*, Warszawa, 1933.

Tarski, A. (1b), 'Der Wahrheitsbegriff in den formalisierten Sprachen', *Studia philosophica* 1 (1936).

Tarski, A. (1c), 'The Concept of Truth in Formalized Languages' in *Logic, Semantics, Metamathematics*, translated by J. H. Woodger, Oxford, 1956.

Tarski, A. (2), 'The Semantic Conception of Truth and the Foundations of Semantics', *Philosophy and Phenomenological Research* 4 (1944).

Whitehead, A. N. (1), *Adventures of Ideas*, Harmondsworth, 1948.

Whitehead, A. N. and Russell, B. (1), *Principia Mathematica*, 2nd. ed., Cambridge, 1947.

REFERENCES

1. Moore [1].
2. Black [1].
3. Strawson [1].
4. Russell [3].
5. Bar-Hillel [1].
6. Cohen [1].
7. A more detailed discussion is given in Beth [4].
8. Russell [1], p. xv.
9. Frege [1].
10. Russell [2].
11. Russell [1], Appendix B.
12. Whitehead and Russell [1].
13. Russell [1], p. xviii.
14. Carnap [1].
15. Tarski [2], Carnap [2].
16. Beth [1]–[3].
17. Leibniz [1], II, xxi, 31; *cf.* Dugas [1], p. 24.
18. Mannoury [1] and [2] – I take this opportunity to draw the attention on the fact that in many respects Mannoury's significs is closely related to British Analytic Philosophy.
19. Strawson [1]; *cf.* Cohen [1].
20. Tarski [1a–c].
21. Stegmüller [1], p. 232.

22. I wish to stress that at the present moment this statement cannot pretend to be more than a hypothesis, which seems to me rather plausible but which is still in need of precise and detailed verification. – On the other hand, the following conclusions concerning the semantic theory of truth do not depend on this hypothesis.
23. Tarski [1a–c], Strawson [1], Stegmüller [1].
24. Rüstow [1]; cf. Beth [4]
25. Whitehead [1], p. 265.
26. Ryle [1]. – Ryle's discussion of the 'Fido'-Fido principle originally appeared in the context of a review of Carnap [2], which was answered in Carnap [3]; cf. Greenwood [1], Stegmüller [1].
27. For a more detailed discussion of the possibility of a nominalistic interpretation of certain formal languages whose structure suggests a strongly platonistic inspiration, cf. Beth [3] and [4].

ROBERT ROGERS

A SURVEY OF FORMAL SEMANTICS

My purpose in this paper is to present an account of certain of the principal results that have been obtained in the field of formal semantics. These results will be stated in what from the logician's point of view is an informal way; that is, in stating them there will be little or no use of a precisely formalized language (though we shall of course speak about such languages). These results will, however, be stated as precisely as is possible within a non-formalized language. Taken together, my statements of these results will constitute what I believe to be a representative survey of what has been accomplished within the field of formal semantics.

The paper will be divided into the following four parts: First, an introductory section in which I give a definition of the term "formal semantics", together with a statement of certain of the leading concepts and distinctions within formal semantics. Second, I shall give statements of two different basic concepts in terms of which much of semantical theory can be developed, and show how the concept of truth can be defined in terms of each of these concepts. These will be the concept of satisfaction, as developed by Alfred Tarski, and the concept of multiple denotation, as recently developed by Richard Martin. Third, I shall present two further concepts in terms of which the so-called *L-concepts*, viz., analyticity and its related concepts, can be defined. These will be the concept of a state-description, as developed by Rudolf Carnap, and the concept of an interpretation, as developed by John G. Kemeny.

Fourth, and finally, I shall present what certain writers have taken to be the significance of formal semantics for problems of ontology.

I

As the term "formal semantics" is understood by those writers I am directing my attention to in this paper, it is the name of a certain kind of systematic inquiry into the problems of meaning and interpretation. More specifically, it is concerned with the problems of meaning in the sense of *cognitive*, or *declarative* meaning, as contrasted with such other types of meaning as emotive or exhortative meaning. Roughly speaking, a *semantical theory* is a theory that provides us with a set of concepts by means of which we can give an account of the meaning of statements, or whole bodies of statements, and of the terms appearing within them. Now a *formalized semantical theory* is one that proceeds with respect to a formalized language. Within such a theory, the semantical problems and concepts being dealt with are relativized to some well-defined language, called the *object-language*. A distinction is made between this language and the language in which the semantical analyses are being carried out, this latter language being called the *meta-language* with respect to that given object-language. The object-language is always an example of what is called a *formalized language*, while the meta-language may or may not itself be a formalized language.

Before I give an account of the nature of formalized languages, let me state why it is that formal semantics has come to use such languages. Principally, it is because of the clarity and precision that are made possible once one relativizes the problems of semantics to such languages. There is about them, and results based upon the use of them, a kind of definiteness which it is impossible to obtain when one is working with an ordinary language. And because, though formalized, such languages have much in common with ordinary languages, and can be made successively to approximate such languages in power of expression, semantic analyses carried out with respect to formalized languages are of interest not only to students of such languages, but also to those who are especially interested in the semantics of ordinary, unformalized languages.

There is in addition a technical type of consideration which argues strongly for the use of formalized languages. This concerns the so-called *semantical paradoxes*. There are a number of such paradoxes known, each of which seems to lead to the impossible conclusion that a contradiction is true. These paradoxes seem to be forced upon us once we decide to carry out semantical investigations within an ordinary language. By carefully setting up a distinction between a formalized object-language and its meta-language, however, we can avoid the known paradoxes. Let me briefly state one such paradox, a version of the well-known paradox of the liar.

Consider the following sentence:

The sentence on page 84, line 1, is not true.

Clearly this sentence is true if and only if the sentence on page 84, line 1, is not true. But the sentence on page 84, line 1, is just this sentence itself. Thus we conclude that the sentence in question is true if and only if it is not true; that is, that it is both true and not true, both true and false. But this is a contradiction. Yet it seems that every step in the above argument is an admissible step within the logic of the ordinary English language. Until that language is more precisely specified than it has been heretofore, the conclusion seems inescapable that anyone who wishes to develop semantical theory within it is committed to this type of contradiction. As we shall illustrate later, however, the semantical paradoxes known to date can be successfully avoided if one has recourse to properly developed formalized languages.

Let us now give an account of what a formalized language is. In order to do this, we make use of the distinction between an *object-language* and a *meta-language*. The language being formalized, and thus being *talked about*, is called the *object-language*; the language within which we formalize this language, that is, the language we *use* for this purpose, is called the *meta-language*. Let us call the object-language "*L*", and the meta-language "*M*". In order to formalize *L*, within *M* we first lay down rules which determine the *syntax* of *L*; and then rules which determine the *semantics* of *L*, and thus provide an interpretation of *L*. As the first of our syntactical rules, we give a full specification of the *primitive signs* of *L*, that is, its alphabet and punctuation signs. Such a rule provides first for those signs that are to be used as punctuation signs and logical connectives; e.g., parentheses, a comma, a sign to play the role of negation, and a sign to play the role of material implication. This rule also specifies which signs are to be used as subject matter variables and constants. Then a *formation rule* is introduced, for the purpose of defining which combinations of these primitive signs of *L* are accepted as well-formed formulas of *L*. Usually this rule is in the form of a recursive definition. When it takes this form, first the most elementary or atomic type of wffs (we use 'wff' as an abbreviation of "well-formed formula") are given a general characterization, and then the various means by which the more complex types of wffs are built up out of simpler wffs are enumerated. Next, definitions of bound and free occurrences of variables are given, and those wffs of *L* (if any) which contain no free occurrences of variables are identified as the *sentences* of *L*. Finally, among one's list of rules determining the syntax of *L* one may include a characterization of certain of the wffs of *L* as *axioms*, and then specify *rules of inference* by means of which one may legitimately infer certain wffs from others. And one may also include rules which permit us to abbreviate certain wffs in certain ways.

If we add no more rules than these syntactical rules, our language *L* is not yet really a language at all, but merely what is called a "calculus", or an uninterpreted system. In order that *L* be truly a *language*, we shall have to provide an interpretation for *L*. This we do by laying down certain *semantical rules* for *L* in our meta- language *M*. There are various ways of doing this, and I shall examine a number of them later, in our next section. Still, each of these ways presupposes that our meta-language *M* has a number of certain well-defined features, and I specify these now. Within *M* we are to give an interpretation of *L*. To do this, within *M* we must first be able to talk about any and all expressions of *L*. We must therefore have within *M* a way of forming *names* for any and all expressions of *L*. The usual means of accomplishing this end is to include within *M* specific names for each of the primitive symbols of *L*, together with an operation sign for concatenation. By putting the concatenation sign between any two signs of *M* which stand for given expressions *a* and *b* of *L*, we form an expression in *M* which stands for the result in *L* of putting the two expressions *a* and *b* together. For example, let "*P*" and "*x*" be two primitive symbols of *L*; let "*pee*" and "*ex*" be their respective names in *M*; and let "$\frown$" be the sign in *M* for concatenation. Then "*pee$\frown$ex*" is an expression in *M* which names the expression "*Px*" of *L*. With such means as I have just described at our disposal, we are able in *M* to refer to, or speak about, all of the expressions within *L*, since all such expressions are formed by concatenating together a finite number of the primitive symbols of *L*.

Second, if we are to interpret the expressions of *L*, we must not only be able to speak about them, but must also be able to speak about whatever they are speaking about. That is, we must be able to say within *M* whatever can be said within *L*; that is, *M* must contain a translation of each of the meaningful expressions of *L*. And, as a final requirement for giving an interpretation of *L*, we must have within *M* some way of relating the expressions of *L* to whatever they are about. This we do by making use of some *specifically semantical term* or *terms*. It is principally at this point that there is considerable room for choice as to how the semantics of *L* is to be given. There are a number of specifically semantical concepts which are known to be adequate to the job. I shall examine some of these in my next section, and show how each of them suffices for the purpose of interpreting an object-language *L*. And I shall show how each of them permits us to give an analysis of the semantical concept of truth. Once we have an analysis of this concept, we are in a position to give an interpretation to each of the sentences of *L*. For with respect to each of the sentences *S* of *L*, we can form within *M* the following sentence: "_____ is true in *L* if and only if...," where in the position marked "_____" we put the name of *S*, and in the position marked "..." we put the translation of *S* into the language *M*. Such a sentence will constitute an interpretation of *S* within *M*, for it tells us within *M* the conditions under which *S* is true. And it provides this interpretation without using any vaguely-defined terms, such as the term "meaning"; in place of such terms, only the precisely-defined and (supposedly) contradiction-free term "true in *L*" is used. Thus it is easy to see why one of the most important tasks of all in interpreting a formalized language *L* is the task of defining within its meta-language the concept of truth in *L*.

II

There are essentially two different ways in which one can introduce semantical terms into a meta-language. First, one may introduce such terms by *defining* them all in terms of the specifically non-semantical terms already available in one's meta-language *M*. These latter terms fall into three distinct groups: (1) the *logical vocabulary* of *M*, including the logical constants (e.g., "not", "or", "all", etc.), and variables; (2) the *syntactical vocabulary*, including names of each of the primitive symbols of the object-language *L*, a sign for concatenation, and syntactical variables ranging over the expressions of *L*; and finally (3) the *translation vocabulary*, which must permit us to translate into *M* all the meaningful expressions of *L*. This means of introducing semantical terms into a meta-language by way of definition was first developed by Tarski, and is probably the most frequently used method for introducing such terms within formal semantics. A strong argument in favor of it is that by defining all one's semantical terms exclusively in terms of non-semantical terms, one has a kind of guarantee that the paradoxes associated with the use of semantical terms will not appear in one's meta-language – supposing, of course, that these paradoxes are not already present in some form in the non-semantical part of that language.

A second way in which one might introduce semantical terms into M is one in which we do not define all such terms in the manner discussed above, but introduce certain of them as *undefined terms* of M, and then lay down axioms governing these terms. We then define the remainder of our semantical terms by means of these primitive terms. Such a procedure is comparable to Peano's treatment of arithmetical terms in his axiomatization of arithmetic, whereas the former procedure is comparable to the manner in which Frege and Russell introduce arithmetical terms into logic; viz., by means of definitional analyses which reduce such terms to terms of logic. Tarski is the originator of this second method also, and has suggested the possibility of introducing the term "true" itself into a meta-language in this manner.

1. I now take up two different bases for semantics. The first of them, by Tarski, illustrates the former of the above two methods of introducing semantical terms, viz., by definition; the second, by Richard Martin, is of the latter type.

Tarski takes as his basic semantical concept the concept of *satisfaction*.[1] The relation of satisfaction is one between objects, or sequences of objects, and wffs; objects, or sequences of objects, are said to *satisfy* wffs. For example, consider the wff "x is a city". This wff is satisfied by anything that is a city, and by nothing else; for example, by New York City, but not by John Jones. Not all wffs are of one free variable, as this one is, however; wffs may contain any finite number of free variables, and in the extreme case of sentences, no such variables at all. Tarski formulates a definition of satisfaction which will cover all cases. In order to achieve this generality, he defines the relation of satisfaction so that it holds not between objects and wffs, or between ordered n-tuples of objects and wffs, but between *infinite sequences of objects* and such formulas. His definition presupposes an enumeration of the variables of L. As examples illustrating his definition, we may say that any infinite sequence that has New York City as its first term satisfies the wff "x_1 is a city", where "x_1" is the first individual variable of L; and any infinite sequence that has Boston,

Mass., as its first term, and Savannah, Ga., as its second, satisfies the wff "x_1 is to the north of x_2", where "x_1" and "x_2" are the first and second individual variables of L, respectively.

In order to state Tarski's definition of the satisfaction relation, we need to give a fairly definite statement of the object language L with respect to which we are defining this relation. I choose for this purpose an example of a *simple, applied, functional calculus of first order;* that is, a language having all of its variables of one type, taken to range over some non-empty domain of individuals, and having predicate constants but no predicate variables. Les us take the primitive sentential connectives of our language L to be the sign "$-$", for negation, and the sign "$\lor$", for disjunction; and let us take the universal quantifier as the sole undefined quantifier. Let us take the individual variables of L to be defined by the following infinite series: 'x_1', 'x_2', 'x_3', . . .; one variable for each positive integer n, the variable with the subscript n counting as the n-th variable. For purposes of simplicity, let us suppose that there are but three predicate constants: "P'", a one-place constant; "P''", a three-place constant; and "P'''", a four-place constant. And for further simplicity, let us suppose that L contains no individual constants. As examples of wffs of L, we have the following: "$P'x_1$", "$P''x_2x_3x_4$", "$P'''x_1x_1x_2x_3$", "$-(x_1)$ $P'x_1$". The last of these wffs is a sentence.

In order to give a recursive definition of "satisfies in L", for this particular L, we consider first the simplest type of wffs, viz., the so-called *atomic* wffs; and then each of the various ways in which one may build up complex wffs from simpler ones. It is in the first part of this definition, in our consideration of the atomic wffs of L, that we give an interpretation to each of the primitive predicate constants of L; for it is here that we specify just what the conditions are under which given sequences of objects satisfy the atomic wffs in which these constants appear. In particular, we shall take the predicate constant "P'" to stand for the property (of an instantaneous event) of occurring within the twentieth century; the constant "P''" to stand for the temporal relation of occurring later than one event and earlier than another; and "P'''" to stand for the relation of two events being equidistant in time with two other events. We take the variables of L to range over instantaneous events.

In my statement of the definition of "satisfies in L", I shall not use the precise, but cumbersome (until considerably abbreviated), syntactical notation wherein we employ names of each of the primitive symbols of L, together with a sign for concatenation. Rather, I use a more convenient version of the quasi-quotes notation, wherein wherever within a context of quasi-quotes we wish to refer to one of the primitive symbols of L we use that symbol itself rather than its name, with the concatenation sign being implicit. Let us use "J", "K" and "N" as syntactical variables ranging over wffs of L. Then "$\ulcorner(K \lor N)\urcorner$", for example, is understood to be a name of the expression in L which results from writing first the left-hand parenthesis, then the wff K, then the disjunction sign "$\lor$", then the wff N, and finally the right-hand parenthesis; while "$\ulcorner P'x_n \urcorner$", for example, is understood to be a name of the expression in L which results from writing first the predicate-constant "P'" of L, then "x" with a subscript n, for some positive integer n. In referring to expressions of L, I use the quasi-quotes notation only when there appears within that notation at least one occurrence of an expression which is not an expression of L; otherwise, I use the usual quotes notation, as in "$"x"$", for example.

I now state the definition of "satisfies (in L)", with respect to our given L, using "f" and "g" as variables ranging over infinite sequences of events:

f is a sequence satisfying K in L if and only if f is a sequence of instantaneous events, K is a wff of L, and one of the following conditions holds: (1) K is $\ulcorner P'x_n \urcorner$, for some positive integer n, and the n-th term of the sequence f is an event occurring within the twentieth century; (2) K is $\ulcorner P''x_m x_n x_o \urcorner$, for some positive integers m, n, and o, and the n-th term of the sequence f temporally precedes the o-th term of f, and temporally succeeds the m-th term; (3) K is $\ulcorner P'''x_m x_n x_o x_p \urcorner$, for some positive integers m, n, o, and p, and the temporal distance between the m-th and the n-th terms of f is equal to the temporal distance between the o-th and the p-th terms of f ; (4) there is a wff J, such that K is $\ulcorner -J \urcorner$, and f does not satisfy J; (5) there are wffs J and N, such that K is $\ulcorner(J \lor N)\urcorner$, and either f satisfies J or f satisfies N ; (6) there is a positive integer n and a wff J, such that K is $\ulcorner(x_n)J\urcorner$, and every infinite sequence of events which differs from f in at most the n-th term satisfies J.

The first three of the conditions in this definition give an interpretation of the primitive predicate constants of L. Thus, condition (1) interprets the predicate constant "P'" so that the wff $\ulcorner P'x_n \urcorner$ means that x_n is an event occurring within the twentieth century. Condition (2) interprets the predicate constant "P''" so that the wff $\ulcorner P''x_m x_n x_o \urcorner$ means that event x_n temporally precedes event x_o and succeeds event x_m. And condition (3) interprets "P'''" so that the wff $\ulcorner P'''x_m x_n x_o x_p \urcorner$ means that the temporal distance between events x_m and x_n is equal to the temporal distance between events x_o and x_p. The fourth and fifth conditions give an interpretation of the two signs "$-$" and "$\lor$" of L, as being the signs for logical negation and disjunction, respectively. And the sixth condition gives an interpretation of the sign $\ulcorner (x_n) \urcorner$, as being the sign for universal quantification on the variable $\ulcorner x_n \urcorner$, for any positive integer n; for if J is satisfied by all sequences differing from f in at most the n-th place (and thus by the sequence f, also), whatever J asserts of x_n must hold true of all events, since any one event appears in the n-th place of some sequence.

Now that we have a definition of "satisfies in L" at hand, we are in a position to define "true in L", for the sentences of L. The definition that Tarski offers of this term is the following explicit one:

K is a true sentence of L if and only if K is a sentence of L, and K is satisfied by every infinite sequence of objects of L.[2]

Let us examine the plausibility of this definition. In the first place, a sentence is defined as a wff with no free variables. Thus whether a given sequence satisfies a given sentence is in no way dependent upon what the terms of that sequence are. If any sequence satisfies a given sentence, all sequences do; and, conversely, if one sequence does not satisfy a given sentence, no sequences do. Thus every sentence is satisfied either by all sequences or by none. Consideration of a specific example will illustrate the decision to identify the true sentences with those that are satisfied by all sequences, and the false sentences with those that are satisfied by none. Consider the following sentence of L: "$-(x_1)P'x_1$". Let f be an arbitrary sequence of instantaneous temporal events. By condition (4) of the definition of "satisfies", f satisfies this sentence if and only if f does not satisfy "$(x_1)P'x_1$". And, by condition (6), f does not satisfy "$(x_1)P'x_1$" if and only if there is some sequence g differing from f in at most the first place which does not satisfy "$P'x_1$". Now by condition (1), to say that g does not satisfy "$P'x_1$" is to say that the first term of g is an event not occurring within the twentieth century. Thus we conclude that our given sentence "$-(x_1)P'x_1$" is satisfied by any arbitrary infinite sequence of events if and only if not every event lies within the twentieth century. But if we take this condition to be what our given sentence asserts, we conclude that for our sentence to be true it must be satisfied by any arbitrary infinite sequence, that is, by all infinite sequences.

The above definition of "true in L" permits us to infer within M that "$-(x_1)P'x_1$" is true in L if and only if not every event lies within the twentieth century. That is, it permits us to infer the following statement: "_____ is true in L if and only if . . .", where the position marked "_____" is occupied by the sentence-name ""$-(x_1)P'x_1$"", and the position marked " . . ." is occupied by the translation in M of the sentence that is named by this sentence-name, viz., by the sentence "Not every event lies within the twentieth century". This result holds in general. For every sentence K of L, Tarski's definition of "true in L" implies the following statement: "_____ is true in L if and only if . . .", where the position marked "_____" is occupied by an expression which names K, and the position marked ". . ." is occupied by a translation of K into M. In addition, we have the result that whatever is true in L is a sentence of L. It is precisely these two results that the so-called *criterion of adequacy* requires of any definition of any term designating the concept of truth in L. A definition of any such term is regarded as an *adequate definition of truth in L* if and only if it meets this criterion. [3] It follows that all adequate definitions of "true in L", for any given L, are equivalent to one another: any sentence which is true in L according to one of these definitions will be true according to all. Thus, for example, on any adequate definition of "true in L" for the particular L I have been considering, the sentence "$-(x_1)P'x_1$" is true in L if and only if not every event lies within the twentieth century. And because any adequate semantic definition of truth has the properties mentioned in the criterion of adequacy, Tarski contends that all such definitions "do justice to the intuitions which adhere to the *classical Aristotelian conception of truth*"; [4] viz., to say of what is that it is not, or of what is not that it is, is false; while to say of what is that it is, or of what is not that it is not, is true.

Notice that the criterion of adequacy is not itself a *definition* of truth. Nor can we regard the infinitely many sentences which follow from any adequate definition of 'true in L' as together constituting a definition of truth in L; for a definition of truth must itself be a (finitely-long) sentence. Indeed, a definition of truth is a kind of "finite product" of these infinitely

91

many sentences, accomplishing in *finitely* many words what they accomplish only in *infinitely* many words.

It is easy to see that from Tarski's definition of truth, and thus from any adequate definition of truth, the following two very important properties of truth follow: every sentence of L is such that either it or its negation is true in L, and no sentence of L is such that both it and its negation are true in L.

Having defined the concept of truth with respect to L, we are now in a position to define a number of other useful semantical concepts with respect to L. Thus, for example, a sentence of L is *false in L* if and only if it is not true in L; one sentence of L *materially implies in L* another if and only if either the first is false in L or the second is true in L; two sentences are *materially equivalent in L* if and only if either both are true in L or both are false in L; and so on.

Let us now see just how it is that the semantical definition of truth in L permits us to avoid the paradox of the liar, which I earlier stated in terms of an unrestricted concept of truth. Here of course we have no such concept, but only one that is relativized to a particular language. And further, here we are requiring that any expression designating the concept of truth with respect to a given language appear not within that language itself, but within its meta-language. Thus the crucial sentence in my earlier statement of the paradox now takes the following form:

The sentence on page 92, line 23, is not true in L, where this sentence — let us call it "A" — appears not within L, but within L's meta-language, M (supposing that within M we can make reference to page 92, line 23). There can, of course, be no question of A's being either true in L or false in L, for A is not even a sentence of L; at most A can be either true, or false, or both, within M. Now for A to be a meaningful sentence within M, the variable implicit within the descriptive phrase "The sentence on page 92, line 23", must be a variable ranging over wffs of L (for "true in L" is defined only with respect to such variables). Thus, the sense of the descriptive phrase appearing within A, together with the criterion of adequacy for "true in M", assures us that on any adequate definition of "true in M", A is true in M only if there is a sentence of L on page 92, line 23, which is not a true sentence of L. But as there is no sentence of L at all at this position, we conclude that A is not a true sentence of M. Thus A is not both true and false in M, but is merely false in M. Thus we see that the distinctions bound up with the semantic definition of truth permit us to escape from the paradox of the liar when stated in the form I have been considering.

2. In order to define the truth-concept for a given language L, Tarski must speak, within its meta-language M, of infinite sequences of the kinds of entities that are discussed within L. This forces him to use as a meta-language M a language which employs variables of higher type than any of the variables appearing within L (supposing that we are restricting ourselves to meta-languages based on the logical theory of types). An alternative approach to semantics, which I shall not examine here in detail, is one that Carnap has examined extensively.[5] Here Carnap takes as basic the semantical concept of *designation*. He interprets the individual and predicate constants of L by specifying just which individuals and properties and relations are designated by those constants. And he argues that we may even go further and take whole sentences of L to be designatory expressions, designating propositions. Within this method of semantical analysis, we speak not of infinite sequences of entities discussed within L, but of the individuals, properties, relations and propositions designated by the designatory expressions of L. This approach, too, forces us to adopt as a meta-language for L one that employs variables of a higher type than that of any of the variables of L; viz., variables ranging over properties, relations, and propositions.

Now it is known that any meta-language in which we can define the concept of truth for a given object-language L must be *essentially richer* than L, roughly in the sense that although M contains an interpretation of L, it is impossible to give an interpretation of M within L.[6] If this requirement were not satisfied, we could introduce the paradox of the liar into L, by first defining "true in L" within M, and then interpreting M within L, thereby obtaining a definition of "true in L" within L itself. The requirement of essential richness is met by both Tarski and Carnap by using as a meta-language for L one that employs variables of higher type than any of the variables of L. Indeed, it is precisely because they have such variables at their disposal that they are able to introduce all of their semantical terms by way of definition. From a "nominalistic" point of view, however, one might be interested in the question whether it is possible to develop a semantical approach that does not force one to use in one's meta-language variables of higher type than those of the object-language being investigated. One might wish to avoid the infinite sequences, classes of classes, positive integers and so on that one is committed to if one does semantics in the manner of Tarski; and one might wish to avoid making reference to the properties and propositions that one is committed to by certain of the methods of Carnap. As a matter of fact, such nominalistic requirements can be satisfied. It is possible to construct a satisfactory approach to semantics, up to the point of defining truth at any rate, which does not require that our meta-language contain variables of higher type than the variables of the object-language, and makes no reference to either infinite sequences or to properties and propositions (supposing, of course, that no such entities are referred to within the language being investigated itself). Within such an approach the requirement that the meta-language M be essentially richer than L is met, not by introducing into M variables of higher type than any of the variables of L, but by introducing into M one or more undefined semantical constants. I shall now examine one example of such an approach; viz., that of Richard Martin.

Martin takes as his semantical primitive the term "Den", standing for a concept of *multiple denotation*.[7] (Strictly speaking, of course, the term "Den", like the term "satisfies", has to be understood as relativized to some language L.) This primitive term is offered as satisfactory for the semantics of any first-order language: that is, any language all of whose variables range over some one class of entities, called the *individuals* of that language. The term "Den" is not introduced into the meta-language by way of definition, but is introduced as an undefined term, with axioms being laid down governing its meaning. The one-place predicate constants of L are the only expressions of L that are said to denote, and they are said to denote not classes, or properties, but severally each of the individuals to which they apply. Thus, for example, the one-place predicate constant "dog" is here said to denote not the class of dogs, or the property of being a dog, but each dog: Rover, Fido, etc. As Martin points out, the concept of multiple denotation is one that was used by Hobbes.

Martin's concept of denotation is meant to be used in connection with a first-order language which employs abstracts. Abstracts are expressions that are obtained by prefacing any wff with one or more instances of an abstraction operator. (For purposes of simplicity, I shall suppose in the following discussion that the languages we are dealing with contain only one-place abstracts.) An abstraction operator binds within any wff to which is applied all free occurrences of whatever variable appears within the operator. If there are no remaining free variables within the abstract, the abstract functions as a one-place precidate constant. Thus, for example, if we apply the abstraction operator "$ə$" to the wff "x is a man" so as to bind the variable "x", we obtain the abstract "$xəx$ is a man", which will function as a one-place predicate constant, denoting severally each man. All such one-place abstracts of L involving no free variables, together with the primitive one-place predicate constants of L, are taken as the *denoting expressions* of L. It is to be noted in particular that the two-or-more-place primitive predicate constants of L are *not* spoken of as denoting anything.

Two types of axioms are laid down in M governing the sign "Den".[8] First, there is a restrictive axiom, to the effect that the only expressions of L which denote are the *one-place predicate constants* of L; that is, the one-place primitive predicate constants, together with the one-place abstracts with no free variables. Second, rather than a single axiom, an *axiom-schema* is laid down, providing for an infinite number of axioms, one in connection with each one-place predicate constant of L. It is these axioms which interpret the predicate constants of L. Roughly speaking, this axiom-schema in effect assures us that the one-place predicate constants of L denote just those entities that satisfy the conditions associated with those constants. Its formulation is as follows, where we let K be any wff of L involving free occurrences of just the variable "x", and let B be the translation of K into the meta-language M (I use "a" as a syntactical variable ranging over expressions of L):

(x) $(a$ Den x if and only if $\ldots x \ldots)$, where either (1) in place of "a" we put the abstract-name "$\ulcorner xəK \urcorner$", and in place of "$\ldots x \ldots$" we put the translation B, or (2) in place of "a" we put the name of any one-place primitive predicate constant of L, and in place of "$\ldots x \ldots$" we put the translation of that primitive predicate constant into M, with the variable "x" in its argument-place.

If we take as L the simple language used earlier in connection with my discussion of the concept of satisfaction (understanding that language now to contain abstracts), this axiom-schema takes the following form (using "K" and "B" as before):

95

(x_1) (a Den x_1 if and only if $\ldots x_1 \ldots$), where either (1) in place of "a" we put "$\ulcorner x_1 \ni K, \urcorner$" and in place of "$\ldots x_1 \ldots$" we put B, or (2) in place of "a" we put "P'" and in place of "$\ldots x_1 \ldots$" we put "x_1 is an instantaneous event lying within the twentieth century".

As examples of the infinitely many axioms provided for by this schema, by (2) we have "(x_1) ("P'" Den x_1 if and only if x_1 is an instantaneous event lying within the twentieth century)"; and by (1) we have "(x_1) ("$x \ni (Ex_2)(Ex_3)P'' x_2x_1x_3$" Den x_1 if and only if there is an event x_2 which precedes x_1 and an event x_3 which succeeds x_1)". Thus we see that this axiom-schema in condition (2) gives the interpretation of the one-place primitive predicate constants of L, and in condition (1) provides for the interpretation of the one-place abstracts of L, and thereby of the two-or-more-place primitive predicate constants of L (since all such constants appear within some one-place abstract, as in my above second example of an axiom on "Den").

Once we have the concept of denotation, we are in a position to define a number of other interesting semantical concepts. Thus, with Martin, we may say that an expression a of L *comprehends* an expression b of L if and only if a and b are both predicate constants of L, and a denotes everything that b does; a *null* predicate constant is one that denotes nothing; a *universal* predicate constant is one that denotes everything; a predicate constant a is the *semantical sum* of two predicate constants b and c if and only if a denotes x if and only if either b denotes x or c denotes x; and so on. [9]

Martin presents a number of ways in which we may define an adequate truth-concept for any first-order language on the basis of the concept of multiple denotation.[10] I present here one of the simplest of these definitions, one that makes use of the logical notion of the prenex normal form of a formula. It is a well-known theorem in logic that corresponding to any given formula K in a first-order language L, there is at least one formula J of L which is logically equivalent to K, and has all of its quantifiers appearing at the beginning of the formula, with the scopes of these quantifiers each extending to the end of the formula. Such a formula J is said to be in *prenex normal form*. Indeed, for any K, there is always a formula J, logically equivalent to K, which is in prenex normal form, and has a universal quantifier as its initial quantifier. Any sentence which is of this special form, i.e., any sentence which is in prenex normal form and has a universal quantifier as its initial quantifier, Martin calls an *atomic universal sentence*. For every sentence K of L, there will be an atomic universal sentence which is logically equivalent to K. Let us speak of the abstract which is formed from an atomic universal sentence by replacing the initial quantifier of that sentence by the abstraction operator involving the same variable as appears in that initial quantifier, as the *associated abstract* of that atomic universal sentence. Thus, for example, the associated abstract of the atomic universal sentence "$(x_1)P'x_1$" is the expression "$x_1 \ni P'x_1$".

Now it is clear that in order for an atomic universal sentence to be true, it is necessary and sufficient that its associated abstract be universal; that is that its associated abstract denote everything in the domain of discourse. Further, every sentence is logically equivalent to some atomic universal sentence. We may, then, define the truth concept for all sentences of any first-order L as follows:

K is true in L if and only if K is a sentence of L, and there is an atomic universal sentence J of L which is logically equivalent to K, such that the associated abstract of J is universal.

Martin is able to show that this definition of truth is adequate, in the sense of the criterion of adequacy, and that it implies that every sentence of L is such that either it or its negation is true in L, and no sentence is such that both it and its negation are true in L.[11] Further, he is able to show how, with the concept of denotation, one can define truth for the very powerful languages of the Zermelo set theory and the simple theory of types.[12]

When we introduce all semantical terms into our meta-language M by way of definition, we have a kind of guarantee that M is consistent if its translational and syntactical parts are consistent. When we introduce one or more semantical terms as undefined primitives, as Martin does, however, a special *relative-consistency* proof is called for, showing that M is consistent, and thus that by introducing our undefined semantical primitives into M we have not thereby introduced any semantical paradoxes into M that were not already present in M in some form. The need for such a proof was first pointed out by Tarski, who suggested the possibility of introducing a sign for truth into the meta-language as an undefined term, and then laying down as axioms governing this sign all those infinitely many formulas described in the statement of the criterion of adequacy; viz., (1) all sentences of the form "_____ is true in L if and only if...", where the position "_____" is occupied by the name of some sentence of L, and the position "... " is occupied by the translation of this sentence into M, together with (2) a sentence to the effect that the term "true in L" applies only to sentences of L.[13] Tarski showed that if the translational and syntactical parts of M are consistent *before* such additions are made, M will also be consistent *after* they are made. And, in a similar manner, Martin shows that the result of adding "Den", together with its infinitely many axioms, to a meta-language M will be consistent if M is consistent before these additions.

III

I turn now to the semantic analysis of an especially important type of concepts, viz., the so-called *L-concepts*. These concepts include as principal examples the concept of L-truth, or analyticity; L-falsity, or self-contradiction; and L-implication, or logical implication. Each of these concepts has, of course, played an important role in much of modern philosophy. The concept of analyticity, for example, makes one of its earliest appearances in a fairly clear form in the writings of Leibniz, in the form of a distinction between necessary and contingent truths; necessary truths being described as those that hold in all possible worlds, and contingent truths being described as those that hold in the actual world, but not in all possible worlds. The various semantic analyses of the concept of analytic truth may indeed be taken to be attempts to define in a precise way Leibniz's notion of a truth's holding in all possible worlds.

In order to give an analysis of the L-concepts, we shall need some semantical concept which permits us to make reference to all the "possible worlds" with respect to a given language L. By a *possible world* with respect to a given language we mean, roughly, a complete state of affairs concerning the individuals within the domain of that language, in so far as that state of affairs can be described by means of the expressions appearing within that language. In this section I shall examine two basic semantical concepts which permit us to speak of all possible worlds. First I shall present Carnap's semantical theory of *state-descriptions*, and show how the concept of a state-description permits us to give an analysis of the L-concepts.[14] Second, I shall present a formulation of a semantical theory in terms of the basic concept of an *interpretation*, as recently developed by John G. Kemeny.

1. The concept of a state-description is one that is meant to be used for the semantic analysis of a language L only when that language possesses, for every individual within the intended domain of discourse, an individual constant designating that individual. Now as the term "language" is usually understood by logicians, no language contains more than a denumerable infinity of signs, and thus no more than a denumerable infinity of individual constants. The method of state-descriptions must be confined, then, to the semantic analysis of those languages that have no more than a denumerable infinity of individuals within their intended domains of discourse. In particular, this method could not be used for the semantic analysis of a first-order language which included among the values of its individual variables the real numbers.

Let us suppose that we have a simple applied language L of first-order, with a finite number of predicate constants, there being at most a denumerable infinity of individuals in the domain of the variables, with L containing an individual constant for each individual in its domain. And let the signs for negation, disjunction and universal quantification be the primitive logical constants of L. We first define an *atomic sentence* of L as an expression consisting of an n-place predicate constant of L followed by n individual constants, not necessarily all distinct. We then define a *state-description* of L as a class of sentences of L which contains for every atomic sentence of L either that atomic sentence or its negation (not both), and no other sentences.

As an example of a state-description, let us suppose that L contains just two individual constants, and two one-place predicate constants. Let these be "a", "b", "P" and "Q", respectively. Then the class consisting of the sentences "Pa", "Pb", "Qa" and "Qb" constitutes one state-description with respect to L. For this particular L, there are sixteen state-descriptions in all.

I now show how to introduce into a meta-language the basic semantical concept of a sentence's holding in a given state-description, by means of the following recursive definition:

K holds in the state-description S of L if and only if K is a sentence of L, S is a state-description of L, and one of the following conditions is

99

satisfied: (1) *K* is an atomic sentence, and *K* is an element of *S*; (2) *K* is the negation of a sentence *J* of *L*, and *J* does not hold in *S*; (3) *K* is a disjunction of two sentences, one of which holds in *S*; (4) *K* is a universal quantification of the wff *J* of *L*, and *J* holds in *S* for every value of the free variable appearing in *J*.

To say, then, that a given sentence holds in a given state-description is to say, roughly, that that sentence would be true if that state-description were true, that is, if all the sentences appearing within that state-description were true. Now every sentence either holds or does not hold in a given state-description. Thus the truth of any one state-description of *L* uniquely determines the truth or falsity of every sentence of *L*. It is in this sense that we may say, provisionally at any rate, that every state-description of *L* determines a possible world with respect to *L*.

Certain sentences of *L* can readily be seen to hold in *all* the state-descriptions of *L*. By condition (2) of the definition of a sentence's holding in a state-description, for any arbitrary sentence *K* of *L*, and any arbitrary state-description *S* of *L*, either K or ⌐–K⌐ holds in *S*. Thus, by condition (3), the disjunction ⌐K ∨ –K⌐ holds in *S*. In general, all the sentences of *L* which are logical theorems of *L*, in the sense of being theorems either within the propositional calculus or quantification theory, will hold in all state-descriptions of *L*.

Now at first Carnap proposed to identify analyticity in a given language with the property of holding in all the state-descriptions of that language. [15] But then it was noticed that certain state-descriptions may not correspond to possible worlds; that is, once a definite interpretation is given to each of the individual and predicate constants of *L*, it may be logically impossible that certain state-descriptions be true, in the sense that all of the sentences appearing within them be true. An example readily shows how this is possible. Suppose that *L* contains "*P*" and "*Q*" as one-place predicate constants, and "*a*" as an individual constant. Suppose further that we intend to interpret *L* so that the sign "*P*" is taken to name the property of being perfectly spherical in shape, and the sign "*Q*" is taken to name the property of being perfectly cubical in shape. Now one or more state-descriptions of *L* (indeed, one quarter of them) will contain both the atomic sentence "*Pa*" and the atomic sentence "*Qa*". But, on the interpretation intended, no such state-descriptions could possibly be true, since no matter what "*a*" is taken to denote, that object *a* cannot be both perfectly spherical and perfectly cubical in shape. Thus no such state-descriptions represent any possible world with respect to *L* under the intended interpretation of *L*.

A number of ways of handling this difficulty have been proposed. I shall consider here that method which makes use of so-called *meaning postulates*.[16] According to this method, whenever we interpret the individual and predicate constants of *L* in such a way that certain of these constants become logically dependent upon others, the logical dependencies of those constants upon one another are to be indicated by laying down certain postulates, the so-called *meaning postulates*. Thus, for example, if *L* contains "*P*" and "*Q*" as one-place predicate constants, and we intend to interpret these constants as designating respectively the property of being perfectly spherical in shape and the property of being perfectly cubical in shape, then we must lay down the following meaning postulate: "$(x)(-Px \lor -Qx)$." Under the intended interpretation, this postulate says that whatever is perfectly spherical in shape is not also perfectly cubical in shape. Now the only state-descriptions of this particular *L* which represent possible worlds are those in which this particular meaning postulate holds. No such state-descriptions contain both "*Pa*" and "*Qa*". In general, for any language *L*, the only state-descriptions with respect to that language which represent possible states are those in which all the meaning postulates forced upon us by the intended interpretation of *L* hold.

We are now in a position to define the L-concepts. Let us agree to include among the meaning postulates of *L* all axioms and definitions of *L*. A sentence of *L* is then said to be *L-true in L*, or *analytic in L*, if and only if that sentence holds in all those state-descriptions of *L* in which the meaning postulates of *L* hold. Similarly, a sentence of *L* is *L-false in L* if and only if its negation holds in all such state-descriptions; and one sentence *J* is said to *L-imply*, or logically entail, another sentence *K* if and only if the material implication from *J* to *K* holds in all such state-descriptions. A sentence which is either L-true or L-false is *L-determinate*; otherwise, *factual*.

We may speak of a state-description in which the meaning postulates hold as an *admissible* state-description; and of the *range* of a sentence as the class of all admissible state-descriptions in which that sentence holds. Employing the concept of range, we may then define an L-true sentence as one whose range is the class of all admissible state-descriptions; an L-false sentence as one whose range is the null class of admissible state-descriptions; and similarly for the remaining L-concepts. The concept of range is, then, a basic one for the theory of the L-concepts, and Carnap also takes it as basic in his construction of a theory of probability.

Notice that in order to define the L-concepts with respect to a given language L it is not necessary that we first give an interpretation of L. If we wish in addition to interpret L, we may now do this, among other ways, with the help of the defined concept of *the true state-description* with respect to L. First, we must state what we take to be the domain of the individual variables of L. Then we must interpret the individual and predicate constants of L. This we may do by making use of any one of the semantical methods discussed in the preceding section, for example. That is, we may interpret these constants by laying down designation rules for them, or by laying down denotation rules, or by making use of the concept of satisfaction. Next we need to define the concept of a true atomic sentence. Once again, this may be done as before. We now define *the true state-description* with respect to L as that state-description which contains all the true atomic sentences of L, together with the negations of all the remaining atomic sentences of L. Finally, we define a sentence as being true in L if and only if it holds in the true state-description with respect to L. Thus we see that in order to define the L-concepts with respect to a language L, it suffices to have in the meta-language of L the concept of a sentence's holding in a given state-description; while in order to give in addition an interpretation to L, we need to add to the meta-language of L some such semantical concept as was considered in the preceding section.

2. It should be noticed that Carnap's concept of a state-description, together with the concept of holding in a state-description, are syntactical concepts, and are not semantical in the strict sense of the term "semantical". That is, they make reference only to the expressions of L, and not to the entities discussed when using L. Thus, on Carnap's analysis the L-concepts turn out to be syntactical concepts. Specifically semantical concepts enter only when the individual and predicate constants are interpreted, as by designation rules, for example. On the analysis we are now to consider, by Kemeny, the L-concepts are defined as genuinely semantical concepts.

The concept of a *model* is a fundamental one within that part of mathematics known as meta-mathematics, that is, the theory of mathematical languages. It is of course not necessary to confine its use to the study of mathematical languages alone; it may be used in the study of languages in general. When so used, the concept of a model becomes one of the fundamental concepts of semantics.

Kemeny makes use of four important concepts in his construction of a semantical system.[17] These are the concepts of a *value-assignment*, a *semi-model*, a *model*, and an *interpretation*. He defines and illustrates these concepts with respect to a vey general type of language, employing type symbols. We shall not follow him in this respect, but shall present an adaptation of his concepts to a simple, applied functional calculus of first order, in which there are no abstracts.

A *value-assignment* is simply an assignment of individuals from the domain of L to each of the individual variables of L; each variable is assigned one individual, and it is not required that distinct variables be assigned distinct individuals. Since any one value-assignment covers all the variables of L, each of the value-assignments of L assigns to each of the free variables of any expression of L some one individual. There are, of course, as many distinct value-assignments as there are distinct ways of assigning values to the individual variables of L, two value-assignments being regarded as distinct if and only if at least one individual variable has one individual assigned to it by one of these value-assignments and a different individual assigned to it by the other.

Kemeny defines a *semi-model* with respect to L as (1) an assignment of a domain of individuals to the individual variables of L, together with (2) an assignment of an interpretation within this domain to each of the individual and predicate constants of L. To each of the individual constants of L, if any, a semi-model assigns an individual within the assigned domain. To each of the one-place predicate constants of L, a semi-model assigns a class of individuals from the assigned domain. Thus, for example, a semi-model might assign the class of dogs as the domain of L; it might then assign the particular dog Fido to the individual constant "a", and the class of Dalmatians to the one-place predicate constant "D". In general, a semi-model of L assigns to each n-place predicate constant of L a class of n-tuples of individuals from the domain assigned to L by that semi-model. And there are as many semi-models of L as there are distinct

103

ways of assigning domains to L and then interpreting the individual and predicate constants of L within those domains.

In contrast with the method of state-descriptions, it is not here required that the domain of individuals assigned to L be at most denumerable. The method of state-descriptions has to impose this requirement in order to be able to characterize each of the possible worlds; if there were a non-denumerable number of entities in the domain of L, no state-description could completely characterize any one of the possible worlds with respect to that domain. As was pointed out earlier, this is because in that case there would not be a sufficient number, viz., a non-denumerable number, of atomic sentences or negations of atomic sentences to give such a characterization as a state-description is supposed to give. But on the method we are now considering, a possible world with respect to a given language is not characterized by a syntactical entity, such as a class of atomic sentences and negations of atomic sentences, but by a genuinely semantical entity, viz., that part of a semi-model in which we make definite assignments to each of the individual and predicate constants of L. Such an assignment of values to these constants of L does the work of a state-description. That is, it determines, (with respect to a given semi-model), whether a given n-place predicate constant applies to a given n-tuple or not. (As we shall see later, an n-place predicate constant applies to a given n-tuple of individuals if and only if that n-tuple is an element of the class of n-tuples assigned to that predicate constant.) The assignment of a class of n-tuples of individuals to a constant, however, does not itself involve the names of those individuals; rather, it involves just those individuals themselves. And in order to *make*, or *give*, the assignment of a class of n-tuples of individuals to a constant, we need not actually name any of the individuals in that class; the class may be given by mentioning its defining property, as when we refer to a certain class as the class of real numbers, for example. The method of models does not, then, as does the method of state-descriptions, presuppose that we have a name for each of the individuals within the domain of L. As a consequence, L may indeed possess a non-denumerable domain.

We need now the concept of a wff's holding in a given semi-model with respect to a given value-assignment. Kemeny defines this concept by recursion, as follows:

K holds in the semi-model M of the language L with respect to the value-

assignment V if and only if K is a wff of L, M is a semi-model of L, V is a value-assignment to the individual variables of L, defined over the domain assigned to L by M, and one of the following conditions is satisfied: (1) K is an n-place atomic wff, and the n-tuple of individuals assigned to the argument expressions of K by the semi-model M and the value-assignment V is an element of the class of n-tuples assigned to the predicate constant of K by the semi-model M; (2) there is a wff J, such that K is $\ulcorner -J \urcorner$, and J does not hold in M with respect to V; (3) there are wffs J and N, such that K is $\ulcorner (J \lor N) \urcorner$, and either J or N holds in M with respect to V; (4) there is an individual variable $\ulcorner \alpha_n \urcorner$ and a wff J, such that K is $\ulcorner (\alpha_n)J \urcorner$, and J holds in M with respect to every value-assignment which is defined over the domain which M assigns to L, and which differs from V in at most its assignment to $\ulcorner \alpha_n \urcorner$.

The intuitive meaning of the above definition is that a wff K holds in a semi-model M with respect to a value-assignment V if and only if that wff would be true if one were to interpret the individual variables and individual and predicate constants within it in accordance with the interpretation put upon them by M and V. The definition very closely resembles that of the earlier definition of the concept of satisfaction. It is, indeed, a generalization of that definition. In the concept of a wff's being satisfied by an infinite sequence, we fix the meaning of the individual and predicate constants of L, and vary the assignments to the individual variables of L by varying the infinite sequences of individuals. In the concept of a wff's holding in (or being satisfied by) a semi-model with respect to a value-assignment, on the other hand, we vary not only the assignments to the individual variables of L, by means of the different value-assignments, but also the meaning of the individual and predicate constants of L, by means of the different semi-models of L. The restricted concept of satisfaction suffices for the definition of truth and its related concepts, with their reference being only to the actual world; while the more general concept (or some equivalent substitute) of holding in a semi-model with respect to a value assignment is needed in order to define the L-concepts, their reference being to all possible worlds.

Kemeny next defines a wff as *valid* in a given semi-model if and if that wff holds in that semi-model with respect to every value-assignment; and *contravalid* in a given semi-model if and only if it holds in that semi-model with respect to no value-assignment. Now a sentence is a wff having no

105

free variables. Thus every sentence is either valid or contravalid in any given semi-model. We may say, informally, that a sentence's being valid in a given semi-model amounts to its being true when understood in accordance with that semi-model.

Next, we define a *model* of a language L as a semi-model of L in which the meaning postulates of L are valid (we again use the term "meaning postulate" in the broad sense, so as to include all axioms and definitions of L among the meaning postulates of L). This definition corresponds to our earlier restriction of attention to those state-descriptions of L in which the meaning postulates of L hold.

Notice now the plausibility of defining an analytic sentence of L as a sentence which is valid in all the models of L. A model of a language L defines a possible world with respect to that language. Thus, for example, suppose that in one model of L we include the dog Rover within the class of individuals assigned to the predicate constant "Br", which we intend to interpret as standing for the class of brown things within the domain of L; while in another model of L we include the dog Rover within the class of things assigned to the predicate constant "Bl", which we intend to interpret as standing for the class of black things within the domain of L. Then the first model represents a possible world in which Rover is brown; while the second model represents a possible world in which Rover is black. There is no possibility, however, that any model assign Rover both to the class of things that are brown (all over) and to the class of things that are black (all over); for no such model would satisfy the meaning postulates.

Now Kemeny writes that at first he intended to take the concept of a model as the basic concept of semantics. But he then noticed that certain results of logic make this impossible; viz., the so-called *incompleteness theorems* of Kurt Gödel. These theorems concern deductive systems, where by a *deductive system* I mean any system which contains an underlying logic together with certain additional constants, and axioms governing these additional constants. Gödel's theorems show that whenever we are dealing with a sufficiently strong deductive system (viz., any system strong enough to contain elementary number theory), that system will be incomplete. Without at this point attempting to give a perfectly general definition of "complete", we may illustrate the meaning of Gödel's incompleteness theorems as follows: Consider a formalization of elementary number theory within a second-order functional calculus. Since we may bind predicate variables within the second-order calculus, our formulation of number theory will contain *sentences* of number theory; that is, it will contain wffs of number theory which contain no free variables. Now Gödel's incompleteness theorems assure us that within our formulation of number theory there will be (infinitely may) pairs of sentences, which are such that in each pair one sentence is the negation of the other, and yet neither sentence is provable within our system. Nor will any addition to our axioms and rules of inference – short of making our system contradictory – so strengthen our system as to permit us to derive one sentence from each such pair of sentences. Our system will be irremediably incomplete.

The significance for semantics of the above result is as follows: Once we give our formulation of number theory its usual interpretation, each of the sentences within that formulation becomes either true or false. Certain of the true sentences within this formulation L, however, will not be provable within L. Now it is known, from further results of logic, that the reason that these sentences are not provable in L is because they are not valid in all of the models of L.[18] Thus, if we were to define an analytic sentence of L as any sentence of L which is valid in all of the models of L, these true but unprovable sentences would not be analytic in our particular L. As they are surely not contradictory, we would have to describe them as synthetic. But such a result would be intuitively unacceptable. Any particular true but unprovable sentence of L will be provable – and thus analytic – in the system L', where L' is obtained from L merely by adding that particular sentence as a further axiom. Thus, if we were to speak of the true but unprovable sentences within our particular formulation of number theory as synthetic, the boundary between the analytic and the synthetic would become perfectly arbitrary. But when we call a wff *analytic*, we mean to say, roughly, that its truth is determined by considerations of meaning alone, as contrasted with factual considerations. Thus the distinction between the analytic and the synthetic is *not* arbitrary, once our sentences are interpreted. Surely, then, if we are to count the provable sentences within some axiomatic approach to number theory as analytic, we must count the true but unprovable sentences as analytic also; for their truth is as much determined by considerations of meaning as is the truth of the provable sentences. Let us suppose that on independent grounds we have decided to regard all the true sentences of number theory – or, indeed, of mathematics in general – as analytic, and the false sentences as contradictory. The conclusion which we must then infer from Gödel's incompleteness theorems is that if we are to obtain a semantic definition of "analytic" which will permit us to record this decision within our language, we shall have to define an analytic wff of any language in some other way than as one which is valid in each of the models of that language.

Let us call those models of any incomplete deductive system in which certain of the analytically true but unprovable sentences of that system are not valid, *incompleteness models* (the term is mine, not Kemeny's). These models clearly interpret incomplete systems in ways they were not meant to be interpreted. It would be desirable if we could eliminate all such models. But they are a consequence of any incomplete system; every incomplete system admits of incompleteness models. Because we are unable in an incomplete system to lay down a set of meaning postulates which captures all the formal relations which hold between our individual and predicate constants once these constants are given their intended interpretations, any set of meaning postulates we *do* lay down will admit of models within which certain of these formal relations are denied. Only if we could complete our formal system, could we exclude all such models.

Now what Kemeny proposes in the view of those results is that the question whether a wff is valid in an incompleteness model be regarded as irrelevant to the question whether that wff is analytic or not. He proposes to define analyticity within a language not in terms of the set of *all* models of that language, but in terms of a certain sub-set of the set of all of its models, which we might informally call the set of *intended* models of that language. The problem we now have to take up is the problem of how to define this set.

Before attempting to define the set of intended models of a language, Kemeny points out that a language may contain constants such that every model we regard as an intended model makes the same assignment to these constants. That is, within the set of intended models the interpretations of these constants are fixed. Kemeny calls all such constants, *logical constants*. (The signs which play the role of *and*, *or* and *all* are special cases of logical constants.) As an example, suppose that our language contains an axiomatic approach to elementary number theory. Then no model will count as an intended model of that language unless it assigns the non-negative integers as a domain to the number variables, and then assigns the number zero to the numeral "0", the number one to the numeral "1", and so on. Now if the intended models all make the same assignments to the number-theoretical constants of our language, then every number-theoretical statement within our language will either be valid in all intended models, or contravalid in all intended models. By identifying analyticity with validity in all intended models, and self-contradiction with contravalidity in all intended models, we obtain the desired result that each of our number-theoretical statements is either analytic or self-contradictory. And of course the situation is precisely similar for any other branch of mathematics, or any branch of knowledge at all. If we require that all intended models make the same assignments to the individual and predicate constants used in our formulation of any area of knowledge, then all the statements within our formulation of that area will be either analytic or self-contradictory. Thus, whether we lay down this requirement for the constants appearing within our formulation of any area of knowledge will depend upon whether we wish to admit any of our statements within that area to be neither analytic nor self-contradictory. Within a formulation of a branch of empirical science, of course, the requirement would not be laid down. That is, here different assignments to the individual and predicate constants would be admitted within the class of intended models. Obviously, in our formulation of any language it is of the utmost importance that we give a complete list of its logical constants; i.e., those constants which receive a fixed interpretation within the class of intended models.

We may now define the class of intended models, or as Kemeny calls them, the class of *interpretations*. We first choose a certain model of L as that model which assigns those meanings to the individual and predicate constants of L which we intend to regard as their "official" meanings; that is, those meanings we assign to these constants when we use L for purposes of communication. Let us call this model "M^*", and let us call the domain it assigns to L, "R". Then by an *interpretation* of L we mean any model of L which (1) assigns R as a domain to L, and (2) differs from M^* at most only in assignments to those constants of L which are not logical constants of L. M^* is itself, of course, an interpretation; indeed, what we might call the "official" interpretation of L. Kemeny is able to show, however, that no incompleteness model is an interpretation.

The above definition of the term "interpretation" may present the appearance of being circular. It is easy to see, however, that it is not circular in any bad sense. When defining a language L, we present a list of constants, which we intend as the logical constants of L. The interpretations of L are then well-defined in terms of this list, together with the domain of the particular model we have chosen as our M^*.

Consider now the class of all models of L, except those models (in the case of incomplete systems) which are incompleteness models of L. Kemeny is able to show that a wff is valid in all interpretations if and only if it is valid in all of these models. The L-concepts may therefore be given satisfactory definitions in terms of the class of interpretations. Kemeny thus proposes the following definitions of these concepts, defining them for wffs in general, rather than for sentences alone: K is *analytically true (L-true) in L* if and only if K is a wff of L which is valid in all of the interpretations of L; and K is *self-contradictory (L-false) in L* if and only if K is a wff of L which is contravalid in all interpretations of L. A wff is *synthetic* if and only if it is neither analytically true nor self-contradictory. One sentence (logically) *implies* a second if and only if the second is valid in all those interpretations in which the first is valid; and two sentences are (logically) *equivalent* if and only if they are valid in the same interpretations. And a language L is said to be *complete* if and only if all its analytically true wffs are provable in L.

(The method of state-descriptions apparently does not require that for the purpose of defining the L-concepts we first define a restricted class of state-descriptions, corresponding to the restricted class of models called "interpretations". This is because state-descriptions themselves do not assign domains to L, as do models. On the method of state-descriptions, the assignment of a domain to L is made independently, and the state-descriptions of L can then be understood as defining possible states within that domain. Thus the problem of unintended domains does not arise.)

In order to define truth and its related concepts for a given language L, we need an interpretation of the individual and predicate constants of L. Such an interpretation is already at hand. The model M^* was understood to be just that model which assigns to these constants those interpretations we intend to regard as their "official" interpretations. The definition of "true in L", for wffs of L, is thus as follows: a wff of L is *true in L* if and

only if that wff is valid in M^*; and a wff is *false in L* if and only if that wff
is contravalid in M^*. These definitions readily yield the desired results
that every sentence of L is either true or false in L, and no sentence is
both. Also, of course, we have the results that whatever is analytically
true is true, and whatever is self-contradictory is false. And Kemeny is
able to show that this definition of "true in L" is adequate in the sense of
the criterion of adequacy.

IV

I now turn to a number of distinct but interrelated topics, which center
in the problem of ontological commitment.

I have earlier remarked that in order to know what a given (declarative) sentence means, it is necessary and sufficient that we be able to state,
within some language we admittedly understand, the conditions under
which that sentence is true. Now being able to state these conditions
implies being able to determine just what entities that sentence explicitly
assumes to exist. Thus we conclude that in order to understand a given
(declarative) sentence, one must be able to determine just what the
ontological commitments of that sentence are. For the purposes of semantic analysis, then, it is of the greatest importance that we possess some
criterion by means of which we can determine just what entities a given
sentence or theory is explicitly committing us to when we assert that
sentence or theory.

W. V. O. Quine has proposed a criterion for determining the ontological commitments of any given body of discourse. [19] Quine first argues that
the mere using of a name or descriptive phrase does not commit one to
the view that there is some entity which is designated by that name or
descriptive phrase. [20] He points out that Russell in his theory of descriptions has shown how to eliminate singular descriptive phrases, that is,
phrases of the form "the so-in-so", from sentences. By means of Russell's
method, we can replace any sentence containing a singular descriptive
phrase by another sentence equivalent to it in which no descriptive
phrases appear. For example, by Russell's method the sentence "The
Bishop of Milan is wealthy", is transformed into the sentence "There is
something x which is a bishop of Milan; everything which is a bishop of
Milan is identical with that thing x; and that thing x is wealthy". Consider now the case where a sentence contains a descriptive phrase which

we would suppose designates nothing; e.g., the sentence "There is no such person as the author of *Principia Mathematica*". One might suppose that the descriptive phrase appearing in this sentence must after all designate *something*, in order for the sentence to be meaningful. With the help of Russell's analysis, however, we can see how such a sentence can be meaningful, and even true, though the descriptive phrase appearing within it designates nothing. On that analysis, our sentence is seen to be equivalent to the sentence "There is no person who is an author of *Principia Mathematica*, and is identical with every person *x* who is an author of *Principia Mathematica*". By examining this latter sentence, which contains no descriptive phrases, we can readily see how the sentence of which it is an analysis can be both meaningful and true, though the descriptive phrase it contains designates nothing. We cannot, then, in general, determine the ontological commitments of any body of discourse by supposing that among these commitments must be included entities designated by the various descriptive phrases appearing in that body of discourse.

Nor, in determining the ontological commitments of a body of discourse, can we in general proceed on the supposition that each of the proper names appearing therein designates something. It is possible for a sentence to be meaningful, and even true, even though that sentence contains proper names which really name nothing. That this is possible may be regarded as but one illustration among many of the fact that being meaningful is one thing, while being a name of something is another. Furthermore, one may, as Quine shows, treat proper names as abbreviations of descriptive phrases, and then eliminate them *via* Russell's analysis. In determining the ontological commitments of any body of discourse, then, we are not entitled to suppose in general either that the descriptive phrases, or that the proper names, appearing therein designate anything whatsoever. They may, or they may not.

Where then are we to look in order to determine what entities we are explicitly committed to whenever we assert a given sentence or theory? In answer to this question, Quine proposes the following *criterion of ontological commitment: "An entity is assumed by a theory if and only if it must be counted among the values of the variables in order that the statements affirmed in the theory be true."* [21] The variables that we use, either explicitly or implicitly, play the role of pronouns, and by their means we express

112

the logical notions of *all* and *some*. It is to them that we are to look when determining ontological commitment, according to Quine, rather than to proper names and descriptive phrases. Thus, for example, when we assert the sentence, "There are five books on the table", we explicitly commit ourselves to the existence of whatever entities must lie within the range of the existential quantifiers implicit within this statement in order for the statement to be true; viz., five books. We do not, however, explicitly commit ourselves to the existence of some entity designated by the term "five". Not even when we assert such a sentence as "Five is less than six" do we so commit ourselves. However, we would explicitly commit ourselves to such an entity were we to assert the sentence "There is a prime number lying between four and six", for example. For here the number five must be included within the range of the existential quantifier appearing within this statement if the statement is to be true. Quantification over numbers is, of course, the usual practice in mathematics. Now it seems that numbers must be thought of as abstract entities. Quine's conclusion is that "Classical mathematics . . . is up to its neck in commitments to an ontology of abstract entities". [22]

The distinction between *nominalistic* and *realistic* languages is one that Quine is able to draw with the help of his criterion, supposing that we are already able to distinguish the concrete and the abstract. [23] Abstract terms may appear in either type language, but may appear as substituents for variables only in realistic languages. Realistic languages contain variables ranging over abstract entities, while nominalistic languages do not. If the nominalist uses variables which seemingly range over abstract entities of any sort, he must either find some way of eliminating them in favor of a more primitive notation in which they do not appear, or renounce nominalism in favor of realism. Here realism, understood as a thesis in the philosophy of science, is defined by Quine as maintaining that a complete expression of all scientific knowledge requires that we use a realistic language, while nominalism is understood as insisting that for this purpose a nominalistic-type language suffices.

It is, of course, one thing to determine what a given language says that there is; and quite another thing to determine what there is. Quine's criterion of ontological commitment is clearly directed primarily towards the first of these two questions. The importance of his, or any, criterion of ontological commitment for the problem of ontology, understood as the problem of what types of entities exist, is indirect, but nonetheless important. In order to significantly raise the question as to what types of entities exist, one will have to do so within a language with respect to which the problem of commitment can be settled. Until we are able to determine what kinds of entities we are explicitly committing ourselves to when we speak, we are in no position to raise ontological problems at all. As Alonzo Church has put it, the relation of the question of ontological commitment to the question of ontology is that of "a necessary-preliminary issue concerning the logic of the matter." [24]

A further respect in which a criterion of ontological commitment bears indirectly on the problem of ontology is as follows. Suppose that employing such a criterion one came to the conclusion that it was possible, at least in principle, to state all of scientific and mathematical knowledge in a nominalistic language; and that, further, it was in principle possible to develop satisfactory theories of the nature of knowledge, morals, art and other areas of philosophy, in a nominalistic language. Presumably, one could not infer, as a logical consequence of such a conclusion, that abstract entities did not exist; but one could, it seems, infer that there was no good reason to believe that they did (supposing, of course, that one was treating the question as to whether they did or did not exist as an objective-type problem, admitting of only one correct answer; and thus not as a problem to be settled by adopting some convention or other). Suppose, however, that one were to conclude that for the purpose of developing some area of knowledge or other, a realistic-type language was needed. Once again, presumably it would not follow, as a logical consequence of this conclusion, that there really were abstract entities. One might, however, care to infer that one now had good reasons for believing that there were. In either case, then, it seems that a criterion of ontological commitment would bear at least indirectly on the problem of ontology itself.

Are there any areas of knowledge in which we are committed to the existence of abstract entities of any sort? I have mentioned that classical mathematics seems to be such an area; at least until a nominalistic interpretation of mathematics is given, the possibility of which seems very doubtful. More germane to the general topic of this paper is the question whether the area of semantics itself is one in which we are committed to abstract entities. We know that the meta-language of a given object-language must be stronger than the object-language itself. Thus, if a given object-language presupposes abstract entities, so must its semantical meta-language. Let us, then, consider the question whether a semantical meta-language must presuppose abstract entities even when its object-language does not. And let us pass over the question whether syntax itself – a part of semantics – is committed to abstract entities. [25] Our question is whether the specifically semantical part of a satisfactory semantical meta-language need commit us to abstract entities.

On Tarski's approach to semantics, *via* the concept of satisfaction, abstract entities are presupposed. Classes, sequences and positive integers all make their appearance. All the abstract entities that Tarski needs may be defined in terms of classes, but classes he must have; and classes, as distinct from heaps, are abstract entities. The method of models, as developed by Kemeny, is ontologically comparable to the method of satisfaction: it, too, requires no other abstract entities than those that can be defined in terms of classes, the commitment to classes themselves, however, being evident. Certain of Carnap's various approaches to semantics presuppose not only such entities as can be defined in terms of classes, but entities which apparently cannot; viz., properties and propositions. Forming an interesting contrast to these writers, on the other hand, Richard Martin, as we have seen, has succeeded in showing how to develop semantics up to the point of defining truth without introducing any abstract entities not already presupposed by the object-language itself. To what extent Martin's particular approach can be extended so as to provide satisfactory answers to such problems as the problems of analyticity, strict synonymity, belief contexts and contrary-to-fact conditionals, remains to be seen. [26]

One who has repeatedly insisted on the importance of keeping one's ontological commitments to a minimum is Quine. Quine has been ready to admit the propriety in semantics of appealing to classes, as we do, for example, when we assign as extensions to general terms the classes of those entities to which these terms apply. [27] He has insisted many times, however, that no good purpose is served in semantics by introducing such abstract entities as are often spoken of as *meanings*; in particular, properties as meanings of general terms, and propositions as meanings of statements. [28] As Quine sees it, in order to explain how an expression is meaningful, it is unnecessary to introduce an entity which that expression is said to have as its meaning. And not only is it redundant and pointless to do so, but many of the very entities which are customarily introduced as meanings (in particular, properties and propositions) are most peculiar, due to the fact that, unlike classes, generally-accepted identity-conditions for them are not yet known. In considering an argument for meanings by a hypothetical person McX, Quine writes as follows: "I feel no reluctance toward refusing to admit meanings, for I do not thereby deny that words and statements are meaningful. McX and I may agree to the letter in our classification of linguistic forms into the meaningful and the meaningless, even though McX construes meaningfulness as the having (in some sense of 'having') of some abstract entity which he calls a meaning, whereas I do not. I remain free to maintain that the fact that a given linguistic utterance is meaningful (or *significant* as I prefer to say so as not to invite hypostasis of meanings as entities) is an ultimate and irreducible matter of fact; or, I may undertake to analyse it in terms directly of what people do in the presence of the linguistic utterance in question and other utterances similar to it . . . The problem of explaining these adjectives 'significant' and 'synonymous' with some degree of clarity and rigor – preferably, as I see it, in terms of behavior – is as difficult as it is important. But the explanatory value of special and irreducible intermediary entities called meanings is surely illusory." [29]

Recently, however, Quine has come to look more favorably upon the introduction of such entities as properties and propositions into semantic analysis. Among the problems of semantics, in one of the more inclusive senses of the term "semantics", is included the problem of providing correct analyses of propositional attitudes, such as knowing and believing, and of what Quine calls "attributary attitudes", such as hunting and fearing. One of the principal problems in connection with the analysis of these attitudes is that of identifying their objects – supposing that they have objects. Certain philosophers have argued that their objects might be taken to be linguistic entities, such as predicates and sentences. Other philosophers have rejected analyses in terms of linguistic entities as too artificial, however, and Quine presently agrees with them. Still, he is apparently at present ready to concede that the attributary and propositional attitudes are relational in nature, and thus demand objects of some sort. If one is to assign objects at all, the natural choices seem to be attributes as objects of such attitudes as hunting and fearing, and propositions as objects of such attitudes as believing and wishing. Quine writes: "Lion-hunting is not, like lion-catching, a transaction between men and individual lions; for it requires no lions. We analyse lion-catching, rabbit-catching, etc. as having a catching relation in common and varying only in the individuals caught; but what of lion-hunting, rabbit-hunting, etc? If any common relation is to be recognized here, the varying objects of the relation must evidently be taken not as individuals but as kinds. Yet not kinds in the sense of classes, for then unicorn-hunting would cease to differ from griffin-hunting. Kinds rather in the sense of attributes.

Some further supposed abstract objects that are like attributes with respect to the identity problem, are the *propositions* – in the sense of entities that somehow correspond to sentences as attributes correspond to predicates. Now if attributes clamor for recognition as objects of the attributary attitudes, so do propositions as objects of the propositional attitudes: believing, wishing, and the rest." [30]

Not that Quine feels that the identity problem has been solved for such abstract entities as properties and propositions: he does not. Nevertheless, he is now ready to admit them into semantic analysis as a curious kind of "half-entity". He writes: "We might keep attributes and propositions after all, but just not try to cope with the problem of their individuation.... Why not just accept them thus, as twilight half-entities to which the identity concept is not to apply?" [31] Abstract entities, then, have recently come to gain favor even with one of their most determined critics.

Among contemporary semanticists, Alonzo Church probably argues most strongly for the need for abstract entities in semantics. Defining the task which any reasonably complete semantical theory must set itself, Church writes as follows: "Let us take it as our purpose to provide an abstract theory of the actual use of language for human communication – not a factual or historical report of what has been observed to take place, but a norm to which we may regard everyday linguistic behavior as an imprecise approximation, in the same way that e.g. elementary (applied) geometry is a norm to which we may regard as imprecise approximations the practical activity of the land-surveyor in laying out a plot of ground, or of the construction foreman as seeing that building plans are followed. We must demand of such a theory that it have a place for all observably informative kinds of communication – including such notoriously troublesome cases as belief statements, modal statements, conditions contrary to fact – or at least that it provide a (theoretically) workable substitute for them. And solutions must be available for puzzles about meaning which may arise, such as the so-called 'paradox of analysis'." [32]

The theory which Church takes to be most satisfactory for accomplishing these objectives, and which, he writes, "seems to recommend itself above others for its relative simplicity, naturalness and explanatory power" [32], is a modification of Frege's semantic theory, which was based on the concepts of sense and denotation. Quine's scruples as regards meanings are on this proposed theory laid aside; meanings are embraced whole-heartedly. Briefly, this modification of Frege's theory, which is due to Church, has as its principal features the following [34]: Every name in the language being investigated is regarded as having both a *denotation* and a *sense*; a name is said to denote its *denotation*, and to *express* its sense. The term "name" is taken in a broad sense, so as to include under it not only individual constants, but also predicates and sentences. Proper names are taken to denote individuals; predicate expressions are taken to denote the classes of *n*-tuples of individuals to which they apply; and sentences are said to denote their own truth-values, viz., truth or falsehood. Senses of names are said to be *concepts* of the denotations of these names. Thus, the sense of an invidimual constant is an *individual-concept*; the sense of a one-place predicate constant is a *class-concept*, which is taken to be a property; and the sense of a sentence is a *truth-value-concept*, taken to be a *proposition*.

There is, further, a second type of meaningful expression under this theory, viz., the class of *forms*. A name is defined as a meaningful expression without free variables; any expression which differs from a name only in possessing free variables at certain of the places where that name possesses constants, is said to be a *form*. Now every variable possesses not only a range, but also a *sense-range*, which is the class of the senses of the admissible substituends for that variable. A form is then said to have not only a *value* for every admissible assignment to its free variables, but also a *sense-value* corresponding to each of its values. Thus, while a name has a unique denotation and a unique sense, corresponding to each form there is on the one hand a whole class of values, and on the other hand a whole class of sense-values. We may say, then, that both names and forms are assigned two types of meaning, in parallel fashion.

In order to handle *non-extensional*, or *oblique*, contexts, – e.g., contexts of the form "believes that . . .", "thinks that . . ." – further distinctions are introduced. [35] Church agrees with Frege that if a name is used in both ordinary and oblique contexts, it does not have the same denotation in the latter contexts as it does in the former; in oblique contexts, the denotation of a name is taken to be its usual sense. Thus, for example, in the sentence "John believes that the world is round", the denotation of the expression "the world is round" is not the truth-value of that expression, but its usual sense; viz., the proposition that the world is round. However, Church regards it as a desideratum – in a formalized system, at least – that every name have but one denotation and one sense. Thus, he proposes that rather than follow Frege, and use a given name both in ordinary and oblique contexts, a second name be chosen to be used in the case of the oblique contexts. That second name will have as its denotation the sense of the first name. This second name will itself have a sense; this sense will have a name; and so on. When applied to the analysis of oblique contexts, Church's semantical approach thus becomes exceedingly complex.

Senses, that is, concepts, are of course abstract in nature. Further, as Church thinks of them, they are as independent of language as are denotations. Church writes: "A concept in this sense is not to be thought of as associated with any particular language or system of notation, since names in different languages may express the same sense (or concept). We suppose that a concept may in some sense exist even if there is no language in actual use that contains a name expressing this concept. And we even wish to admit a non-denumerable infinity of concepts – thus more concepts than there can be names to express in any one actual language."[36]

That this theory involves a considerable commitment to abstract entities is evident. It is Church's opinion, however, that the extent to which a theory is committed to such entities is but one of the criteria by which its worth is to be judged. Others include workability and generality. Taking all these considerations together, Church contends that his theory, though surely open to possible correction or modification, is worth investigation. Complex though it is, "the problems which give rise to the proposal are difficult and a simpler theory is not known to be possible."[37]

University of Colorado, Boulder, Colorado, U.S.A.

ROBERT ROGERS

REFERENCES

1. See Alfred Tarski, 'The Concept of Truth in Formalized Languages', and 'The Establishment of Scientific Semantics' both in his *Logic, Semantics, Metamathematics,* Oxford, 1956, pp. 152–278 and 401–408; also 'The Semantic Conception of Truth and the Foundations of Semantics', *Philosophy and Phenomenological Research* 4 (1944) 341–376.
2. Tarski, A., *Logic, Semantics, Metamathematics,* p. 195.
3. Tarski, A., *Ibid.,* pp. 187–188.
4. Tarski, A., 'The Semantic Conception of Truth and the Foundations of Semantics', *Philosophy and Phenomenological Research* 4 (1944) 342.
5. Rudolf Carnap, *Introduction to Semantics,* Part B., Harvard University Press, Cambridge, Mass., 1948.
6. See Alfred Tarski, *Logic, Semantics, Metamathematics* pp. 247–254.
7. Richard Martin, *Truth and Denotation,* Chicago, 1958.
8. *Ibid.* pp. 108–110.
9. *Ibid.* pp. 104–108.
10. *Ibid.* pp. 115–119.
11. *Ibid.* pp. 119–122.
12. *Ibid.* Chapter VI.
13. Alfred Tarski, *Logic, Semantics, Metamathematics* pp. 255–263.
14. See Rudolf Carnap, *Introduction to Semantics,* Part C; and *Logical Foundations of Probability,* Chicago, 1950, pp. 70–89.
15. Rudolf Carnap, *Introduction to Semantics,* pp. 134–138.
16. See Rudolf Carnap, 'Meaning Postulates', *Philosophical Studies* 3 (1952) 65–73. Carnap here ascribes the method of meaning postulates to John G. Kemeny.
17. John G. Kemeny, 'A New Approach to Semantics', *The Journal of Symbolic Logic* 21 (1956) 1–27 and 149–161.
18. *Ibid.* Theorem 27. See Leon Henkin, 'Completeness in the Theory of Types', *The Journal of Symbolic Logic* 15 (1950) 81–91. See also Alonzo Church, *An Introduction to Mathematical Logic,* Vol. I, Princeton, 1956, pp. 307–315.
19. See W. V. O. Quine, 'Designation and Existence', *Journal of Philosophy* 36 (1939) 701–709; 'On What There Is', *Review of Metaphysics* 2 (1948) 21–38; 'Logic and the Reification of Universals' in *From a Logical Point of View,* Cambridge, Mass., 1953, pp. 102–129.
20. W. V. O. Quine, *Review of Metaphysics* 2 (1948) 25ff.
21. W. V. O. Quine, *From a Logical Point of View,* p. 103.
22. W. V. O. Quine, *Review of Metaphysics* 2 (1948) 32.
23. W. V. O. Quine, *Journal of Philosophy* 36 (1939) 708.
24. Alonzo Church, 'Ontological Commitment', *Journal of Philosophy* 55 (1958) 1008.
25. See Nelson Goodman and W. V. O. Quine, 'Steps Towards a Constructive Nominalism', *Journal of Symbolic Logic* 12 (1947) 105–122, and Richard Martin, *op. cit.,* Chapters XI and XII. See also R. M. Martin and J. H. Woodger, 'Toward an Inscriptional Semantics', *The Journal of Symbolic Logic* 16 (1951) 193–203.
26. See Richard Martin, *The Notion of Analytic Truth,* University of Pennsylvania Press, Philadelphia, 1959.
27. W. V. O. Quine, 'Semantics and Abstract Objects', *Proceedings of the American Academy of Arts and Sciences* 80 (1951) 94–95.

28. See *Ibid.*, pp. 91–94; Also W. V. O. Quine, 'Two Dogmas of Empiricism', *Philosophical Review* **60** (1951) 22.

29. W. V. O. Quine, *Review of Metaphysics* **2** (1948) 30–31.

30. W. V. O. Quine, 'Semantics and Abstract Objects', *Proceedings and Addresses of the American Philosophical Association* **31** (1957–1958) 19.

31. Ibid. p. 20.

32. Alonzo Church, 'The Need for Abstract Entities in Semantic Analysis', *Proceedings of the American Academy of Arts and Sciences* **80** (1951) 100–101.

33. *Loc. cit.*

34. Ibid. 101–104.

35. A. Church, 'A Review of Quine', *The Journal of Symbolic Logic* **8** (1943) 45–47.

36. A. Church, 'A Formulation of the Logic of Sense and Denotation', *Structure, Meaning and Method* (Ed. by P. Henle), Liberal Arts Press, 1951, p. 11.

37. A. Church, *Proceedings of the American Academy of Arts and Sciences* **80** (1951) 104.

JOHN G. KEMENY

ANALYTICITY VERSUS FUZZINESS *

I. THE PHILOSOPHY OF FUZZINESS

The last decade has produced a long series of articles criticizing a variety of precise definitions in philosophy on the grounds that they do not correspond to distinctions made in ordinary usage. We are told that all distinctions in ordinary languages are fuzzy. There are no sharp borderlines to be found in nature, and all attempts to impose them by the analytical philosopher are bound to lead to violations of ordinary usage. These points have been made, most eloquently, by a number of distinguished philosophers.

Though it is readily admitted that an adequate philosophy must account for the fuzziness found in every-day discourse, there are two key questions that may be asked: Can philosophy gain by proposing precise distinctions to replace the fuzzy concepts found in every-day discourse? Is it methodologically sound to formulate these definitions in terms of a formalized language? The purpose of this paper is to answer both of these questions in the affirmative. I shall try to maintain that such definitions, even in artificial languages, may lead to fruitful philosophical speculation; and that more insight can be gained about ordinary languages from this point of view than by an approach that stops at the level of fuzziness.

The raging debate of the last decade concerning the analytic-synthetic classification is perhaps the outstanding example of the great fight just mentioned. It will therefore serve as the principal illustration in terms of which I shall make my methodological points. It gives me special pleasure to dedicate this discussion to Professor Woodger, who has contributed much to the philosophy of precision.

II. THE GRAND DEBATE

During the years 1950–1956, at least twelve major books and articles

* The author is indebted to T. J. Duggan for constructive criticism of an earlier version of this paper.

122

were devoted to pros and cons on the analytic-synthetic debate. Since that time even more articles have appeared, but it seems fair to say that the major points on both sides are already contained in the earlier articles. Without a doubt Quine and White performed a major service in launching what may prove to be the most important losing battle in the history of modern philosophy. Perhaps now, after the passage of a few years, we are in a position to evaluate the merits of disputants on both sides unemotionally. I shall attempt to do this in the present section, as far as Quine's views are concerned. I shall take up White's position in the next section.

Quine, in [2], denounced what he considered two unjustifiable dogmas in empiricist philosophy. These views of his were reinforced in his book [8]. The two works together created a major stir amongst philosophers all over the world.

Specifically, Quine argued that the sharp distinction between analytic and synthetic statements is indefensible. He also stated that the belief that individual statements, especially in science, can be verified or falsified in isolation is a closely related indefensible dogma. His method of argument was to show, to his own satisfaction, that a variety of proposed definitions were unacceptable in principle, and to argue that any other type of definition of the term 'analytic' was doomed to failure.

His major effort was to demonstrate that there is a class of inter-definable terms such as 'analytic', 'synonym', 'intentional language', and 'semantic rule', which cannot be defined in terms of concepts outside the set. Since Quine feels that all of these concepts are too nebulous to have a precise meaning to him, he thought that an empiricist had no way of defining any of these terms to Quine's satisfaction, without involving himself in a vicious circle. He therefore concluded that all these terms should be deleted from the language of philosophy.

His second dogma is related to the first one, if one is willing to accept the verifiability theory of meaning. If the meaning of a statement is to be found in terms of the method of verification, then presumably an analytically true statement is one that will be verified no matter what the facts turn out to be. However, this very definition presupposes that it is possible to verify statements in isolation, a dogma that Quine emphatically rejects.

Quine's strongest arguments concern actual examples taken from

JOHN G. KEMENY

everyday discourse. He sheds doubt on the acceptability of the analytic-synthetic distinction by producing many troublesome examples where we would be puzzled as to whether a given statement is analytic or synthetic. He also shows that similar puzzles exist in the semi-formalized languages practiced by scientists.

Hence, Quine argues that we are confronted with a series of terms manufactured by philosophers, which they cannot make clear to their fellow philosophers. We should therefore stop pretending that we are talking in terms of precise and justifiable distinctions. His position, if generally accepted, would of course have radical effects on many branches of philosophy.

The earliest criticism of Quine's position is found in Mates' [3]. He argues for the necessity of circularity in many types of philosophical definitions. It must certainly be granted that any definition of the dictionary-type must eventually lead to circularity. If we insist on defining new words in terms of ones not previously used, then the only alternative would be an infinite regress. In a finite language, our escape must be either circularity or eventual reduction to the level where we can point in place of defining. But it is questionable whether any high-level abstraction could possibly be defined entirely in terms of pointings. Therefore it is reasonable to maintain that the precise definition of any set of related abstract concepts must involve a certain degree of circularity.

Mates argues that such definitions are by no means useless. The listener may have acquired a certain degree of familiarity with various terms, just from the way they are used, and pointing out inter-relations between these unclearly understood words may make the meaning of each one of them clearer.

Mates also disputes Quine's insistence that, if a concept is clear, one should be able to tell in practice where it is applicable. He cites convincing examples where the opposite seems to be the case. For example, a mathematician certainly claims to have a clear-cut understanding of what it means that a certain statement is a theorem of a given system of arithmetic. However, in practice it may be hard to decide whether the so-called Fermat's Last Theorem is indeed a member of this class. On the other hand, Mates points out that there is a fairly strong intuitive concept underlying each of these philosophers' terms, and therefore the problem confronting us is one of explication, of making precise a fuzzy, intuitive

124

concept. Such a task may be very difficult, but it is never, in principle, impossible.

Martin, in [5], questions Quine's desire to give a definition for analyt-ic-in-*L* for ordinary languages *L*. He points out that it is unreasonable to ask for such precise definitions concerning a language that is very un-precisely formulated. Therefore he maintains that a definition that would be suitable for formalized languages *L* should be all that any philosopher demands.

But even in this approach it is unreasonable to require a single definition that would be applicable to all formalized languages. Indeed, the definition of truth proposed by Tarski in [19] can only be formulated for a given language *L* in a semantic metalanguage that is stronger than *L*. Tarski has demonstrated conclusively that any attempt to escape from an infinite hierarchy of stronger and stronger languages is bound to lead to semantic paradoxes. Martin criticizes Quine for placing a requirement on the defenders of analyticity that is more stringent than that which is satisfied by the Tarski definition, which is acceptable to Quine. In conclu-sion he states that, while Quine and White may have shown inadequacies in definitions proposed so far – for example, those proposed by Carnap – they have in no sense destroyed either the goal or the basic approach proposed by Carnap.

My own contributions at this stage consisted primarily of reviews of previously-mentioned articles in [7] and [9]. I tried to underline some of the major criticisms made of Quine's position. In addition, I felt it impor-tant to point out the necessity of *understanding* a given language before one tries to define any semantic concepts concerning this language. I believe that I showed in these reviews that, if one is willing to accept Quine's criticism of analytic truth, one must also reject the sharp distinction between true and false, to which Quine was willing to ad-here. The following example may make my earlier points somewhat clearer.

Let us try to think ourselves into the position of a person who has neither seen nor heard of swans of any other color but white. He would be quite likely to make the assertion "All swans are white". Indeed, a number of people might reach agreement on the truth of this particular assertion, granted that they are ignorant of the existence of certain similar animals in Australia. Let us now suppose that, as a surprise, the famous

125

black swan of Australia is introduced to them. Each of these men is likely to make one of two assertions. One may say, in surprise, "I thought that all swans were white, but I now see that there are black swans". But another could equally reasonably say "Isn't it surprising that there are animals in Australia which are just like swans, but are black? Perhaps we ought to think up a new name that would cover both types of animals". I would be very much tempted to say that for the latter person the original saying was analytically true, while for the former person the assertion was synthetic and was eventually disproved.

The interesting point in this example is that the type of information needed, once the facts are agreed upon, is exactly the same for a decision as to the analyticity of the statement as for a decision as to its truth. That is, the statement in question, as used by a given person, is analytic and true if he chooses to use the word 'swan' in such a way as to *require* the animal to be white. If, on the other hand, he uses this word to describe an animal of certain general appearance and biological make-up, not specifying its color, then the assertion is clearly synthetic, and is certainly falsified once black swans are produced. All the arguments suggested by Quine and White which shed doubt on the sharp distinction between analytic and synthetic are applicable to this particular example. Therefore I can take each of their arguments and use them equally strongly to question the true-false dichotomy.

My reviews also contained a suggestion for the improvement of previously proposed definitions of analyticity. This suggestion originated in my [4] as a proposal for strengthening Carnap's approach to inductive logic. It was there noted that Carnap's restriction to a language in which atomic sentences are independent prohibited the philosopher from applying the definition to languages containing any one of a number of interesting relational predicates. It was argued, however, that this restriction was quite unnecessary. Carnap originally introduced it to prevent the introduction of analytically false state-descriptions. My counter-proposal was to introduce restrictions on the use of extra-logical constants by means of special semantic axioms (since called *meaning postulates* by Carnap). Instead of placing a restriction on the nature of the language, one could restrict state-descriptions to be maximal conjunctions of atomic sentences which are consistent with all our meaning postulates.

126

This same paper contains an even more fundamental suggestion for revision of Carnap's approach. It is proposed that the syntactic tool of state-descriptions be abandoned in favor of work in a metalanguage. In this case the concept of a state-description may be replaced by that of a model of a logical system. By making it part of the definition of a model that all logical, as well as extra-logical axioms must be true in each model, one acquires a highly flexible and powerful tool for the formulation of a variety of semantic concepts.

It was suggested in [7] that a complete understanding of a given language requires knowledge of its meaning postulates. These serve for extra-logical terms analogously to the way that the logical axioms serve for logical constants. Once these constants are known, Carnap's approach to semantics can be extended to very powerful languages and many of the known difficulties can be overcome. The resulting definition would presumably be a modern explication of Leibniz's concept of "true in all possible worlds". A "possible world" of a modern semanticist would be represented by a model of the language L.

An excellent sketch of the way this program could be carried out for simple languages was given by Carnap in [6]. However, later developments convinced me that the forceful criticism of Quine and White could not be answered short of an actual precise definition for a class of rich formalized languages. I therefore proceeded to give such a definition, basing my work heavily on a variety of ideas due to Carnap, Church and Tarski. I shall give a brief account of this work in Section IV.

Perhaps the most forceful criticism of Quine's position came from Grise and Strawson [11]. Their position is not to defend the analytic-synthetic distinction, but rather to show that Quine has failed to shed doubt on the possibility of making such a distinction. Apparently the authors were unaware of the various articles discussed above, and hence a good part of their article repeats some of the same points. I shall therefore restrict myself to points that are new.

They question Quine as to whether the terms under consideration are really only philosophers' words. After all, terms like 'synonymous' have close counterparts in ordinary language such as 'means the same as'. They are willing to admit that Quine has shed serious doubts on the acceptability of definitions proposed so far, but they argue that this is irrelevant to the question as to whether the distinction can ever be made. They present as strong evidence the fact that people in ordinary conversation *can* reach remarkably good agreement on such questions as to whether two terms do or do not have the same meaning. Anyone who would try to convince the man on the street that he does not know what he is talking about in such cases would have to present arguments different in nature from those presented by Quine.

Grise and Strawson contribute a most useful criterion for telling, in certain everyday situations, whether a given statement is analytic or not. They argue that in a dispute over synthetic truth, the argument can be reduced eventually to questions of fact, which are either unanswered at the moment or about whose answer there is dispute. In other words, such disagreements end in some mutual expression of disbelief. On the other hand, questions of disagreement concerning analytic statements eventually lead to a conviction that we do not know what the other person is talking about. Thus, if we dispute about the number of students who attended a given lecture, we are presumably arguing that one or the other person has made an error in judgment or a miscalculation in collecting his facts. However, if someone honestly denies the statement "All tall men are tall", and he convinces us that he has neither misunderstood nor is he joking, we are forced to say that we do not know what he is talking about. This difference in attitude certainly furnishes a partial criterion for recognizing analytic statements in everyday languages.

They also argue that Quine's claim, that no statement in an ordinary language is permanently immune from revision, does not establish that there are no analytic statements in ordinary discourse. They point out that there are two different ways in which someone would give up a claim. He may admit error, and hence revise his assertion, or he may decide to shift the way he uses certain terms. Since this distinction is readily admitted in every-day disputes, it constitues strong evidence for the reasonableness of the analytic-synthetic distinction.

III. MORTON WHITE ON ANALYTICITY

I should like to consider White's position, in [12], at some length. This is the most detailed formulation of the criticisms of the analytic-synthetic dispute, and the book appeared since my last contribution to this field.

One of the four parts of the book is devoted entirely to the analytic-synthetic question. Of this, Chapter XII carries on the sacred war of the nominalist against universals, and need not concern us at the present time.

Chapter VIII takes up a definition that Quine has indicated would be acceptable to him, if the terms in the definition could be made precise. This is the definition of an analytically true statement as one that results from a logical truth by substituting synonymous terms for synonyms. The definition is made plausible by such often-used examples as "All bachelors are unmarried", which results from the statement "All unmarried men are unmarried" by substituting 'bachelor' for its synonymous phrase 'unmarried man'. This statement is an instance of $(x)[(Px \ \& \ Qx) \rightarrow Px]$, and hence is logically true. However, it is argued with great vehemence, and at considerable length, that the very concept of synonymy is just as much in need of clarification as analyticity, and therefore this approach is as dubious as all the others proposed for the explication of analyticity.

I find this discussion particularly illuminating in view of the fact that the proposed definition of analyticity is obviously unacceptable, even to the most ardent advocate of such distinctions. Consider the following very simple example. The assertion "No tree is taller than itself" would certainly be classified as analytically true by anyone who is willing to make such a classification. Yet there is no conceivable way that this particular statement could be obtained by substituting synonym for synonym in a logically true statement. If we try to establish this as a substitution instance of a logical truth, it would presumably have to result from the schema

$$(x) \sim (xRx) \, .$$

However, this schema does not hold for an arbitrary relation R, and hence is not logically true, though it is presumably a consequence of meaning postulates for a reflexive relation like 'taller than'. This example is perhaps the simplest of hundreds of different types that could be manufactured to illustrate that, while substituting synonyms for synonyms in logically true staements will yield analytically true statements, it is not true that all analytically true statements can be so obtained.

It is ironic that the major effort extended by both Quine and White in their criticisms of various proposed definitions of analyticity should be spent on a definition that can so easily be shown to be unacceptable. It points out the danger of conducting these discussions on the level of ordinary languages and, in particular, in a framework that is essentially aristotelian in nature.

This brings us to Chapter IX, where White considers proposed solutions of the problem in terms of what he calls artificial languages. Since the solution that I am proposing in the present article uses this approach, it will be important for me to consider his criticisms.

White seems to be prepared to concede that for a given formalized language, given enough arbitrariness in the definition, one can succeed in formulating a definition of analytically true statements. However, he argues that the result is only the ritual of labeling sentences as analytic or synthetic, and he seriously questions that anything useful is accomplished by it.

Indeed, if one could take seriously White's version of how Carnap proposes to solve the problem, it would be somewhat hard to answer this criticism. We are given the impression that Carnap proposes for each language to give a list of (presumably infinitely many) analytic statements. He would then say that a statement of the language is analytic if it appears on this list. This distortion of the use of meaning postulates is exactly analogous to a description of a formalized logical system as a catalogue of all the theorems. Certainly the proposal made by Carnap and myself was to give as short a list of meaning postulates as we possibly could, which would enable the deduction of the analytically true statements in our formalized language.

One can therefore give two types of answers to White's objections. Let us modify one of his own examples. Suppose that language L_0 contains the meaning postulate "Men, and only men, are rational animals". Let us further suppose that another language L_1 contains a meaning postulate "Men, and only men, are featherless bipeds". Let us now imagine that, on a planet outside the solar system, we discover a feathered quadruped which scores 147 on a standard IQ test. Someone makes the claim that this animal is a man, but this claim is immediately hotly disputed. The specific availability of meaning postulates would allow us to settle it in either of the two given languages. Clearly the statement is true-in-L_0, but false-in-L_1.

Secondly, we could argue that if we had a language without meaning postulates, certain disputes could never be settled. It is quite unclear as to whether the newly-discovered animal is to be classified as a man or not in ordinary English. A poll of men on the street would no doubt result in a wide divergence of opinions, and the same may be the case in a poll of competent biologists. The disagreement would not be one of facts, since we all agree that the newly-discovered animal is a feathered quadruped which is entirely rational. Our disagreement is as to the "correct" usage of the word 'man'. In other words, at this stage we would be forced to reach agreement on a meaning postulate, in order to settle a dispute.

Indeed, throughout his entire discussion, one is left with the impression that White considers English to be a single, reasonably-clear language. He considers questions as to whether a given statement is true-in-English as a meaningful and unambiguous question which, however, is very difficult to answer in many specific instances. I should like to take the liberty to carry this position to its logical, but ludicrous extreme.

It so happens that the sentence "Sir Walter Scott is the author of *Waverly*" is a perfectly meaningful sentence, not only in English, but also in Laputian. However, in Laputian the sentence has the same meaning as the English sentence "Some men are immortal". Therefore, one is forced to admit that the sentence "Sir Walter Scott is the author of *Waverly*" is a sentence true-in-English, but false-in-Laputian. I presume that White would at this point ask me why I raise such a trivial point as the fact that just because two people uttered the very same words in two different languages, we should therefore require that both assertions should be true or both assertions should be false. However, it seems to me that the same remark is applicable to White's examples from English. English-speaking people often utter the very same sounds with entirely different meanings, and therefore the only reasonable question is whether a given sentence uttered by a given man at a given period of his life it analytic or synthetic, and true or false. If one wishes to avoid this tremendously cumbersome method of classification, one will have to insist on a language whose syntactic and semantic structure is perfectly precise, which either means a completely artificial recasting of English or work in terms of a formalized language.

But White raises the point that even for formalized languages the solution is far from adequate. He ties the question of analyticity to that of *a priori* knowledge and states that, according to the best of his knowledge, there is no logic that is "sufficient for the deduction of all *a priori* knowledge". However, this objection seems pointless in view of Gödel's results. As a matter of fact, we know that there is no logical system that is sufficient for deducing all logical truths, or even deducing all logical truths concerning whole numbers. The existence of these results of Gödel seem to be a clear warning to the philosopher that reasonable sounding questions formulated in terms of ordinary language may turn out to be completely unreasonable, looked at from a more rigorous point of view. I think that most logicians would agree that Gödel's results demonstrate that the requirement of deduction of all logical truths is unreasonable, not only in formalized languages, but in any language utilized by human beings. We are forced into the position of dealing only with a limited class of statements in all such considerations, and to a weakening of the requirement of deducibility. This will be considered in the next section.

Again, White raises the question as to whether the claimed distinction is so fundamental, when it is prefectly reasonable to establish an analytic statement by observation rather than by reasoning about it. But one can't help wonder whether this is really a correct description of the situation. Suppose that we make a tally of 100 students in a given course and find that 60 of them are males and 40 of them are females. It might then be claimed that the count of females constituted an empirical check of the analytic assertion that $100 - 60 = 40$. However, let us suppose that our tally showed that we had a total of 100 students, 60 of whom were males and 42 were females. Would we then be equally willing to accept this as an empirical disproof of the same analytic statement? I should like to assert categorically that no one, whether speaking English or Laputian, either of the ordinary or of the formalized variety, would ever accept such a disproof. Or, if by any chance an English-speaking person accepted such disproof (Laputians are excellent mathematicians), he could certainly be persuaded by pure reason that he has made a mistake. This very difference in attitude seems to be one of the best tests of the fundamental nature of the distinction in question.

One of the most interesting questions raised by White is whether the thesis held by many empiricists "Analytic and only analytic statements can be established *a priori*" is itself an analytic or synthetic assertion. I will assume that the analytic-synthetic classification relates to the content of statements, while the *a priori-a posteriori* classification refers to the way human beings establish statements.

On reflection, it seems to me that this question can best be answered by separating the two parts of the thesis. It appears quite clear that the assertion "All analytic statements can be established *a priori*" is itself analytically true. Presumably this is the reason why we have had few, if any, serious proposals for the existence of analytic *a posteriori* knowledge. However, the converse assertion, which may equivalently be formulated as "All synthetic statements must be established *a posteriori*", appears to me to be synthetic in nature. While I believe it to be true, on the basis of the best available knowledge, I could conceive of human beings of different constitution for whom this assertion would be false. Let us suppose that some children now alive grew up with the uncanny ability to predict such facts as the mean temperature in New York City for a given day, a year in advance, by simply closing their eyes and thinking about the question hard enough. If sufficient evidence accumulated that these predictions were invariably – or almost invariably – accurate, one might be forced into the position of admitting that these human beings had the capacity of establishing synthetic statements *a priori*.

However, some of these remarks have diverted us from the main stream of the discussion. Let us recapitulate three major questions confronting us:

(1) Can one give a precise explication for the concept of analytic statements for a formalized language? I shall try to answer this question in the affirmative in the next section.

(2) Is the information obtained from such a definition useful to a person whose interest centers around ordinary languages? I shall again try to answer this in the affirmative, in Section V.

(3) Is this type of approach – of giving a precise definition in terms of formalized languages and then applying it to ordinary languages – philosophically fruitful? Some arguments have already been presented in favor of this position. I shall try to argue more generally, in Section VI, that in all similar cases it *is* fruitful to work for a precise explication and to view the fuzziness of ordinary languages from this higher point of view.

IV. ANALYTICITY IN A FORMALIZED LANGUAGE

Let us first consider the possibility of giving a clear and reasonable definition of analyticity for a fully formalized language. Building on the foundations laid down by Carnap, Church, and Tarski, I proposed such a defini-

133

tion in [10]. An excellent summary of this definition, as well as other work in formalized semantics, will be found in the article by Robert Rogers, on page 38 ff. I shall, therefore, restrict myself to a very brief sketch, to make the present article self-contained.

The definition may be said to produce a modern counterpart of Leibniz's "true in all possible worlds". A model of a formalized language L is a means of describing a "possible world" in the metalanguage of L. Roughly speaking, a model assigns a denotation to each term of L, allowing us to interpret (in an extensional rather than intensional sense) each statement of L. This is a somewhat extended use of the term 'denotation', in which each syntactic constituent of a sentence has a denotation. The sentence itself denotes either 'true' or 'false'. A statement of L is said to be *valid* if it denotes 'true' no matter what denotation is assigned to its free variables.

The concept of validity, when applied to sentences (i.e. statements without free variables), is an exact analogue of truth as defined by Tarski, only it may be relative to an arbitrary model. Indeed, one model assigns to each term its denotation in the actual world, and we may say – in agreement with Tarski – that a sentence is *true* if it is valid in this actual-world-model. The sentence is *analytically true* if it is valid in all models.

The denotation of a logical constant is fixed once and for all, but the denotation of a term like 'man' varies. In some models (e.g. the actual-world-model) the term denotes a class all of whose members are animals under ten feet tall, but in other models there may be members of the denoted class that are twenty feet tall. Hence the statement "all men are under ten feet tall' is valid in the actual-world-model, but not in all models; hence it is true, but not analytically true.

The definition briefly sketched so far is adequate only for fairly simple languages. We must require that the denotation of logical constants be the same in all models; if the logical axioms are strong enough to characterize these constants, then each model – which must make all axioms valid – will assign the intended denotation to these constants. However, we now know from the work of Henkin, in [17], that sufficiently strong languages are always incomplete in the sense that they admit some models not intended by the creators of the language. This leads to a restricted class of models, which I have called interpretations, in which all logical constants have the intended denotations. A sentence is analytic if it has the same truth-value in all interpretations; it is analytically true if it is valid always, and it is self-contradictory if it is always contravalid.

This definition was given for a very strong and rich family of languages, presumably adequate for ordinary philosophical discussions as well as for most of mathematics. It can be adapted to a variety of languages with only minor modifications. Once the formal definitions are available, a long list of significant theorems can be proved in the metalanguage.

For example, if we define a *logical sentence* as a sentence all of whose constants are logical, then it is easy to show that it must have the same truth-value in all interpretations (since the denotations of the constants do not change). Hence all logical sentences are analytic. Under a reasonable definition of synonymy, it can also be shown that a sentence resulting from an analytically true sentence by substituting synonyms for synonyms is analytically true. Hence all sentences resulting from logical truths by substitution of synonyms are analytically true; but it is easy to show that in all non-trivial languages the converse is false (see Section III above).

As a matter of fact, even deeper insight can be gained. By passing to a meta-metalanguage, conditions of adequacy for various definitions can be considered. For example, Carnap proposed the condition that in our metalanguage the statement "The sentence S of L is analytically true" should be true if and only if "The sentence S of L is true" is analytically true. This can be demonstrated for the proposed definition of analyticity. One can lay down further requirements for various concepts, e.g. that they be invariant under translations. It can be shown that this holds for analytic truth, but it does not hold – in general – for logical truth. We thus find that rigorous analysis demonstrates that our intuition misleads us when we feel that 'logically true' is a sounder concept than 'analytically true'.

We may conclude that it *is* possible to give rigorous and reasonable definitions of analyticity for formalized languages. I do not claim that my own is in any sense the best possible definition. I sincerely hope that others will be able to improve on it. But it establishes the *possibility* of a sharp and useful distinction in formal languages.

Once a precise criterion is accepted it leads to a new unified viewpoint for a variety of problems. For example, definitions – mathematical or other – may simply be viewed as meaning postulates. This will divorce various pragmatic considerations from semantic problems. Again, the status of statements as axioms or theorems is completely arbitrary, but their status as analytically true is not. Problems of logical truth, synonymy translation, etc., can all be studied within the same framework.

But does this help us in our study of ordinary languages?

V. LANGUAGES ORDINARY AND ARTIFICIAL

The literature of the last twenty years contains a wide variety of criticisms of attempts to solve philosophical problems in terms of artificial languages. The attack on the employment of formalized systems is led by a group of philosophers who seem to be intoxicated by the sound of their ordinary language.

Without a doubt the philosophical school of language-analysis has made several major contributions to modern philosophy. It has amply demonstrated the therapeutic value of carefully analyzing what a given term would mean in ordinary discourse. For example, a criticism of a proposed justification of induction on the grounds that the word 'justify' is used in a sense never ordinarily employed, is a novel and excellent philosophical practice.

Again, the practice of attempting to translate a given assertion from one ordinary language to another is a good safeguard against drawing conclusions depending on accidental features of an individual language. Let us take for granted from here on that concern with the ordinary use of words is an important part of the task of the philosopher.

However, the basic issue is whether language analysis is the only legitimate task for a philosopher, and whether philosophical analysis must be carried out in terms of an ordinary language. Considerable insight may be gained by applying language analysis to the term 'artificial language'.

It is interesting to note the refusal of many philosophers to refer to certain constructs by the term 'formalized language', and that they insist on using 'artificial language'. In what sense of the word is 'artificial' used? The normal connotation of artificial is an antonym of 'natural', rather than of 'ordinary'. In such contrast it carries an implication that one language was created by nature, while another language is man-made. But surely no language comes about except as an artifact, as a conscious creation of mankind. What, then, are the real differences between ordinary and artificial languages?

136

Ordinary languages have been with us since the beginning of historic time, while artificial languages are essentially the creation of the last century. But the age of a given language is irrelevant to its employment for philosophical purposes. Again, ordinary languages are presumably the gradual creations of a large group of people, while artificial languages have been created by a single person or by small groups. Yet it is hard to see how this is relevant to the employment of such languages for philosophical purposes.

A very noticeable difference is that most artificial languages are formulated in terms of symbols rather than ordinary words. However, words themselves are symbols, though more familiar-appearing than those employed in symbolic languages. In spite of this apparent strangeness, most students learn a symbolic language much more rapidly than a foreign language. The employment of newly-manufactured symbols is, after all, an inessential feature of formalized languages. Carefully selected words from the English language could be substituted for symbols without destroying the essence of a formalized language. This practice was avoided presumably because it was thought easier to teach precise meanings for entirely new symbols than to make precise the meaning of previously ambiguous ones.

I have reluctantly come to the conclusion that many philosophers believe that ordinary language contains within it, hidden, "the wisdom of the ages" in some mysterious way, and that this wisdom cannot be reproduced in formalized languages. I should hate to subscribe to such a principle unless it were meant in a sense in which ordinary languages also contain within themselves all the foolishness of the ages. The key question appears to be not so much whether we are to employ ordinary languages as a basis for our philosophical discussions, but rather, whether we are forced to employ them in their illogical, vague, and ambiguous form. Is the philosopher to be a slave to the linguistic accidents of 10,000 years?

Philosophers must face up to the fact that there are certain important problems that cannot be discussed except in terms of a language free of vagueness and ambiguity, and with a clearly-understood structure. For example, when Quine discusses the problem of ontological commitment, he does this not in terms of ordinary language, but in terms of a "refinement of scientific language". Such a language is presumably a segment of ordinary language which has been freed of its most obvious vagueness and ambiguity, and to which a portion of mathematics has been added, to enable the scientist to carry on technical discussions with ease. For many purposes it is not even important whether anyone has ever constructed such a language, that is completely satisfactory. We can certainly conceive of such a language, and that is adequate for philosophical discussions. However, once such a language is constructed, or even imagined, it turns out to be a language artificial in any sense of that word in which it is applicable to formalized languages. Once one agrees to a discussion of refinements of scientific languages, one might as well agree to take as the basic object-language a formalized language. The latter has the advantage that its structure is explicitly stated and that it is, in a clearly understandable sense, simpler than a refinement of ordinary language. It also has the advantage that we now know how to construct stronger and stronger formalized languages, and can presumably construct any language that may be needed for the most abstruse philosophical consideration. That such a hierarchy of stronger and stronger languages is essential for semantic considerations is clearly demonstrated by the work of Tarski.

There are many other advantages in dealing with formalized languages. For example, when one is confronted with complete freedom in constructing one's language, one may discover certain accidental features that are common to all ordinary languages. While the simple method of translation from one language to a different one may show that philosophical beliefs are relative to accidents of a given language, the method will not free philosophy from accidents common to all human languages. After all, there are only a few hundred languages constructed by mankind, and probably only a few dozen fundamentally different in form. Granted the basic similarity of all men, one must assume that there are many accidental features in common to all languages.

A perfect example of results achieved in terms of formalized languages, which would have been completely impossible for ordinary languages, are the previously mentioned theorems of Gödel concerning the essential incompleteness of all logical systems. It sounds entirely reasonable to demand that one construct basic principles of logic or mathematics from which one can deduce all logical truths, or at least a certain fraction of them. The most eminent mathematicians were convinced that it is possible to construct a set of axioms for whole numbers from which all properties of these numbers are deducible. Yet Gödel demonstrated that it is in principle impossible to accomplish this goal in a man-made language free of inconsistencies. His very approach would have been impossible in terms of ordinary languages. We thus see a dramatic demonstration that a principle that sounds perfectly reasonable when formulated in vague, everyday languages, can be shown to be unreasonable from a more precise point of view.

VI. PROPOSALS FOR PHILOSOPHICAL PROGRESS

I have already admitted that in the early stages of the analysis of a philosophical problem the approach of the ordinary-language-school is most fruitful. It leads to a clarification of the issue puzzling us. It may even result in the startling discovery that the puzzle dissolves, once the language in which it is stated is cleared up. But I firmly deny that all philosophical problems are of this type.

An important class of problems belongs to the area in which the role of the philosopher is to explicate, to make intuitive concepts clear. Here the philosopher may play a more positive role; instead of dissolving a pseudo-problem, he may provide us with a sharper tool for the analysis and classification of knowledge. This task must by its nature go beyond the realm of the ordinary-language-philosopher, since the explicator is called upon to improve an ordinary language.

I should like to propose four methodological principles to guide us in disputes where we find ordinary language too vague for philosophy.

First of all, the disputants must *agree on a language* to be studied, whose logic and syntax are fully understood, and which is free of vagueness and ambiguity. This language may be a reconstruction of an ordinary language, a refinement of a scientific language, or it may be taken as fully formalized. Part of the understanding of the language is specifying all its meaning postulates, whether these concern logical constants or extralogical constants. It has now been admitted that it is legitimate to specify the so-called logical axioms governing a given formalized system. But these are no more and no less than postulates specifying the way in which logical constants are to be used. The extralogical meaning postulates proposed by Carnap and myself serve precisely the same purpose for extralogical constants. No one would dream of studying a logical system, in which a certain constant stands for "or", without specifying whether this is used in the exclusive or inclusive sense. Indeed, some of the proposed logical axioms may turn out to be self-contradictions if interpreted in the wrong sense. Why is it any less reasonable to require knowledge of whether "man" is used in a sense including all rational animals or in a more restricted sense? It would appear to me that philosophical studies conducted in terms of incompletely understood languages must eventually lead to chaos.

Secondly, I would like to suggest that the basic task of such philosophical analysis is the *explication of intuitive concepts*. While I would gladly admit that such analysis should start with a careful consideration of the various ways a given word is used in ordinary discourse, and of the different possible meanings of this word, I would like to suggest that it is the duty of the philosopher to go beyond this stage. Once we have understood how ordinary discourse uses certain concepts to classify and clarify knowledge, we may come to a proposal for improving this method of classification. The criteria for a reasonable explication have been spelled out many times, notably by Carnap. We try to hit upon one reasonably clear meaning of a term and arrive at an exact concept that agrees with the original one whenever its meaning is clear. It is precisely in the cases of ambiguity or vagueness that the philosopher must be given the freedom to legislate.

The third principle concerns the *judgment of proposed explications*. Of course we would wish to know that the concept proposed corresponds reasonably with the way a term is used in ordinary language. But this cannot be an overriding consideration. After all, if a proposed concept fails this test, it only means that it is an explication of a concept other than the one originally proposed. Therefore we may argue that some other term should be used for describing it, but the explication need not necessarily be abandoned. I propose that the only legitimate criterion for the criticism of a philosophical explication is the criterion of fruitfulness. It should be judged as to whether it helps us to organize knowledge, as to whether it provides insight into ordinary discourse as well as into the proceedings of science, and it should be judged in terms of the kind of results that may be established about it.

I suggest that the criteria for judging an explication are in part similar to the criteria by which a mathematician judges the attractiveness of an axiom system. An axiom system may be justified either by its usefulness in terms of applications or because of the aesthetic attraction of the theorems that can be proved in it. One might also claim that the real attractiveness of an axiom system to the mathematician may be measured in terms of the theorems a mathematician believes to hold in the system, but is unable to prove. Similarly, a philosophical concept may be judged partly in terms of the results one can establish about them, and partly in terms of the entirely new problems that are suggested by a study of this concept.

Finally, I would like to propose an *attitude towards explications.* While it is very frequently the case that first attempts at explication have serious faults, usually much more progress can be made by suggesting improvements in the proposed explication than by throwing one's hands up in horror at the sight of the shortcomings. While constructive criticism, pointing out the failings of a proposed definition can be most helpful, these should *never* be used to block further attempts at an improved explication.

Two outstanding examples of such explications are the various proposed criteria of the positivists for an empiricist criterion of meaning, and the proposed explication of scientific explanation by Hempel and Oppenheim in [16]. In both cases the discussion started with, or eventually was carried on in terms of a supposed semiformalized refinement of science. Many examples were considered, and an attempt was made to generalize the principles of meaningfulness, or of a reasonable scientific explanation, from these examples. In each case the proposed criterion went beyond ordinary usage and legislated as to what would be a reasonable, precise concept corresponding to our intuitive ideas. The fruitfulness of these attempts is amply demonstrated by the long list of extremely interesting and varied philosophical papers that have arisen as a result of these proposed definitions.

I think that each of the authors involved would now admit that no criterion yet proposed is entirely adequate as an empiricist definition of meaningfulness. One must also admit that the Hempel and Oppenheim explication of the concept of explanation had certain shortcomings. But the very fact that the definitions existed has led philosophers to ask questions that were never previously asked. The most fruitful discussion has concentrated on relatively small, precisely formulable points, in which a proposed definition failed to meet intuitive criteria of adequacy, and where there was hope of arriving at improvement with a reasonable expenditure of effort. Insofar as one can ever speak of progress in philosophy, I would feel that these are two areas in which twentieth century philosophy has made major progress.

In other cases explications were fairly generally ignored by the ordinary-language-school. Aside from frequent papers proving *a priori* that inductive logic cannot be made precise, and that the concept of simplicity must forever remain vague, the work of Carnap on inductive logic (beginning with [14] in 1950) and my own efforts to explicate simplicity (in [18] in 1953), were conveniently overlooked.

Let us now attempt to apply the proposed criteria to the debate on the analytic-synthetic distinction. If the question is formulated as "Is there any sharp distinction between analytic and synthetic statements?" the problem appears to be an ontological question. But surely it is not that. It concerns a concept that is used in one form or another in connection with ordinary discourse, and that has played a fundamental role in the philosophies of a number of distinguished thinkers. If we can agree to the first principle I have proposed, most of the criticisms leveled against the concept will automatically disappear. The basic criticisms of Quine, White, and their followers then appear as difficulties in applying a proposed explication to actual instances in everyday or scientific discourse, and not as difficulties in principle in formulating a definition. They thus appear to be pragmatic rather than semantic in nature.

If we agree to the principle that the philosopher must try to explicate important intuitive concepts, then we should pay serious attention to the various proposals that have been made for the explication of the concept of analytic and synthetic sentences. According to my third principle, such a proposed explication, if roughly in agreement with ordinary usage, should be judged on the basis of fruitfulness. Let us now consider in what sense the proposed definitions are fruitful.

Presumably no one will deny the fruitfulness of these concepts when they are applied to formalized languages. It enables one not only to answer traditional questions, and to prove precisely a number of principles that were proposed earlier concerning analyticity, but it also leads to the formulation of new and fascinating problems. For example, the deep and difficult problem of the senses in which one translates from one language to another may be formulated in these terms. Some discussion of this problem will be found in [10]. It seems also to occur invariably that such precise formulations point out the unreasonableness of assertions generally believed previously. For example, the inadequacy of defining analytic statements as resulting from logical truths by substituting synonyms for synonyms, and the general questionability of the concept of logical truth, are two such by-products of my formal definition of analyticity.

Next let us consider the fruitfulness of applying these concepts to the study of science. Here we run into our first difficulties, in that scientists do not make precise which of their various proposed hypotheses are analytic in nature and which are synthetic. This seems to be an area in which the philosopher of science might make a useful contribution. He would certainly arrive at the conclusion that the scientist has a considerable amount of freedom in labeling some of his hypotheses as meaning postulates. On the other hand, he does not have complete freedom. Pointing out various options may be very useful to the scientist when he is confronted with unfavorable evidence forcing him to change his hypotheses. I would also argue that there are some clearcut examples in science of statements that were presented as if they were synthetic propositions, which more careful analysis will reveal to be analytic. For example, I would propose Newton's Law, $F = ma$, as an outstanding example of this. Again, the study of instances in which propositions that were previously analytic were abandoned, shed considerable insight as to the procedure of scientists. It is fascinating to study the circumstances under which a scientist decides to use old words with entirely new meanings. Good examples are the change of the term 'velocity' from a scalar concept to a vector concept, and the gradual broadening of the concept of energy.

Let us finally consider whether such an approach gives useful insight into ordinary discourse. It seems to me that it does. Quine can certainly make a strong case for the claim that it is nearly impossible to tell whether "All swans are white" is analytic or synthetic as used by some one person in normal conversation. (Assuming that such assertions are ever made in normal conversations.) One may stop here, and arrive at a completely negative view of the role of philosophy, which is the impression left with the reader at the end of Part 2 of White's [12]. Instead we could note that this is an instance in which much sensible conversation is possible even though no decision has been made as to the exact meaning of 'swan'. A child could, and often does, grow up with a quite vague notion of what animals it would apply the term 'swan' to. If it never hears of black swans, it may never be forced to make a decision. But the fact that a decision may be forced leads to the interesting question of how ordinary languages slowly make vague concepts more precise. This seems to me to be a much more constructive point of view.

The principal alternative, proposed by White in [1], is to consider analyticity a matter of degree in ordinary languages, depending on how reluctant a person is to give up a belief. But it is hard to see how such a position could lead to fruitful philosophy. We know of thousands of people in the Middle Ages who refused to abandon the belief that the earth is flat, even in the face of overwhelming evidence. Would we therefore propose to call this an analytic or nearly analytic principle? It would appear much more fruitful to take the position of Hempel in [15]: "The interpretive rules governing the *application* of such models to concrete contexts are always more or less vague; but such vagueness may not require the replacement, *within* the theoretical model, of a sharp dichotomy by a schema permitting gradual transitions. . . . A decision as to whether such a change within the theoretical model is called for would seem to depend, in each case, upon the severity of the vagueness involved, the adequacy of the theory involving the dichotomy, and the availability of a more adequate theoretical model embodying the proposed gradualization."

The fourth principle is, therefore, crucial. I would hold that any fairly reasonable precise explication is preferable to a vacuum. It should be abandoned only in favor of a superior proposal.

I would suggest that there is an escape from Gilbert Ryle's celebrated pessimism in identifying philosophy with analysis of meaning in ordinary languages. The escape route is provided by explications, by assigning to the philosopher the role of improving on ordinary discourse, by devising better concepts for the organization and clarification of knowledge. This task is vastly more difficult than that proposed by Ryle, for example, by requiring knowledge of the powerful techniques devised by twentieth century mathematical logicians. But we know that the easiest procedure, which in this case would qualify the man on the street as a potential philosopher, is rarely the most fruitful. At the price of a greater effort the philosopher can retain not only his traditional task of watchdog over other thinkers, but also the role of the system-builder.

Dartmouth College, Hanover, New Hampshire, U.S.A.

BIBLIOGRAPHY

The 12 basic items in chronological order

[1] Morton G. White, 'The Analytic and Synthetic: an Untenable Dualism' in *John Dewey: Philosopher of Science and Freedom*, The Dial Press, New York, 1950, pp. 316–330.

[2] W. V. O. Quine, 'Two Dogmas of Empiricism', *The Philosophical Review* **60** (1951) 20–43.

[3] Benson Mates, 'Analytic Sentences', *The Philosophical Review* **60** (1951) 525–534.

[4] John G. Kemeny, 'Extension of the Methods of Inductive Logic', *Philosophical Studies* **3** (1952) 38–42.

[5] R. M. Martin, 'On 'Analytic'', *Philosophical Studies* **3** (1952) 42–47.

[6] Rudolf Carnap, 'Meaning Postulates', *Philosophical Studies* **3** (1952) 65–73.

[7] John G. Kemeny, Reviews of [2], [3], and [5], *The Journal of Symbolic Logic* **17** (1952) 281–284.

[8] W. V. O. Quine, *From a Logical Point of View*, Harvard University Press, Cambridge, Mass., 1953.

[9] John G. Kemeny, Review of [8], *The Journal of Symbolic Logic* **19** (1954) 134–138.

[10] John G. Kemeny, 'A New Approach to Semantics', *The Journal of Symbolic Logic* **21** (1956) 1–27 and 149–161.

[11] H. P. Grice and P. F. Strawson, 'In Defense of a Dogma', *The Philosophical Review* **65** (1956) 141–158.

[12] Morton G. White, *Toward Reunion in Philosophy*, Harvard University Press, Cambridge, Mass., 1956.

Additional Bibliography

[13] Rudolf Carnap, *Introduction to Semantics*, Harvard University Press, Cambridge, Mass., 1948.

[14] Rudolf Carnap, *Logical Foundations of Probability*, The University of Chicago Press, Chicago, Ill., 1950.

[15] Carl G. Hempel, Review of [1], *The Journal of Symbolic Logic* **16** (1951) 210–211.

[16] Carl G. Hempel and Paul Oppenheim. Studies in the Logic of Explanation', *Philosophy of Science* **15** (1948) 135–175.

[17] Leon Henkin, 'Completeness in the Theory of Types', *The Journal of Symbolic Logic* **15** (1950) 81–91.

[18] John G. Kemeny, 'The Use of Simplicity in Induction', *The Philosophical Review* **62** (1953) 391–408.

[19] Alfred Tarski, 'Der Wahrheitsbegriff in den formalisierten Sprachen', *Studia Philosophica* **1** (1936) 261–405.

R. M. MARTIN

TOWARD A LOGIC OF INTENSIONS *

Logical syntax and semantics constitute a very central part of modern logical theory. Syntax is concerned exclusively with the signs or expressions of a language and their interconnections. In semantics, on the other hand, we are interested not only in signs and their interconnections but also in the relationships between signs and the objects which they designate or denote or stand for in one way or another. Such a semantics is a *denotational* or *designational* semantics. A second branch of semantical theory is concerned not only with denotation but with the *meaning* or *intension* of expressions. Such a theory is sometimes called an *intensional* semantics. We know a good deal about denotational semantics, thanks to the work of Carnap, Kotarbiński, Tarski, and others. In a sense, denotational semantics may now be regarded as a completed body of theory. The study of intensions, however, is in its infancy, and indeed it can be said safely that at the present time we have no fully satisfactory semantical theory of intensions at all.

Mathematicians who concern themselves with semantical matters are on the whole, it would seem, uninterested in matters of *ontology*. The fundamental questions as to what objects there are or are not somehow fail to attract them. The attention of the mathematician focusses primarily upon mathematical structure, and his intellectual delight arises in part from seeing that a given theory exhibits such and such a structure, from seeing how one structure is "modelled" in another, or in exhibiting some new structure and showing how it relates to previously studied ones. But in all of this the mathematician is satisfied so long as he has some "entities" or "objects" or "sets" to work with, and he does not inquire into their inner character or ontological status.

The philosophical logician, on the other hand, is more sensitive to matters of ontology, and to him the fundamental problem as to what

* Supported by a grant from the National Science Foundation, Grant No G9737. This paper contains a preliminary version of some of the material in the author's *Intension and Decision, a Philosophical Study*, Prentice-Hall, New York, to appear.

146

kind or kinds of entities there are actually may be of fundamental interest. He will not be satisfied with being told merely that such and such entities exhibit such and such a mathematical structure. He will wish to inquire more deeply into what these entities are, how they relate to other entities, how they are known or identified in experience, what role they play within knowledge generally, and so on. Also he will wish to ask whether the kind of entity dealt with is *sui generis* or whether it is in some suitable sense *reducible* to or *constructible* in terms of other perhaps more fundamental kinds of entities. If he takes the *sui generis* view, he has of course the enormously difficult task of characterizing the new kind of entity *ab initio*. If he takes the reductionist or constructionist view, his task may be no less formidable, but here he will explicitly show how the new kind of entity may be characterised in terms of entities already studied. There are many types of entities in our universe, and there should be no more dreamt of in one's philosophy than there are actually. The reductionist is free to use such entities as are already available and no more. Those who uphold the *sui generis* view are more liable to introduce a new kind of entity unnecessarily than to explore carefully interconnections between or among entities already at hand.

Suppose a biologist wishes to introduce the term "entelechy" into the language of a certain area of biology. Suppose he introduces this as a primitive and finds that he needs a new kind of entity with variables ranging over them. Also he gives axioms characterizing the new primitive and governing the new kind of variable. Of course these axioms must to some extent at least interrelate the new primitive and variables with the antecedent biological theory in a satisfactory way. The reductionist, on the other hand, will introduce the term "entelechy" if at all only on the basis of definitions and without new entities as values for variables. Entelechies or whatever would then emerge merely as special kinds of entities already available. The difficulties in developing satisfactorily either of these methods suggests that entelechies are not suitable objects for scientific study.[1]

Two important theories concerning the intension of terms have been put forward in recent years – the theories of Carnap and Church.[2] Both of these are built to some extent upon foundations suggested by Frege. Neither of them can be said to have been formulated with much concern for ontological economy. Both of them regard intensions as *sui generis*.

Both admit all kinds of new entities that are perhaps not actually needed. The philosophic logician sensitive to matters of ontology will wish to formulate a more economical theory, probably of a reductionist kind, in which propositions, individual, class, and relational concepts and the like take their place among entities already available. The entities already available will be the individuals, classes, and relations of the kind normally admitted in the sciences and presupposed to some extent at least by our everyday speech and behaviour. The kinds of entities admitted in Church's theory, like entelechies, seem to have no legitimate place in science at all.

Another approach to the theory of intension has recently been sketched by Leonard in his *Principles of Right Reason*.[3] This approach is clearly an important one and should be more fully developed.

Let us outline in this paper first a kind of "rational reconstruction" of Leonard's theory. We shall do this on the basis of Russell's *simplified type theory* and within an wholly *extensional semantical meta-language*. We may then go on to *generalize* this theory in certain important ways. Finally we append to this semantics a kind of *quantitive pragmatics*, gaining therewith a general theory in which semantical and pragmatical intensions of various sorts may be systematically defined and interrelated. A detailed development of this whole theory will not be given, but merely a brief and tentative sketch.

Leonard speaks of objects and of their "characteristics" but it is not altogether clear how we are to construe characteristics. Let us identify them with *classes* of objects in the sense of Russell's type theory. Perhaps such an identification violates some of Leonard's intent. But the only difference presumably between classes and characteristics is the condition of identity: two classes are identical if and only if their membership coincides, whereas if two attributes or characteristics are identical then every object which has one has the other (but not necessarily conversely). This difference between classes and characteristics would not seem to be sufficient to destroy the main thread of Leonard's theory, as we shall see.

Let us presuppose a semantical meta-language based on *designation*. The details of its formulation need not concern us here, and will be clear enough as we proceed. We let "x", "y", etc., be the variables (of type 1) for individuals, "F", "G", etc., the class variables (of type 2), and "κ" and "λ" the variables (of type 3) for classes of classes. (For the moment we need no variables for classes of higher type or for relations.) We let "Des" stand for the relation of designation, and "PredCon" for the syntactical notion of being a primitive or defined predicate or class constant (of type 2). "a", "b", etc., will be the variables of the meta-language taking expressions of the object-language as values. "a Des F" thus states that the expression a designates the class F, and "PredCon a" states that the expression a is a primitive or defined class constant (of type 2). In addition to these notations we use "$\vee$", "$\sim$", " $\cdot$ ", "$\supset$", and "$\equiv$" as the usual logical connectives, and "(x)", "(Ex)", "(F)". "(EF)", etc., as the quantifiers. [4]

We first give a few definitions. Following Leonard, we may say that a class F is *common* to members of a class G if and only if every member of G is a member of F. Thus

$$\text{``}F \text{ Com } G\text{''} \quad \text{abbreviates} \quad \text{``}(x)(Gx \supset Fx)\text{''}.$$

And a class F is *peculiar* to x where Fx.

$$\text{``}F \text{ Pclr } x\text{''} \quad \text{is an alternative notation for} \quad \text{``}Fx\text{''}.$$

A class F is *peculiar to members of G or to G simpliciter* if and only if every member of F is a member of G.

$$\text{``}F \text{ Pclr } G\text{''} \quad \text{abbreviates} \quad \text{``}(x)(Fx \supset Gx)\text{''}.$$

A class of classes κ, on the other hand, is *jointly peculiar* to an object x, provided x is a member of *every* member of κ.

$$\text{``}\kappa \text{ JPclr } x\text{''} \quad \text{abbreviates} \quad \text{``}(F)(\kappa F \supset Fx)\text{''}.$$

To say that κ is jointly peculiar to the individual x is to say essentially that x is a member of the *product* of κ (in the sense of *40.01, *Principia Mathematica*.) This notion may be extended to a *class of objects* as follows. We say that κ is *jointly peculiar* to the class of objects F if and only if every object to which κ is jointly peculiar is a member of F.

$$\text{``}\kappa \text{ JPclr } F\text{''} \quad \text{abbreviates} \quad \text{``}(x)(\kappa \text{ JPclr } x \supset Fx)\text{''}.$$

All of the foregoing definitions can be given within the object-language and employ no notions of syntax or semantics. The notion of being an *intension*, however, involves the semantical notion of designation. We introduce first what Leonard calls the *total contingent intension* for predicate (i.e., class) constants. We say that κ is the total contingent intension of a if and only if a is a predicate constant and there exists a G such that

(i) a Des G and (ii) κ is the class of all classes common to G.

"κ TotContInt a"   abbreviates   "(PredCon $a \cdot$ (EG)(a Des $G \cdot$
$$\kappa = F(F \text{ Com } G))\text{"}^5$$

Note that for any predicate constant a there exists one and only one total contingent intension κ of a. The proof utilizes the uniqueness law for designation that if a Des F and a Des G then $F = G$.[6]

Actually there are several kinds of intensions to be introduced, all distinguishable from each other through their membership. Leonard regards it as a defect of the older logic – as indeed it is of most theories of intension hitherto formulated – that they speak of *the* intension of a term, as though there were only one. The fact is that these theories have not discriminated the many different kinds of intension. The reason is probably two-fold. No clear condition under which two intensions are the same is usually given. Also intensions are usually regarded as *sui generis* and hence how they involve or consist of or are generated out of other kinds of entities is not considered.

In general, an intension of a predicate constant is, for Leonard, any class of classes κ such that κ is jointly peculiar to the extension of that constant and its members are common to it. Thus

"κ Int a"   may abbreviate   "(PredCon $a \cdot$ (EF)(a Des $F \cdot$ (G)
$$(\kappa G \supset G \text{ Com } F) \cdot \kappa \text{ JPclr } F))\text{"}.$$

Every particular kind of intension of a predicate constant must, then, be shown to be an intension in this sense. Given any predicate constant a, there exists at least one class of classes κ such that κ is an intension of a. But uniqueness here of course cannot be proved.

An example or two will help to clarify these notions. Let a be the predicate "triangle". Common to the class of triangles are the classes of entities (i) lying in a plane, (ii) being bounded, (iii) having straight sides, and (iv) having three sides. Also the set of classes (i)–(iv) is jointly peculiar to the class of triangles. In Euclidean geometry of three dimensions all triangles have the characteristics (i)–(iv), and every object which has the characteristics (i)–(iv) is a Euclidean triangle. Thus the class of classes κ consisting of the characteristics (i)–(iv) is an intension of "triangle". Every characteristic in this class is common to all triangles and the class is jointly peculiar to the class of triangles. But κ does not constitute a total contingent intension of "triangle". There are other characteristics common to triangles that are not included, e.g. the characteristic of having interior angles summing 360° or less. Only if we add to κ *all* the classes common to triangles, do we gain the total contingent intension.

Leonard points out that the set of all characteristics common to the extension of a term is always jointly peculiar to the extension of that term. Clearly then we have that

$$\vdash \hat{F}(F \text{ Com } G) \text{ JPclr } G. [7]$$

The extension of a class term is here identified with the class itself, so that this formula seems adequately to express the suggested law. Its proof is immediate by noting that

$$\vdash (x)((F)(F \text{ Com } G \supset Fx) \supset Gx).$$

Also we note that if κ is the TotContInt of a and a Des F, then κ is jointly peculiar to F. And thus clearly if κ is the TotContInt of a it is also an Int of a.

Next Leonard introduces the notion of the *total strict intension* of a term, and for this reference to the notion of *necessity* is needed. The total strict intension of a term is that intension consisting of all and only the necessary members of its total contingent intension. Let us construe "necessary" here in the sense of the predicate "Anlytc" of the author's *The Notion of Analytic Truth* or *Toward a Systematic Pragmatics* (Chap. V). [8] The definition of this predicate is intended to explicate the notion of analytic or logical truth in essentially the senses of Hilbert-Bernays, Tarski, and Carnap. The analytic or logically true sentences here include just those sentences true in virtue of their truth-functional or quantificational structure or in virtue of the theory of identity. There is nothing mysterious about this notion, when it and its semantical foundations are properly understood.

"Anlytc a" is to express that the sentence a is analytic. Suppose a Des F and b Des G. Then the formula "$(x)(Fx \supset Gx)$" has a certain *structural descriptive* name in the semantical meta-language, i.e., in this instance "$(ex\ qu\ (a \cap ex\ hrsh\ b \cap ex))$". "$ex$" is the structural description of "x", "$hrsh$" of "$\supset$", and "$\cap$" is the sign of concatenation. [9]

The notion of being a total strict intension of a predicate constant may now be introduced as follows.

"κ TotStrInt a" abbreviates "(PredCon $a \cdot (EF)(a$ Des $F \cdot$ $\kappa = G(Eb)(b$ Des $G \cdot$ Anlytc $(ex\ qu\ (a \cap ex\ hrsh\ b \cap ex))))$)".

Clearly every total strict intension is included in the total contingent intension.

In a similar way a notion of total *synthetic* intension may be introduced. The total contingent intension is then clearly the logical sum of the total strict and synthetic intensions. But these latter are not in general mutually exclusive, even though no synthetic sentence is analytic.

Leonard notes that the total strict intension of a term (predicate constant) is infinitely large – assuming presumably an infinity of individuals. The proof may be adapted here roughly as follows. Let F_1 be a class in the total strict intension of a predicate constant a, and let F_2 be any non-null class not included in F_1. Then the logical sum $(F_1 \cup F_2)$ of the two classes F_1 and F_2 is also a member of the total strict intension of a and is distinct from F_1. If in fact there is an infinity of individuals, then given any F_1 in the total strict intension of a there is an infinity of classes of the kind F_2 distinct from F_1. But if there is only a finite number of individuals, this proof fails, because, given any F_1 there is at most a finite number of classes F_2 of the kind described.

It is not difficult to extend Leonard's treatment to class constants of higher logical type. Also relations may be handled as classes of higher type, by means of the well-known devices (due to Wiener and Kuratowski) of regarding relations as classes of ordered couples. For constants of each type, their intensions are of next higher type, as is clear from the foregoing. But how are we to handle intensions of *individual* constants? – in Carnap's terminology, how are we to introduce *individual concepts*? Leonard does not consider this problem. Perhaps for him, as for Mill, there are no intensions of individual constants.

One way of introducing intensions of individual constants suggests itself as follows, in terms of an alternative type of definition. Let us consider *unit* classes, i.e., classes with only one individual as a member. Let $\{x\}$ be the class whose only member is x. For a unit class constant designating the class whose only member is x the total contingent intension is κ where $\kappa = \hat{F}(F$ Com $\{x\})$, in other words, where $\kappa = \hat{F}(Fx)$, i.e., the class of all classes of which x is a member. We might then let

"κ TotContInt a"   abbreviate   "(InCon $a \cdot$ (Ex) (a Des $x \cdot$
$$\kappa = \hat{F}(Fx))",$$

where "InCon a" states that the expression a is a primitive or defined individual constant.

Whitehead in his lectures during his last years on the faculty at Harvard used to call the class $\hat{F}(Fx)$ the *essence* of x. It seems not unreasonable to identify such a class of classes with the total contingent intension of an individual constant designating x. In fact, following this suggestion we could perhaps identify the total contingent intension of a class constant a with the essence of the class designated by a as an alternative to Leonard's definition. And similarly for the total contingent intensions of class constants of higher type.

We might also consider introducing the total strict intensions of a constant a in the manner of the Whitehead definition, as suitable *subsets* of the essence of the class designated by a. The total strict intension of a, where a Des F, would consist of just those classes of classes of which F is analytically a member, so to speak.[10] In general, it might be thought, the total strict intension, in this Whiteheadian sense, is closer to what we intuitively would wish to call the *meaning* or intension of the constant designating that class F, than is the total strict intension in Leonard's sense. As constituents of the meaning of a we should wish to include perhaps not just the classes in which F is analytically contained, so to speak, as with Leonard, but all of the properties analytically possessed by F, i.e., all of the classes of classes of which F is analytically a member. The Leonard definition leaves out too much, it might be thought, being too closely patterned as it is on the logic of class inclusion. The meaning of a term is determined by the whole bundle of the analytic properties of its extension, it could be maintained, and not just by the analytic superclasses of that extension, as for Leonard.

Technically speaking, the Whiteheadian definitions have the advantage of being uniformly the same for constants of all types, including individual constants. Leonard's definitions, on the other hand, have the advantage, if such it be, of treating intensions as only one type higher than their terms. On the basis of the Whiteheadian definitions, intensions are two types higher than their terms. Where the whole ontology of types is at one's disposal, the Leonard advantage would seem slight. On the other hand, if one's ontology is to be restricted to only a few lower types, then this advantage might well be significant.

Let us consider Frege's example concerning the morning star and the evening star on the basis of the Whiteheadian type of definition. According to Frege the two phrases "morning star" and "evening star" designate the same object but have different meanings – they have the same *Bedeutung* but different *Sinne*. Within some suitable system, let "Sx" read "x is a star", "Mx", "x shines in the morning", and "Ex" "x shines in the evening". (We disregard here all irrelevant astronomical matters, all matters concerned with the flow of time, precise definitions of "morning", "evening", etc., as is customary, whether justly or not.) We suppose also that "ms" and "es" are distinct individual constants defined in such a way that

$$\text{"ms} = (\imath x)(Mx \cdot Sx)\text{"}$$

and

$$\text{"es} = (\imath x)(Ex \cdot Sx)\text{"}$$

are both analytic sentences. (The notation "$(\imath x)(Mx \cdot Sx)$" is that of Russell's theory of singular descriptions. Descriptions we may assume here to be part of the *primitive* notation of the object-language.) The total contingent intensions of "ms" and "es" are identical, because as a matter of astronomical fact ms = es. The total strict intensions differ, however, because we can find a member of the total strict intension of "ms" not a member of that of "es". Clearly "Mms" is analytic, whereas "Ems" is not, and hence M is a member of the total strict intension of "ms" but not of that of "es".[11]

Two individual constants are said to be *synonymous* provided the sentence which says that the object designated by one is identical with the object designated by the other is itself analytic.

"a Syn b"   abbreviates   "(InCon a. ·InCon b · Anlytc ($a \cap id \cap b$))",

where "id" is the structural description of "$=$". The synonymy relations for constants of higher type may be introduced analogously.

Two individual constants may be said to be *strictly co-intensive* provided there exists a κ such that κ is the total strict intension of both.

"a StrCoInt b"   abbreviates   "(InCon a · InCon b · (Eκ) (κ TotStrInt a · κ TotStrInt b))".

And similarly for constants of higher type.

Clearly synonymous constants are strictly co-intensive and conversely.

We may also introduce relations of *synthetic* and *contingent* co-inten-siveness if desired.

Note that the various intensions here are in fact non-linguistic enti-ties. It has frequently been suggested that intensions might be regarded as classes or virtual classes of *expressions*.[12] For example, the total strict intension of a constant *a* might be regarded as the (virtual) class of co-intensive or synonymous constants. The disadvantage of this type of definition might be that it appears to rely too heavily on merely notational features of the language at hand. As against this disadvantage it should be pointed out that an *interpretation* of the language is pre-supposed. A truth-concept is presupposed by the definition of "Anlytc". Strictly then the intension of a constant regarded as a virtual class of co-intensive or synonymous constants depends fundamentally upon this interpretation. The intension of *a* is the virtual class of constants synony-mous to *a in the given interpretation*.[13]

In spite of this argument, it may well be better for certain purposes to regard intensions as classes of non-linguistic objects, as in the theory here. Suppose we wish to *translate* a constant belonging to one language *L* into another language *L'*. The relation of *being a translation of* is itself a relation within a *comparative semantical meta-language* containing both *L* and *L'* in some way as parts. Let us suppose that *L* and *L'* are given a uniform calligraphy and suppose the theory above is formulated in such a way that within it we can compare them systematically. Let *a* be some individual constant in *L* and *a'* one in *L'*. Thus *a* has a certain total strict intension associated with it, as does *a'*. Suppose the strict intension of *a* is in fact the same class of classes as the strict intension of *a'*. We should then regard *a* as *translatable* into *a'* and conversely. Within this kind of theory, we should be able to formulate an exact comparative semantics in which different systems could be compared not only with respect to their syntax and designational semantics, but with respect to their intensional semantics as well.

The metalanguages sketched here are of course Platonic in two senses: they contain abstract objects as values for variables, and expressions are regarded as shapes or sign-designs. But no doubt they could be reformu-lated in such a way as to regard expressions as inscriptions rather than shapes. They would then still remain Platonic in the first sense, being explicitly based on type theory.

It would be of interest to see how a meta-language of this general kind could be formulated without presupposing the type distinction, using some suitable set theory (such as that of Zermelo or Von Neumann and Bernays) in its place.

For a full semiotical analysis of language semantics must be augmented with a *pragmatics*. In pragmatics we consider not only the expressions of a language, their syntax and designational (and intensional) semantics, but also their relations to the *users* of language. If pragmatical factors are admitted, the semantical theory of intensions may be supplemented with a theory of *pragmatical intensions*. In *Toward a Systematic Pragmatics* tentative foundations for a very narrow kind of theory were laid down based upon a relation of *acceptance*. The basic idiom there is "*X* Acpt *a, t*", read "person *X* accepts sentence *a* at time *t*". The relation there is merely classificatory. *X* either accepts *a* at *t* or not. Let us consider now the possibility of constructing along somewhat similar lines a *comparative* and even *quantitative* pragmatics.

Let us consider a relation of *preference* as between sentences of the object-language. Ordinarily preference is construed as a relation not between sentences but between economic goods, commodities, and the like. Of the various ways of construing preference the one as between sentences may be the fundamental one in terms of which the others may be definable.

More specifically let

$$\text{``}X \text{ Prfrs } a,b,t\text{''}$$

be significant primitively to express that *X* prefers *a* to *b* at time *t*. We consider an experimental situation now as follows. The experimenter *E* himself knows the object-language, its syntactical and semantical rules, and perhaps also its inductive logic in Carnap's sense.[14] Further, we suppose that *E* conducts his observations and experiments under controlled laboratory conditions. Suppose *E* presents the subject *X* with two sentences *a* and *b* of the object-language at time *t*. And suppose he asks *X* which of the sentences *X* regards as the *more probable*, or in which he *more firmly believes*, or whether he is more willing to accept one than the other, or whether he is more willing to *base his actions* on one than the other. E then himself accepts and records a sentence of the meta-language to the effect that X at time t indicates preference for sentence a over sentence b, if such is the case. Strictly E gives X the *choice* if the two sentences and records which of the sentences X has chosen. Several hours later, or the next day or month, E might repeat the experiment on X and get the same choice in response. E would reasonably infer, there being no evidence to the contrary, that X would have made the same choice at any time between. In other words, E concludes that X *prefers* a to b during the whole time interval. And conversely, if E knows that X prefers a to b during some time interval, he is then presumably willing to infer that X would choose a to b at any time during that interval.

Perhaps E would wish to establish that X prefers a to b at t in some other way, in particular, by observing whether X *behaves* or *acts* as though he regarded a as more likely than b at t. If so, E would devise an experiment or series of experiments so that in principle given any two sentences he could decide which X would or does actually choose.

In experimental practice, E might use both of these methods, letting the results of one verify and corroborate the results of the other. Rarely are observations made in isolation. Rather a whole cluster of them are made almost simultaneously and upon the basis of a conceptual background. And an experiment involves not only the cluster of observations and the conceptual background, but usually various operations or actions on the part of the experimenter as well. To give a full description of the whole experiment E would make in order himself to accept a sentence of the form "X Prfrs a,b,t" would thus involve a good deal. For present purposes, in fact, we should not wish to settle once and for all the exact character of the experiments needed. Leaving open the matter of detail, we shall have greater freedom in interpreting the theory in different ways and for different purposes as needed.

Clearly the experimenter E will not wish to impose restrictions of any kind upon X's preferences. He will wish rather to await the results of observation and experiment before stating what X's preferences are or that X's preferences exhibit such and such a ranking or structure. Thus any general axioms or rules to be laid down should be extremely weak. In fact we need only two general rules governing "Prfrs", stipulating that if X Prfrs a,b,t then a and b are distinct sentences, and that if X Prfrs a,b,t

then X Prfrs a,b,t_1 for every momentary t_1 which is a part of t, and conversely.[15]

In terms of "Prfrs" the relation of *indifference*, symbolized by "Indiff", may be defined. If X neither prefers a to b at some momentary time t_1 nor b to a at t_1, X then is presumably indifferent as between a and b at t_1. This notion we may extend to non-momentary times t by requiring that X be indifferent as between a and b at every momentary part of t. It is then readily provable that if X is indifferent as between a and b at t then X neither prefers a to b at t nor b to a at t, that if X is indifferent as between a and b at t he is then indifferent as between b and a at t, and for any sentence a of the language X is indifferent as between a and a itself at t.

Note that the transitivity (more precisely, a quasi-transitivity) of Prfrs has not been postulated, i.e., we do not postulate that if X prefers a to b at t and b to c at t, he then also prefers a to c at t. Of course this circumstance may obtain in special cases. Also transitivity (more precisely, quasi-transitivity) for Indiff is not assumed, but may obtain in special cases. Nor is it postulated that either X prefers a to b at t, or b to a at t, or else that he is indifferent as between them at t. We might require that only one of these three conditions hold for any X at t. But this again we should not wish to postulate in general.

The four conditions mentioned are often taken by economists, and social scientists generally, as constituting conditions of *rational* behaviour. If X's preferences are in accord with these conditions, his preferences are regarded as rational. Using this customary terminology, we may introduce the notion of a *rational preference ranking*.[16]

In view of the symmetry condition concerning indifference, however, the actual definition may be given without the fourth condition mentioned, i.e., that only one of the three conditions holds. But in place of this fourth condition we should have that if X Prfrs a,b,t then it is not the case that X Prfrs b,a,t.

More specifically, we may say that a given virtual class of sentences P exhibits a rational preference ranking relative to X and t as follows.

"P RPR X,t" abbreviates "$((Ea)(Eb)(Ec)(Pa \cdot Pb \cdot Pc \cdot \sim a = b \cdot \sim a = c \cdot \sim b = c) \cdot (a)(b)(c)(Pa \cdot Pb \cdot Pc \cdot X$ Prfrs $a,b,t \cdot X$ Prfrs $b,c,t \cdot \sim a = c : \supset : X$ Prfrs $a,c,t) \cdot (a)(b)(c)(Pa \cdot Pb \cdot Pc \cdot X$ Indiff $a,b,t \cdot X$ Indiff $b,c,t : \supset : X$ Indiff $a,c,t) \cdot (a)(b)(Pa \cdot Pb : \supset :$

X Prfrs a,b,t $\lor$ X Prfrs b,a,t $\lor$ X Indiff a,b,t) $\cdot$ $(a)(b)(Pa \cdot Pb \cdot$
X Prfrs a,b,t : $\supset$: $\sim$ X Prfrs b,a,t))".

Rational preference rankings are by no means the only kind of rankings that are of interest to the logician. There are also rankings concerning X's preferences as between sentences containing in various ways the logical constants "$\sim$" or "$\lor$". Such rankings we may call *normal* preference rankings. These are, however, of no immediate interest for the theory of intensions.

A natural extension of the theory of preference is provided by modern decision theory, stemming from the theory of utility in Von Neumann and Morgenstern's *Theory of Games and Economic Behaviour*.[17] This theory has attracted wide attention on the part of social scientists and mathematicians. But unfortunately philosophers have shown little interest in it and seem to have been unaware of its philosophical relevance. The reason lies in part that attention has been focussed exclusively on the utility of economic goods or commodities or concrete objects. Let us attempt apply this theory now to the *sentences* of a well-knit language-system.

We consider a new primitive "Eq" in contexts of the form

$$\text{"}X \, \text{Eq}_e \, a,b,c,t,\alpha\text{"}.$$

This may be read "person X *equates* a with the *combination* of sentences b and c at time t to the degree α on evidence e".

An exact experimental meaning to "Eq" must of course be given and this is not easy to do. Much contemporary research is being devoted to what is essentially this and allied problems. Very roughly we may think of "Eq" as being operationally characterized as follows. Given three distinct sentences a, b, and c of the object-language, suppose the experimenter E discovers that X does not prefer a to b, nor c to a. What E now wishes is to able to discuss X's preference, or rather his indifference, not only as between sentences, but as between sentences and "combinations" or "alternatives" of sentences with stated probabilities or degrees of confirmation on suitable edivende e. The sentence e here is presumed to be the total available relevant evidence available at the time to X or to E or to both. The sentences in the combination b and c are chosen by E in such a way that their degrees of confirmation on evidence e sum to 1, so that if b has α as its degree of confirmation on evidence e, c will have $(1-\alpha)$.

Perhaps X is an expert in inductive logic, is familiar with the notion of confirmation, and, being presented with a sentence of the object-language, can compute its degree of confirmation on given evidence quickly and correctly. But this we need not assume. If X is familiar with the concept of degree of confirmation but is not skilled at computing numerical values, E may tell him the degrees of confirmation of the alternatives involved. But this we need not assume either. In fact, X may be completely ignorant of the notion of confirmation, relying on E to supply the numerical measure.

For E to decide in a special case whether a given sentence of the form "X Eq_e a,b,c,t,α" holds or not, he may devise some suitable experiment or set of operations. He may *ask* X whether he, X, equates a with the combination of b and c. E may observe X's behaviour, providing it is relevant, to decide. He may do both of these, as well as use other data about X's behaviour, including any relevant preferences or indifferences previously established.

Instead of seeking a strictly operational meaning for "Eq", perhaps we should regard the relation here as a "theoretical construct", in the sense of Carnap and Hempel.

The notion of confirmation is used here to provide the necessary numerical foothold. More specifically, where X does not prefer a to b, nor c to a, the degree of confirmation of b on the total available evidence e is taken as the *numerical ratio for his preference of a over c to that of b over c*. That this is reasonable may be seen as follows.

Scientists in their official parlance provide perhaps ideal subjects for this kind of an application of decision theory. Scientists presumably have a proclivity for sentences highly confirmed on the available evidence as over against falsehoods or sentences with only low confirmation, and their preferences as between sentences should reflect this proclivity to some extent. At any event it is not unreasonable in the present context to assume this. The higher the degree of confirmation of b on the available evidence the greater is the scientist's "satisfaction" with b, so to speak.

Let us speak informally for the moment of the satisfaction attached to b as the utility of b (relative of course to some person X at t), and let it be designated by "$u(b)$". If person or scientist X equates a with the combination of b and c, then his satisfaction with a, $u(a)$, should equal the sum of (i) his satisfaction with b, $u(b)$, times its likelihood of being true on

160

the available evidence and (ii) his satisfaction from c, $u(c)$, times its likelihood of being true on the available evidence. This circumstance is described by the following "natural" equation, where α is the degree of confirmation of b on e, $(1-\alpha)$ of c on e.

$$u(a) = (\alpha \cdot u(b)) + ((1-\alpha) \cdot u(c)).$$

Solving for α, provided $u(b) - u(c) \neq 0$, we have that

$$\alpha = (u(a) - u(c))/(u(b) - u(c)).$$

Hence, as suggested, α may be taken as the ratio for X's preference of a over c to that of b over c.

It has been remarked that X himself need not be familiar with the notion of confirmation. It is E who uses this notion to estimate the ratio of X's preferences. If X seeks to maximize truth on available evidence, so to speak, then his preferences will reflect to some extent the "rational" ratios provided by confirmation theory.

The use of degrees of confirmation in this way provides the effect of an *interval measure* of preference. That such a measure gives rise to a system of numerical measurement for utility has frequently been noted in the literature of economics, and apparently was first observed by Pareto in his *Manuel d'Économie Politique*.

Concerning "Eq" we assume only very weak axioms analogous to those for "Prfrs" or "Acpt".

In terms of "Eq" some interesting types of patterns may be defined, concerned with the "logically correct" or "normal" use of the logical constants "$\sim$", "$\vee$" and so on. Also the notion of a *rational preference pattern* may be introduced essentially as in decision theory.

Consider again a suitable virtual class of sentences P and suppose that it exhibits a rational preference ranking relative to X and t. X's preferences might exhibit certain other features also. Suppose, for example, that X Eq_e b,a,c,t,α, where a, b, and c are all in P, and that X is indifferent as between a and d where d is in P. Then very likely X Eq_e b,d,c,t,α also. Similarly if X Eq_e b,a,c,t,α and X Indiff b,d,t then X Eq_e d,a,c,t,α. Again we should not require that these conditions hold of X's preferences. But it is not unreasonable to think that these and similar conditions would obtain for many P, X, and t.

161

Let us define directly the notion that P exhibits a rational preference pattern relative to e, X and t.[18]

"P RPP$_e$ X,t" abbreviates "(P RPR $X,t \cdot$ Sent $e \cdot \sim$LFls $e \cdot$ $(\alpha)(\beta)(a)(b)(c)(d)(Pa \cdot Pb \cdot Pc \cdot Pd \cdot \sim a=b \cdot \sim a=c \cdot \sim a=d \cdot$ $\sim b=c \cdot \sim b=d \cdot \sim c=d \cdot \sim a=e \cdot \sim b=e \cdot \sim c=e \cdot \sim d=e : \supset :$ (X Indiff $a,b,t \cdot X$ Indiff $a,c,t : \supset : X$ Eq$_e$ $b,a,c,t,\alpha) \cdot (X$ Eq$_e$ $b,a,c,t,\alpha \cdot$ X Indiff $a,d,t : \supset : X$ Eq$_e$ $b,d,c,t,\alpha) \cdot (X$ Eq$_e$ $b,a,c,t,\alpha \cdot X$ Indiff b,d,t $: \supset : X$ Eq$_e$ $d,a,c,t,\alpha) \cdot (X$ Eq$_e$ $b,a,c,t,\alpha \cdot X$ Indiff $c,d,t : \supset : X$ Eq$_e$ $b,a,d,t,\alpha) \cdot (X$ Prfrs $a,c,t : \supset : X$ Indiff $b,c,t \equiv X$ Eq$_e$ $b,a,c,t,0) \cdot$ (X Prfrs $a,c,t : \supset : X$ Indiff $a,b,t \equiv X$ Eq$_e$ $b,a,c,t,1) \cdot (X$ Prfrs $a,c,t \cdot$ $\sim X$ Prfrs $b,a,t \cdot \sim X$ Prfrs $c,b,t : \supset : (\mathrm{E}\gamma)(X$ Eq$_e$ $b,a,c,t,\gamma \cdot (\delta)(X$ Eq$_e$ $b,a,c,t,\delta \supset \delta = \gamma))) \cdot (X$ Prfrs $a,b,t \cdot X$ Prfrs $b,c,t \cdot X$ Prfrs $c,d,t \cdot$ $\sim \alpha = 0 \cdot \sim \beta = 1 : \supset : (X$ Eq$_e$ $b,a,d,t,\alpha \cdot X$ Eq$_e$ $c,a,d,t,\beta : \equiv :$ X Eq$_e$ $c,b,d,t,\beta/\alpha \cdot X$ Eq$_e$ $b,a,c,t,(\alpha-\beta)/(1-\beta)) \cdot (X$ Eq$_e$ $b,a,d,t,\alpha \cdot$ X Eq$_e$ $c,b,d,t,\beta/\alpha : \equiv : X$ Eq$_e$ $c,a,d,t,\beta \cdot X$ Eq$_e$ $b,a,c,t,(\alpha-\beta)/(1-\beta)) \cdot$ (X Eq$_e$ $b,a,d,t,\alpha \cdot X$ Eq$_e$ $b,a,c,t,(\alpha-\beta)/(1-\beta) : \equiv : X$ Eq$_e$ $c,a,d,t,\beta \cdot$ X Eq$_e$ $c,b,d,t,\beta/\alpha))))$".

It should be noted that the evidence e here is in effect taken as the whole evidence required to give the appropriate degrees of confirmation to members of P, the whole evidence which E or X wish to take into account in the given context. In the interesting and important uses of this definition, e presumably will be a long conjunction stating the relevant observational and other evidence, including perhaps some general scientific laws or hypotheses.

We should note that Prfrs and Indiff are not here directly governed by the evidence e. Rational preference might be thought, however, to be governed by "rational" degrees of confirmation on given evidence. X rationally prefers a to b at time t *on evidence* e provided he estimates at time t the degree of confirmation of a on e to be greater than that of b on e. In place of "Prfrs" we might now write "Prfrs$_e$" to stand for this kind of relation. Perhaps in the foregoing definition we should replace "Prfrs" throughout by "Prfrs$_e$", and hence "Indiff" by a suitably defined "Indiff$_e$". This is a nicety we need not consider further here.

Let us attempt now to set up a metric for the measurement of *degrees of acceptance*. The fundamental theorem here is an adaptation of that of Von Neumann and Morgenstern. Following them in essentials, we note that real numbers may be assigned to the sentences of a suitable virtual class P of sentences in such a way as to preserve and exhibit the structure of a rational preference pattern. This assignment or metric will have the usual characteristics of an interval scale of measurement that it is unique once an interval and point of origin have been chosen. Let $\Phi_{P^e}(a,X,t)$ be the number assigned by such a metric to the sentence a of P (relative to X, t, and e).

Suppose that a is in P and that P RPP X,t. We might then say that X accepts a of P at t to the degree α if and only if $\varphi_{P^e}(a,X,t) = \alpha$. This would be to identify outright the degrees of acceptance of a sentence with its Φ_{P^e} value. Different virtual classes of sentences would then ordinarily have different scales of values. One such class might have a scale ranging from -20 to $+35$, whereas another might have a scale ranging from $\frac{1}{3}$ to $\frac{2}{3}$. In order to bring a certain uniformity to the scales for different virtual classes of sentences, and therewith to facilitate their comparison, it will be convenient to restrict the scales for degrees of acceptance to values between and including 0 and 1.

For this we need the notion of being the *maximum* value of Φ_{P^e} for X at t and of being the *minimum* value of Φ_{P^e} for X at t. We let

"α Max Φ_{P^e},X,t" abbreviate "$(P \text{ RPP}_e\ X,t \cdot (Ea)(Pa \cdot \Phi_{P^e}(a,X,t) = \alpha) \cdot (b)(Pb \supset \Phi_{P^e}(b,X,t) \leqq \alpha))$",

and

"α Min Φ_{P^e},X,t" abbreviate "$(P \text{ RPP}_e\ X,t \cdot (Ea)(Pa \cdot \Phi_{P^e}(a,X,t) = \alpha) \cdot (b)(Pb \supset \Phi_{P^e}(b,X,t) \geqq \alpha))$".

We may now define the notion that X accepts a of P at t to the degree α, with $0 \leqq \alpha \leqq 1$, as follows.

"$X \text{ Acpt}_e\ a,P,t,\alpha$" abbreviates "$(Pa \cdot (E\beta)(E\gamma)(\beta \text{ Max } \Phi_{P^e},X,t \cdot \gamma \text{ Min } \Phi_{P^e},X,t \cdot \alpha = (\Phi_{P^e}(a,X,t)-\gamma)/(\beta-\gamma)))$".

Still further types of patterns may be distinguished now in terms of this relation of acceptance to such and such degree, e.g., patterns of normalcy for various logical constants.

Now at last we can return to the theory of intensions. On the basis of the relation of acceptance to such and such a degree, further types of intension may be introduced, akin to what traditionally have been called *subjective intensions.* We limit the discussion to just four types, although there are many more to be distinguished in a fuller discussion.

Subjective intensions must clearly be relative to the person – the use of "subjective" is intended to suggest this. Also they clearly should be relative to time. The subjective intension of a for X at t_1 may well differ from that at some distinct t_2. Subjective intensions must also be relative to a virtual class P of sentences in which the experimenter E is interested. This relativization to P provides a factor of experimental control, as we have in effect already noted.

Let a be an InCon of the object-language and let P have as members a good many sentences containing a. These may be regarded as ascribing properties to the object designated by a. Consider just those sentences of this kind which person X accepts at time t to degree α. The class of all such properties (or classes) is the total *contingent* subjective intension of a of degree α relative to X, P, and t. We let

$$\text{"}\kappa \text{ TotContSubjInt}_1{}^e\ a,X,P,t,\alpha\text{"} \quad \text{abbreviate} \quad \text{"(InCon } a \cdot \kappa =$$
$$F(Eb)(b \text{ Des } F \cdot X \text{ Acpt}_e\ (b \cap a),P,t,\alpha))\text{".}$$

The condition "Tr $(b\cap a)$" might also be added here, but even though $(b\cap a)$ were not true we might still wish to include it in X's subjective intension of a. (The subscript "1" here is to indicate that this definition introduces merely a *first* kind of total contingent subjective intension, a second to follow.)

In a similar way we may introduce total *strict* subjective intensions of degree α.

$$\text{"}\kappa \text{ TotStrSubjInt}_1{}^e\ a,X,P,t,\alpha\text{"} \quad \text{abbreviates} \quad \text{"(InCon } a \cdot$$
$$\kappa = F(Eb)(b \text{ Des } F \cdot \text{Anlytc } (b \cap a) \cdot X \text{ Acpt}_e\ (b \cap a),P,t,\alpha))\text{".}$$

These definitions may be generalized to provide subjective intensions (in the first sense) for constants of higher type. Note that these definitions are of the Whiteheadian kind. For constants of higher type, subjective intensions as provided by definitions along the lines of Leonard may also be introduced.

All of the subjective intensions considered thus far are of degree *exactly* α. The sentences of P determining the members of the subjective intensions are all accepted to the degree exactly α. A very natural extension of these definitions is to include relevant sentences accepted to degree α *or greater*. We add to the subjective intensions of degree exactly α those properties or classes, to speak loosely, determined by appropriate sentences accepted to a degree greater than α. A subjective intension in this *second* sense we speak of as being of degree α *and upwards*. Here also we shall have total contingent and strict subjective intensions of degree α and upwards to distinguish. First we let

"κ TotContSubjInt$_2{}^e$ a,X,P,t,α"   abbreviate   "(InCon a · (Eb)(PredCon b · X Acpt$_e$ ($b \cap a$),P,t,α) · ~ (Eb)(Eβ)($\beta < \alpha$ · PredCon b · X Acpt$_e$ ($b \cap a$),P,t,β) · $\kappa = F$(Eb)(Eβ)(b Des F · $\beta \geqq \alpha$ · PredCon b · X Acpt$_e$ ($b \cap a$),P,t,β))".

A class of classes κ is the total strict contingent subjective intension of degree α and upwards of an individual constant a relative to X, P, and t, if it consists of all appropriate classes determined by sentences of P to degree β for $\beta \geqq \alpha$, provided P contains at least one sentence of the form ($b \cap a$) where b is a predicate constant accepted by X to degree α and no such sentence accepted by X to a lower degree.

The formal definition for total strict and synthetic subjective intensions may be given similarly, with the necessary changes. We may go on to characterize *intersubjective, intertemporal*, and *objective* intensions, somewhat in the manner of *Toward a Systematic Pragmatics*.[19] Also various relations of subjective, intersubjective, intertemporal, and objective *co-intensiveness* may be introduced. Intersubjective intensions should be of especial interest for the study of subjective intensions of *social groups*.

The foregoing brief sketch of a semantical and pragmatical theory of intensions is tentative and exploratory and many details may no doubt be improved upon in various ways. This kind of approach nonetheless is thought to possess some advantages over the Fregean alternatives. The intensions introduced are not *sui generis*, all being traced back to the more fundamental kinds of entities available in the theory of types. The meta-language, like most if not all languages of interest for the sciences, is an extensional language and its logical structure can be made fully explicit. Intension theory need no longer be regarded as a second-best

R. M. MARTIN

theory about shadowy and mysterious pseudo-entities, but when properly developed should be able to take its rightful place among other legitimate branches of logic and empirical science.

New York University, New York, U.S.A.

REFERENCES

1. Cf. Carl G. Hempel, *Fundamentals of Concept Formation in Empirical Science* (*International Encyclopedia of Unified Science,* Vol. II, No. 7), University of Chicago Press, Chicago, 1952, p. 14 and *passim.*
2. R. Carnap, *Meaning and Necessity,* 2nd ed., University of Chicago Press, Chicago, 1956; and A. Church, ' A Formulation of the Logic of Sense and Denotation' in *Structure, Method, and Meaning, Essays in Honour of Henry M. Sheffer,* Liberal Arts Press, New York, 1951, pp. 3–24 and 'The Need for Abstract Entities in Semantic Analysis,' *Proceedings of the American Academy of Arts and Sciences* **80** (1951) 100–112.
3. Holt, New York, 1957, pp. 234–254.
4. Cf. the author's *Truth and Denotation, A Study in Semantical Theory,* University of Chicago Press, Chicago; University of Toronto Press, Toronto and Routledge and Kegan Paul, London, 1958, pp. 151–159.
5. The notation "$F(F \text{ Com } G)$" reads: the class of all classes F such that F is common to G. The circumflex, often used as the notation for class abstraction, omitted here and troughout.
6. Cf. *DesR3, Truth and Denotation,* p. 168.
7. The "⊢" in the so-called assertion sign applying to the symbolic context following it and indicating that that context is a theorem in the language at hand.
8. *The Notion of Analytic Truth,* University of Pennsylvania Press, Philadelphia and Oxford University Press, London, 1959; and *Toward a Systematic Pragmatics* (*Studies in Logic and the Foundations of Mathematics*), North-Holland Publishing Co., Amsterdam, 1959. On page 28 of the former, line 22, please insert "$\sim (Ed')d'$ Occ$\hat{b}$ c" after the first "b". Also please note that the first and third definitions on page 78 of the latter are slightly too broad but may easily be emended.
9. For a full discussion of structural descriptions and concatenation see *Truth and Denotation,* pp. 70–90 and pp. 156–157.
10. Note that the total strict intension of a in this sense would be essentially what is called the *absolute quasi-intension* of a, as defined in *Toward a Systematic Pragmatics,* p. 88.
11. It is to assure that "Mms" is analytic that descriptions are taken here as primitive.
12. Cf. B. Russell, *An Inquiry into Meaning and Truth,* Norton, New York, 1940, p. 209; R. Carnap, *Meaning and Necessity,* p. 152; W. V. Quine, 'Notes on Existence and Necessity', *Journal of Philosophy* **40** (1943) 120; and *Toward a Systematic Pragmatics,* pp. 91–92. Concerning virtual classes, see *Truth and Denotation,* pp. 49–52 and p. 106 f.
13. Cf. *Truth and Denotation,* pp. 278–281, for an analogous argument.
14. We shall utilize informally some suitable adaptation of Carnap's theory of confir-

mation, as applied to the object-language at hand. See R. Carnap, *Logical Foundations of Probability*, University of Chicago Press, Chicago, 1950.

15. The notion of being a moment is definable within the underlying theory of time. See *Toward a Systematic Pragmatics*, pp. 36–37. Cf. also J. H. Woodger, *The Technique of Theory Construction* (*International Encyclopedia of Unified Science*, Vol. II, No. 5), University of Chicago Press, Chicago, 1939, pp. 32–33 and *The Axiomatic Method in Biology*, Cambridge University Press, London, 1937, p. 56 ff.

16. Cf. especially D. Davidson, J. C. C. McKinsey, and P. Suppes, 'Outlines of a Formal Theory of Value I', *Philosophy of Science* 22 (1955) 140–160.

17. 2nd ed., Princeton University Press, Princeton, 1947.

18. Cf. Davidson, etc., *op. cit.*, p. 156. A sentence is LFls *(logically false)* if and only if its negation is analytic.

19. Pp. 47–49.

PART II

LOGICAL ANALYSIS OF THEORY STRUCTURE

KARL R. POPPER

CREATIVE AND NON-CREATIVE DEFINITIONS
IN THE CALCULUS OF PROBABILITY

It was in 1935, in Paris, at a congress for 'scientific philosophy', that I met Woodger first – also Bertrand Russell, Susan Stebbing and Freddy Ayer. Woodger read a paper which I still remember because it not only paid just tribute to *Principia Mathematica* but described in frank and moving personal terms its impact upon the author of the paper.

When shortly afterwards I had the great pleasure of staying with the Woodgers at Tanhurst, their home in Epsom Downs, and Socrates (as he allowed me to call him) had shown me his as yet unfinished book which was to become *The Axiomatic Method in Biology*, he asked me for an axiom system for probability theory to be used in his book. Questioned whether he wanted a more or less complete system or just as much as was needed for the purpose of his book, he replied that he merely wanted a minimum system permitting him to justify his inferences. The task was very easy; but my quite trivial suggestions were most generously acknowledged when the book finally appeared.[1]

It may be perhaps a fitting though a very modest contribution to a volume which is to be dedicated to Socrates if I now present here some of my latest efforts in the same field – the axiomatic treatment of elementary probability theory.

My aim in this paper is to present some recent slight improvements of an axiom system of which I have published several versions previously[2], and to use this system as the basis for a kind of case study of what are known as *creative definitions*.[3]

A definition *d* is called '*creative*' if and only if in its presence otherwise underivable thereoms become derivable which are written solely in the primitive terms of the system (or which contain neither the expression defined in *d*, nor any expression defined with its help); otherwise it is called '*non-creative*' (or 'normal' or 'proper' or 'purely verbal').

It is easily seen that the creativity of a term within a deductive system entails its independence. The opposite is not the case, however (as shown, for example, by a system of positively identical conditionals of the propositional calculus to which intuitionist negation is added by way of a definition; to see this one has only to develop a little some arguments given in Section VIII below).

After having presented my new axiom system for probability I shall first point out a creative definition in it. Next I shall establish the relativity of its creativity: I shall show that creativeness, like independence, is a property whose possession by a formula (the definition formula) depends on the rest of the axiom system. Finally I shall try to describe a simple method of avoiding creative definitions, or of eliminating their creativity.

I. GENERAL ASSUMPTIONS AND NOTATION

We assume a theory of real numbers in which the usual operations – addition, subtraction, multiplication, division, powers, and roots – are available. *RN* is the name of the class of real numbers and we assume that the expression

$$r \, \varepsilon \, RN$$

is a true formula of our system provided r is a real number.

Two fundamental concepts will be implicity defined by our system of postulates and axioms:

(i) S, a class of some elements for whose names we use lower case italics from the beginning of the alphabet: 'a, b ε S' will express that a and b are elements of S.

(ii) A two termed numerical measure-function of the elements of S, denoted by '$p(a, b)$' or '$p(c, d)$' etc. (to be read 'the probability of c given d'). In addition to these primitive terms the following defined terms play a fundamental role:

(iii) The operation of forming the product-element ab (to be read 'a-and-b') of two elements a and b of S, expressed by the concatenation of the names of these elements.

(iv) The operation of forming the complement-element $-a$ (to be read 'non-a') of an element a of S, expressed by placing the sign '$-$' in front of the name of that element.

The atomic statements of our system are equations or inequalities such as '$p(a, a) = 1$' or '$p(a, a) \neq 0$' or '$p(ab, c) \leqslant p(a, c)$' (to be read 'the probability of a-and-b, given c, is smaller than or equal to that of a, given c'). Molecular sentences are obtained from atomic ones in the usual way by using methods of the lower functional calculus. '(a)' and '(Ea)' will be used, respectively, to mean 'for all elements a in S' and 'there is at least one element a in S'. '$\leftrightarrow$', '$\rightarrow$', '&', '$\lor$' express as usual the biconditional, conditional, conjunction and disjunction respectively. Open sentences will be used in the usual way.

II. POSTULATES AND AXIOMS

Postulates

PA $a, b, \varepsilon\, S \rightarrow p(a, b)\, \varepsilon\, RN$
PB $a, b, \varepsilon\, S \rightarrow ab\, \varepsilon\, S$
PC $a\, \varepsilon\, S \rightarrow -a\, \varepsilon\, S$

Axioms

A $(Ea)(Eb)\, p(a, b) \neq 1$
B $((d)\, p(ab, d) = p(c, d)) \leftrightarrow (e)(f)(p(a, b) \leqslant p(c, b)\, \&\, p(a, e) \geqslant$
         $\geqslant p(c, e) \leqslant p(b, c)\, \&\, ((p(b, e) \leqslant p(f, e)\, \&\, p(b,f)) \geqslant p(f,f) \leqslant$
         $\leqslant p(e,f)) \rightarrow (p(a,f)\, p(b, e) = p(c, e))))$
C $p(-a, b) = p(b, b) - p(a, b) \leftrightarrow (Ec)\, p(b, b) \neq p(c, b)$

Axiom A, it should be noted, is weaker than those axioms which demand the existence of at least two different probabilities such as, for example,

$$(a)(b)(Ec)(Ed)\, p(a, b) \neq p(c, d)$$

or

$$(Ea)(Eb)\, p(a, a) \neq p(a, b)$$

Axiom B is a little complicated, although of course as simple as I can make it. (It replaces some still more complicated formulae I have used previously.)[4]

     Axiom C may be brought into a form more like B, by writing

C $((c)\, p(-a, c) = p(b, c)) \leftrightarrow (d)(p(a, d) + p(b, d) = p(d, d) \leftrightarrow$
         $\leftrightarrow (Ee)\, p(d, d) \neq p(e, d))$

I have introduced C before. (It is called Cd in [3], p. 42.)

III. DERIVATIONS

In order to establish the adequacy of the system I shall derive my old system of [2], pp. 332 and 349 from the system given here.

Substituting c/ab (i.e., replacing c by ab) in B we get

(1) $\qquad p(a, b) \leqslant p(ab, b)$ $\hfill$ B

(2) $\qquad p(ab, e) \leqslant p(a, e)$ $\hfill$ B

(3) $\qquad p(ab, e) \leqslant p(b, ab)$

(4) $\qquad \big(p(b, e) \leqslant p(f, e) \,\&\, p(b, f) \geqslant p(f, f) \leqslant p(e, f)\big) \to p(ab, e) =$
$\qquad = p(a, f)\, p(b, e)$ $\hfill$ B

(5) $\qquad p(a, b) = p(ab, b)$ $\hfill$ 1, 2

By e/b and f/b in 4, we get

(6) $\qquad p(a, b) = p(ab, b) = p(a, b)\, p(b, b)$ $\hfill$ 5, 4

(7) $\qquad p(a, b) \neq 0 \to p(b, b) = 1$ $\hfill$ 6

(8) $\qquad (b)\big(((a)\,p(a, b) = 0) \to (\text{E}c)\, 0 \neq p(c, b)\big)$ $\hfill$ C

(9) $\qquad (b)\,(\text{E}a)\, p(a, b) \neq 0$ $\hfill$ 8

(10) $\qquad p(a, a) = p(b, b) = 1$ $\hfill$ 7, 9

By e/c and f/d, in 4 we get

(11) $\qquad \big(p(b, c) \leqslant p(d, c) \,\&\, p(b, d) \geqslant 1 \leqslant p(c, d)\big) \to p(ab, c) =$
$\qquad = p(a, d)\, p(b, c)$ $\hfill$ 4, 10

and by c/b, d/c, in 11,

(12) $\qquad p(b, c) \geqslant 1 \leqslant p(c, b) \to p(a, b) = p(a, c)$ $\hfill$ 11, 10, 5

(13) $\qquad \big((c)\,p(a, c) = p(b, c)\big) \to p(a, b) = p(b, b) = p(a, a) =$
$\qquad = p(b, a) = 1$ $\hfill$ Instantiation; 10

From this and a/d, b/a, c/b in 12, we get

(14) $\qquad \big((c)\,p(a, c) = p(b, c)\big) \to p(d, a) = p(d, b)$ $\hfill$ 12, 13

By a/b, b/c, and e/bc in (1), (2), and (3), we get

(15) $\qquad p(b, c) \leqslant p(bc, c)$ $\hfill$ 1

| (16) | $1 = p(bc, bc) \leqslant p(b, bc)$ | 10, 2 |
| (17) | $1 = p(bc, bc) \leqslant p(c, bc)$ | 10,3 |

and thus by d/bc in 11, and detachment (or *modus ponens*)

| (18) | $p(ab, c) = p(a, bc)\,p(b, c)$ | 15, 16, 17, 11 |

We have now derived all the axioms of my old 'standard system' of [2] pp. 332 and 349 as follows (I am calling here the last of these axioms C1 in order to distinguish it from our C of which it is obtained as an immediate consequence by b/a, c/b, a/c in C):

A 1	$(Ec)\,(Ed)\,p(a, b) \neq p(c, d)$	A, 10
A 2	$((c)\,p(a, c) = p(b, c)) \rightarrow p(d, a) = p(d, b)$	14
A 3	$p(a, a) = p(b, b)$	10
B 1	$p(ab, c) \leqslant p(a, c)$	2
B 2	$p(ab, c) = p(a, bc)\,p(b, c)$	18
C 1	$p(a, a) \neq p(b, a) \rightarrow p(a, a) = p(c, a) + p(-c, a)$	C

In [2] this system has been proved to be not only independent, but to remain so in the presence of the formula

| (D=) | $a = b \leftrightarrow (c)\,p(a, c) = p(b, c)$ |

and of the assumption that S is a Boolean algebra. (This assumption is demonstrable in the presence of D=.)

IV. A BONA FIDE DEFINITION

In our system the following (somewhat unusual) formulae are derivable as immediate consequences of 16 and 17:

(19)	$p(a, a(-a)) = 1$	16
(20)	$p(-a, a(-a)) = 1$	17
(21)	$(Eb)\,p(a, b) = 1 = p(-a, b)$	19, 20
(22)	$(Eb)\,p(a, b) + p(-a, b) \neq 1 = p(b, b)$	21, 10
(23)	$(Eb)\,p(-a, b) \neq p(b, b) - p(a, b)$	22

The element b whose existence in S is here proved is $b = a(-a) = c(-c)$ We may call this element the empty or the self-contradictory element. By C it is easily shown that

175

(24) $b = c(-c) \rightarrow p(b, b) = p(a, b) = 1$ 19, 20, C, 10, D=

for every element a in S.

Accordingly,

(Ec) $p(c, b) \neq 1$

is a necessary and sufficient condition for b not to be empty.

This situation makes it clear that in our system it would be impossible to define the complement, $-a$, of a, simply by the complementarity of their probabilities with respect to 1, that is by

(−) $p(-a, b) = 1 - p(a, b),$

without demanding that b is not empty, or in other words, without postulating (Ec) $p(c, b) \neq 1$ or the equivalent formula

(Ec) $p(b, b) \neq p(c, b)$

We thus see that if we intend to define the complement $-a$ by some formula like (−), we are led, in our system, to some formula like our C:

C $p(-a, b) = p(b, b) - p(a, b) \leftrightarrow$ (Ec) $p(b, b) \neq p(c, b)$

This formula expresses quite directly our intention to determine the status of the complement $-a$ of a by saying that these two have, for every b which is not empty, probabilities complementary to each other with respect to $1 = p(b, b)$.

We can therefore describe C as a *bona fide* definition of the complement $-a$ in terms of the probability function $p(x, y)$.

It is, of course, easy to lay down formal rules for definitions such that C becomes a definition from a formal point of view. But I have used this simple example in order to establish that C has, intuitively or *bona fide*, the character of a definition.

In the presence of D=, we may replace C for example by

C= $-a = b \leftrightarrow (c) \, (p(a, c) + p(b, c) = p(c, c) \leftrightarrow$ (Ed) $p(c, c) \neq$
        $\neq p(d, c))$

which, upon substituting $b/-a$, is at once seen to be the same as C.

The idea of a *bona fide* definition is a completely intuitive and informal one, and the whole point is perhaps not very important; but it seems to me clear that not all definitions are *bona fide*. Take for example our axiom B which has the form of a definition and compare it with the following much simpler formula which has the same definitional form as B yet which explains in effect the product ab as the widest or weakest element at least as narrow or at least as strong as both, the element a and the element b.

BF $\quad ((d)\, p(ab, d) = p(c, d)) \leftrightarrow (e)\, (f)\, (p(a, e) \geqslant p(c, e) \leqslant$
$\quad \leqslant p(b, e)\ \&\ (p(a, f) \geqslant p(e, f) \leqslant p(b, f) \rightarrow p(c, f) \geqslant$
$\quad p(e, f)))$

Now what I am trying to suggest here, quite informally, is that BF may be considered as a *bona fide* definition of the product, while B should not be so considered. The reason is that B does much more than explain the bare meaning of the product; it actually lays down that B2, the general multiplication theorem, holds for the product. And while B1, the monotony law which is also implied by B, follows from BF, the general multiplication theorem B2 does not follow from BF. This is connected with the fact, it seems, that the general multiplication theorem has (intuitively) the character of an axiom. Thus B, though it may have the same definitional form as BF and though it may be used to eliminate the defined term, may be taken as an example of a definition which is an axiom rather than a *bona fide* definition.

The definitional character of BF becomes particularly clear when it is realised that in any otherwise adequate axiom system strong enough to allow us to deduce the general multiplication theorem in the presence of BF, BF can be shown to be a definition of the product.

In order to show even more clearly that C is a *bona fide* definition, I wish to present an alternative and weaker definition which takes $-a$ not as the complement, or the classical negative, of a, but rather as the *intuitionist negative* in the sence of Brouwer. $-a$ is then the widest element narrow enough to form together with a an empty product-element $a(-a)$. This idea can be expressed by

$(\text{Neg}^{int})\quad -a \geqslant b \leftrightarrow (c)\, c \geqslant ab,$

where '$\geqslant$' is defined by

$\quad a \geqslant b \leftrightarrow ab = b$

In the primitive terms of our system, this can be expressed by

177

$$a \geqslant b \leftrightarrow (c)\, p(a, c) \geqslant p(b, c)$$

We can express Neg^{int} in the terms of our system as follows:

$\mathbf{C}_{int}$ $\quad ((c)\, p(-a, c) \geqslant p(b, c)) \leftrightarrow (d)\, (p(ab, d) \neq 0 \leftrightarrow (e)\, p(ab, d) = p(e, d)))$

Here we may finally eliminate the product 'ab'; for example by writing [5]

$\mathbf{C}^{int}$ $\quad ((c)\, p(-a, c) \geqslant p(b, c)) \leftrightarrow (c)\, (d)\, (p(a, d) \geqslant p(c, d) \leqslant$
$\quad \leqslant p(b, d) \rightarrow (p(c, d) \neq 0 \leftrightarrow (e)\, p(c, d) = p(e, d)))$

The fact that we can replace C by a definition which gives the defined term a somewhat different meaning makes the *bona fide* definitional character of C even more obvious. (All these are, of course, merely intuitive or *ad hominem* arguments.)

V. A CREATIVE DEFINITION

Definitions in a formal system are as a rule considered as mere abbreviations. Quine, for example, after introducing certain definitions into his system of mathematical logic, comments upon them as follows:

'Such conventions of abbreviation are called formal definitions... To define a sign formally is to adopt a shorthand for some form of notation already at hand ... To define a sign is to show how to avoid it.' [6]

This description fits our case: whenever symbols of complements, say '$-a$' and '$-b$', occur in some of our theorems, they can be eliminated by the simple though clumsy method of (i) replacing them by some variables, say x and y, not occurring in the theorem in question, and (ii) adding with the help of '$\leftrightarrow$' the conjunction of their defining formulae, that is to say, adding, in our case, an expression like

$$\leftrightarrow (c)\, (p(a, c) + p(x, c) = p(c, c)\, \&\, p(b, c) + p(y, c) = p(c, c) \leftrightarrow (\mathrm{E}d)\, p(c, c) \neq p(d, c))$$

Thus our definition C is indeed an abbreviating convention. [7]

But is it no more than this? Certainly it is. The fact that it is not *merely* an abbreviating convention can be easily established. *For a number of important theorems in which the sign of complementation does not occur fail to be demonstrable in the absence of our definition C.*

It is of course quite trivial that many of those theorems in which the sign of complementation does occur fail to be derivable in the absence of the Axiom C. Yet many such theorems remain derivable by substitution, as long as we have the *postulate* PC at our disposal; for example, we can obtain, without using C, $p(a, a) = p(-a, -a) = p(a(-a), a(-a))$ if we have A3 of our old standard system at our disposal.

But it is a fact that we cannot in our system A, B, and C, derive

(*) $p(a, a) = 1$ 10

without using C; and (*) does not contain the sign of complementation which is defined by C.

Other important complement-free theorems which are nevertheless dependent upon C are: A1, the theorem that establishes the existence of at least two probabilistically different elements in S; the theorem asserting that some probabities equal zero, and that all fall within the interval between zero and one; and the various addition theorems. Of special interest is the important yet little known 'principle of redundancy'

(25) $p(b, c) = 1 \rightarrow p(ab, c) = p(a, bc) = p(a, c)$

which depends essentially on C^8; also a corollary of this principle

(26) $p(a, c) = 1 \rightarrow (p(b, c) \neq 0 \rightarrow p(a, bc) = 1)$ 25 (subst.); B2

which may be called the 'principle of stability'.

It will be seen that these are among the most fundamental theorems of our calculus. Our system would therefore be utterly inadequate without C or something like it, such as C^{int}. Thus C, in spite of being a *bona fide* abbreviative definition, is very much more than this: it is a *creative definition*, and forms an essential part of our system.

The fact that the complement-free theorems here mentioned are dependent upon C can very easily be established by showing that (a) they are derivable in our system, and (b) they are independent of the rest of the system, that is, of the system without C. (a) has been shown here for some of these theorems; for the others it has been shown in [2]. (b) can be shown by the usual methods of proving independence; for example, the independence of (*) – that is, 10 – can be shown by the simple model that assumes $p(a, b) = 0$ for all elements a and b of S.

This model, however, satisfies (25), i.e. the principle of redundancy.

179

KARL R. POPPER

It is therefore perhaps of some interest to show, with the help of a very simple model, that the principle of redundancy cannot be derived without C or C1.

Take the following model S: S consists of four elements, 0, 1, 2 and 3. We define $-a$ by $-a = a + (-1)^a$, so that $-0 = 1$; $-1 = 0$; $-2 = 3$; $-3 = 2$. We define $2.3 = 3.2 = 0$ and $a.a = a$; in all other cases ab is equal to the arithmetical product of a and b. We further define: $p(a, b) = 0 \leftrightarrow ab = 0 \neq b$; in all other cases $p(a, b) = 1$

This model (which is a Boolean Algebra) may be summed up in the form of matrices:

ab	0	1	2	3	$-a$
0	0	0	0	0	1
1	0	1	2	3	0
2	0	2	2	0	3
3	0	3	0	3	2

$p(a,b)$	0	1	2	3
0	1	0	0	0
1	1	1	1	1
2	1	1	1	0
3	1	1	0	1

We can see at a glance that A1, A2, and A3 are satisfied – the latter even in the strong form

$$p(a, a) = p(b, b) = 1 \geqslant p(a, b) \geqslant 0.$$

B1 and B2 will also be found to be satisfied, and even the laws of commutation and association; also $p(a, b) \leqslant p(ab, b)$, and $p(a, ab) = p(b, ab) = 1$. Thus B is satisfied.

But 25 and 26 are not satisfied by $a = 3$, $b = 2$, $c = 1$; and since 25 and 26 follow from the satisfied axioms if we add C or C1, it is clear that C and C1 cannot be satisfied. They fail for $a = 3$, $-a = 2$, $b = 1$.

Yet both C^{int} and $p(a, b) + p(-a, b) = 1 \leftrightarrow p(a, b) + p(a, b) \neq 2$ are satisfied. (Thus neither of them can fully replace C or C1.)

Thus C is a creative definition.

Once we have seen this point, it will be clear that, at least from a formal point of view, B is also a creative definition (though its *bona fide* character may be questioned: it is clearly so contrived as to yield the desired results; but this may be said of the most ordinary definitions in formal systems). As such it is, if possible, even more important and essential for our calculus than C. And of course, should we omit both B and C from our system as 'merely abbreviating devices', then nothing

180

whatever remains of the calculus; A is utterly powerless without the assistance of B and C although the role it plays in the presence of B and C is quite important.

VI. THE RELATIVITY OF CREATIVENESS

An interesting point is that the (Boolean) sum of a and b, which we may write '$a + b$', can be defined in our system by

D+ $\quad p(a + b, d) = p(c, d) \leftrightarrow p(c, d) = p(a, d) + p(b, d) -$
$\quad\quad - p(ab, d),$

or by

$D_+^=$ $\quad a + b = c \leftrightarrow (d)\, p(a, d) + p(b, d) = p(ab, d) + p(c, d)$

Both these definitions of '$a + b$' are non-creative – although they establish the fact that *probability is an additive measure function*; a fact which has often been considered to be decisive for probability theory. These definitions of '$a + b$' are, indeed, *mere* instruments of abbreviation: everything that can be said, and proved, with their help can also be said, and proved, without their help, simply by writing '$-((-a)(-b))$' in place of '$a + b$'.

The question arises whether or not we can construct an equivalent (and independent) axiom system of probability theory in which D+ is creative, or a system in which B or C would be non-creative. In other words, is creativeness an absolute property, either of the defining formula, or of the term to be defined, or is creativeness a relative property, depending upon our choice of the other primitive formulae?

As might be expected, the answer to this question is that *creativeness is relative* to the rest of the axioms chosen. Thus we can construct independent systems in which (a) the term '$+$' is defined by a creative definition, other than D+, and in which (b) the formula D+ becomes creative. I shall show this by giving as an example an independent system consisting of the formulae A and C, and in addition, two further creative definitions DS and DP; that is to say, a definition DS of the sum, $a + b$, in primitive terms, obtained by eliminating the product ab from D+ by the method sketched in Section V, above; and a definition DP of the product ab in terms of the sum, $a + b$. Here the formula DP is identical with D+. (This is possible because in D+ sum and product occur symmetrically.) Our system thus consists (in addition to PA, PB, PC, and P+, i.e. $a, b \varepsilon S \rightarrow a + b \varepsilon S$) of the following formulae:

DS $\quad ((d) p(a + b, d) = p(a, d) + p(b, d) - p(c, d)) \leftrightarrow$
$\leftrightarrow (d) (e) (p(a, b) \leqslant p(c, b) \& p(a, d) \geqslant p(c, d) \leqslant$
$\leqslant p(b, c) \& ((p(b, d) \leqslant p(e, d) \& p(b, e) \geqslant p(e, e) \leqslant$
$\leqslant p(d, e)) \rightarrow p(a, e) p(b, d) = p(c, d)))$

DP $\quad p(ab, c) = p(a, c) + p(b, c) - p(a + b, c)$

That this system is independent, and that all three definitions, C, DS, and DP, are creative, can be shown very easily.

(a) First, we can obtain from DP and DS (c/ab), by detachment, formulae 1 to 4 of Section III above, and therefore the rest of our system; for example formula 18 is the same as B2, the general theorem of multiplication.

(b) That this cannot be done without DS is proved by the following model: take first any 'normal' model (a Boolean algebra with normal product, sum, and complement) writing '$q(a, b)$' instead of '$p(a, b)$', and define then $p(a, b) = 2q(a, b)$. One easily sees that A, C, and DP are satisfied, but not, for example, (18) (that is, B2); moreover, $p(a, a) = 1$ and $p(a, b) \leqslant 1$ are likewise not satisfied. Thus DS is clearly needed for the derivation of these formulae: it is creative, and also independent.

It is surprising to find that DP is here creative also; for our system *seems* to contain, even without DP, both sum and complement, and thus the means of defining the product. But this is not the case; for in the absence of DP our system *need not contain the normal or ordinary sum* (because we have no assurance that an element with the properties ascribed by DS to the element c always exists; only DP – together with PB, of course – ensures the existence of such an element c).

To show all this with the help of a very simple model, we take a two-element Boolean algebra, $S = \{0,1\}$, defining ab and $-a$ normally, as the numerical product of a and b, and as $1 - a$, respectively. We next define $q(a, b) = 1$ unless $a = 0$ and $b = 1$ in which case $q(a, b) = 0$. Thus we have a 'normal' model. Now we define $a + b$ and $p(a, b)$ in a non-normal way as follows $a + b = |a - b|$; and $p(a, b) = 2q(a, b)$. As a result, DS is satisfied, since both sides of the equivalence become false at the same time; that is to say, for $a = b = 1$. This proves that DP (which is not satisfied) is needed for the derivation of, for example, $p(a, b) \leqslant 1$. DP is therefore creative.

182

The situation revealed by this model is peculiar indeed: as our model shows, DS, although it is a definition, does not determine the meaning of the defined term unless we also have DP – in spite of the fact that DS *'precedes'* DP (that is to say, DP uses in its definiens the term defined by DS). One may express this by saying that although DS *formally* precedes DP, it is preceded by DP in some other sense (a sense which is not easily expressed in a formal way); for DS was obtained from another definition i.e. from D+ (which uses the term defined in DP) by way of eliminating this term (with the help of B).

We can of course apply the same elimination technique to DP, eliminating from it, with the help of DS, the sign of the sum, '+'. Let us call the resulting rather complicated formula 'DP*e*' (for 'DP after elimination'). In a system that uses DP*e* instead of B, *both* definitions of the sum, D+ and DS, turn out to be creative. (Thus D+ may be creative even if used as a definition of the sum rather than the product.)

Returning to the system containing A, C, DS, and DP, this system is *a fortiori* independent since the definitions DS and DP are creative. (The creativeness of C is unchanged, as can be easily seen.) If we now add B, then both DS and DP become non-creative, and one of them becomes redundant.

Because of DS this system is complicated and not 'natural'; in fact, the calculus of relative probabilities $p(a, b)$ is more naturally constructed by taking the product and the complement as basic rather than, say, the sum and the complement. In this lack of symmetry, the classical probability calculus differs from the propositional calculus. The lack of symmetry is due to the fact that the general multiplication theorem B2 is not derivable from the complementation and addition theorems, while the general addition theorem is derivable from the (classical) complementation and multiplication theorems, as shown in appendix *v of [2]. This explains the need for a complex DS if we wish it to be creative. However this may be, DS and DP establish our thesis of the relativity of creativeness, and they do so in a case which is somewhat unfavourable to this thesis.

VII. NON-CREATIVE AND CREATIVE, VERBAL AND REAL DEFINITIONS

It seems to me that the old distinction between verbal and real definitions has something to do with our distinction between non-creative and creative definitions. The intuitive idea of a 'verbal' definition is, no doubt, that its function is *merely* that of an abbreviation, while a 'real' definition is supposed, somehow or other, to enrich our 'body of knowledge'. It seems not an unlikely surmise that some of those who believed in 'real' definitions had some inkling of the existence of creative definitions, and that they were led by this to insist that there are definitions which are not merely verbal abbreviations.

Perhaps the reason for the uneasiness felt by some people in accepting recursive definitions as 'proper' definitions can partly be explained by the fact that they were creative in systems which were not sufficiently rich. Their creativity may even be transferred to the explicit definitions which, as we know from Hilbert and Bernays [10], can replace them.[9]

If there is anything in this surmise, then our result might help to dispel certain views about these so-called 'real' definitions. For it shows that whether or not a definition is *merely* abbreviative depends upon the rest of our system. In fact, it is clear that a sharp distinction between *merely* abbreviative (or normal or verbal or non-creative) definitions on the one side, and creative definitions on the other side, can only be made if we have before us an axiomatised system, which is rarely enough the case outside mathematics and Woodgerian biology. Elsewhere, the distinction breaks down; and if we try to replace it by a distinction between some less definite ideas, we would have to insist that this distinction is also relative – relative to what we may vaguely call 'the present state of our knowledge', or perhaps better, 'our present assumptions'.

VIII. CREATIVE NON-DEFINITIONAL AXIOMS

The idea of creativeness or non-creativeness may also be applied to axioms which do not have the form of definitions. (In fact, the idea will appear more familiar in this context because of its close relation to the idea of independence.)

This may be seen from the axioms B1 and B2 of my old standard system, here derived in Section III above. In this system they replace B, and are therefore just as creative as B. Such important laws as $p(a, a) = 1$ cannot be derived in this system without them. More particularly, $0 \leqslant p(a, b)$ cannot be derived without B1, and $p(ab) \leqslant 1$ cannot be derived without B2, as may be seen from the independence proofs given in [2].

As this example shows, the idea of creativeness can be applied to axioms which introduce some new term, or function, or operation, as do B1 and B2; and it can be applied either to single axioms or groups of axioms.

To take another and more familiar example, we may consider the following axiom system for the propositional calculus which is part of the system of Hilbert and Bernays.[10] It consists of (i) three formulae for the conditional, and (ii) three formulae for negation.

(i)   (1) $A \to (B \to A)$

         (2) $(A \to (A \to B)) \to (A \to B)$,

         (3) $(A \to B) \to ((B \to C) \to (A \to C))$.

(ii)   (1) $(A \to B) \to (-B \to -A)$

         (2) $A \to -(-A)$

         (3) $-(-A) \to A$

These two groups of axioms are complete in the sense that they allow the derivation, by substitution and detachment, of all logically true formulae (tautologies) of the propositional calculus which can be expressed in terms of the conditional and of negation. Roughly speaking, the first three formulae 'define' the conditional, and the last three 'define' negation.

Now the conjunction of the last three of these formulae – the three formulae for negation, including the third, that is, the 'second law of double negation' – may be described as 'creative' in a sense which is very nearly equal to our earlier usage, while the conjunction of the first two formulae for negation may be described as 'non-creative'.

The situation is well known: from group (i) alone, the axioms for the conditional, we can derive a certain set of theorems, the so-called 'positive identical conditionals', which form a well-defined sub-set of the class of all logically true formulae expressible solely in terms of the conditional. If we add to the axioms of group (i) the first two axioms of group (ii), then the new theorems which become derivable are only

185

theorems which contain in addition to the sign of the conditional also the sign of negation: the set of derivable formulae expressible in terms of the conditional alone is not extended. But when we add the last axiom, the second law of double negation, the set of derivable formulae expressible in terms of the conditional is vastly extended: we can now derive *all* the logically true formulae which can be expressed in terms of the conditional.

Again, the creativity of the second law of double negation in the presence of axioms 1 and 2 of group (ii) – and therefore that of 1 and 2 in the presence of 3 – is relative to the rest of the system. For example, if we replace the second formula of group (i) by the formula (Peirce's Law)

$$((A \to B) \to A) \to A$$

then the second law of double negation becomes non-creative, because now all the logically true formulae expressible in terms of the conditional are derivable from (i).

It may be mentioned in this connection that the non-creativeness of certain formulae – for example the so-called 'ε-formula' of Hilbert-Bernays[11] – may prove crucially important in certain meta-mathematical considerations.

IX. THE ELIMINATION OF THE CREATIVENESS OF A DEFINITION

Let us now turn back to the probability calculus and to our first example, the axiom C. This, as we have seen, may be taken to be a *bona fide* definition of the complement. This creative definition

C $\qquad p(-a, b) = p(b, b) - p(a, b) \leftrightarrow (Ec)\, p(b, b) \neq p(c, b)$

may easily be rendered non-creative if we add to our axioms (of group A) the following existential axiom

AC $\qquad (a)\, (Eb)\, (c)\, (p(b, c) = p(c, c) - p(a, c) \leftrightarrow (Ed)\, p(c, c) \neq$
$\qquad \neq p(d, c))$

In the presence of this existential axiom AC, our formerly creative definition C becomes non-creative. But it in no way becomes redundant; rather, it becomes a perfectly normal definition whose sole function is to abbreviate, and to allow us to eliminate our abbreviation. (Similarly, by introducing a far more complicated existential formula AB, corresponding to our axiom B, we may introduce the definition of the product by the *bona fide* definition BF, mentioned above, which in the presence of AB is easily proved non-creative.)

It also becomes clear now why the definition of the sum D+ is non-creative with respect to the axiom system A, B, C: we can prove in this system the formula

A+ $\qquad (a)\,(b)\,(Ec)\,(d)\,p(c, d) = p(a, d) + p(b, d) - p(ab, d)$

that is to say the existence, for all elements a and b in S, of an element c in S which has all the properties of the sum, $a + b$.

The situation seems to be exactly the same with all definitions by which we introduce operators or functions which form from elements of S some other elements of S. In all these cases the creativeness of a definition can be eliminated by adding to our axioms a requirement demanding the existence of an element which has exactly the properties of the element defined by the previously creative definition.

This observation sheds, I think, some light upon the whole question of creativeness: a definition is creative if, in addition to its normal function of abbreviating, it also functions (perhaps in conjunction with an existential postulate such as our PC of Section II) as an existential axiom.

However, this does not mean that we can, in all cases, actually eliminate the creativeness of any given creative definition. In a system like ours, based on the lower functional calculus, we have only the means of expressing the demand that certain elements of S exist. We cannot, for example, express explicitly the demand that certain relations between these elements exist.

Instead, ordinary, non-existential axioms, or an ordinary definition, such as our D=,

D= $\qquad a = b \leftrightarrow (c)\,p(a, c) = p(b, c)$

have to be used.

Now let us assume that we have a system like A, B, and D=. In this system D= will be non-creative. It will, however, at once become creative if we add further

C= $\qquad -a = b \leftrightarrow (c)\,(p(a, c) + p(b, c) = p(c, c) \leftrightarrow (Ed)\,p(c, c) \neq$
$\qquad\qquad \neq p(d, c))$

or else

$$AC= \quad (a)\,(Eb)\,(-a = b \leftrightarrow (c)\,(p(a,\,c) + p(b,\,c) = p(c,\,c) \leftrightarrow$$
$$\leftrightarrow (Ed)\,p(c,\,c) \neq p(d,\,c)))$$

This last formula, AC=, is interesting because it shows that the addition of a new *existential axiom* may turn a non-creative formula, such as D=, into a creative one. But as far as our present problem is concerned – the elimination of the creativeness of a definition – we see that with the means of expression at our disposal in the lower functional calculus we cannot eliminate the creativity of D= which results from adopting AC=. Although this affects the applicability or the technical significance of the method of eliminating the creativeness of a definition, it does not, I think, affect the philosophical significance of our analysis. Whether or not we can *express*, in our formalism, an existential postulate, it appears that one of the essential factors which make a definition creative is that it serves the double purpose of abbreviating *and* postulating existence.

X. CONCLUDING REMARKS

It might be said against my assertion that C= or AC= makes D= creative that I should have added [12] to my definition of the creativity of *d* in Section I a clause excluding the use not only of *definitions* employing terms defined by or with the help of *d*, but also of *axioms* employing such terms But I do not believe that the addition of such a clause (or the equivalent requirement that all the axioms should be written in primitive terms) would suffice to exclude the possibility of turning a proper or non-creative definition of a relation symbol (such as D=), into a creative one, by way of adding new *axioms* (or else new *axiom schemata*).

Let us assume that predicate variables are admitted in our system and especially *two-termed relational variables* such as '*R*', and atomic sentences such as '*aRb*', while universal or existential operators such as '(*R*)' or '(E*R*)' are not admitted. And let us assume that our axioms are, as before, A, B, and the definition D= (which in this system is non-creative). We then add, in place of C=, the following axiom CR which is obtained from C= by the elimination method used in Section VI above, and which is written entirely in terms of the primitive symbols of the system and therefore not 'preceded' by D= :

CR $\quad (a)(b)(-aRb \leftrightarrow (c)(p(a,c)+p(b,c)=p(c,c) \leftrightarrow (Ed)p(c,c) \neq$
$\quad\quad \neq p(d,c))) \leftrightarrow (a)(b)(aRb \leftrightarrow (c)p(a,c)=p(b,c))$

It is easily seen that the addition of this axiom CR makes D= creative; for in the presence of D=, but not otherwise, C and C1 can be obtained from CR.

In any case it should be stressed that there can be no general effective criterion of creativity or non-creativity. As a consequence, if we decide that we wish to admit only non-creative definitions, then we may not know, and we may be unable to find out, whether a certain proposed definition d is an admissible or proper definition of our system.

Also, it is clear from what has been said about CR that questions about creativeness can hardly be seriously discussed unless the logical basis of the system is specified with sufficient explicitness to allow us to decide such questions as, for example, whether or not predicate variables (or else *axiom schemata*) are admitted.[13]

University of London, London, England

NOTES AND REFERENCES

1. Cf. Woodger [1], p. 52.
2. Cf. [2], appendices *iv and *v, with the literature there cited, and [3], pp. 41 f.
3. I wish to thank here Dr. J. Agassi, Dr. I. Lakatos and Dr. C. Lejewski for interesting discussions of some problems connected with creative definitions. I gather from Dr. Lejewski and from P. Suppes [4] p. 153, that Leśniewski was the first to appreciate and to state the problem. Dr. Lakatos has drawn my attention to the historical background of the problem of creative definitions: Gergonne [5] seems to have been the first to put axioms and definitions on the same level, as it were, by introducing the concept of implicit definitions. Padoa seems to be correct when he claims to be the first to have seen the problem of independence of terms [6]. See also Suppes [4], p. 170, and his references to A. Tarski [7] and J. C. C. McKinsey [8]. A brief discussion is to be found in P. Suppes, *op. cit*, pp. 152 ff.
4. See, especially, formula BD on p. 336 of [2] and the much simplified formula BD on p. 41 of [3].
5. We obtain from C^{int} (by $b/-a$)

$1'$ $p(a,d) \geqslant p(c,d) \leqslant p(-a,d) \rightarrow (p(c,d) \neq 0 \leftrightarrow (e)p(c,d)=p(e,d))$ $\quad\quad C^{int}$
$2'$ $p(a(-a),d) \neq 0 \leftrightarrow (e)p(a(-a),d)=p(e,d)$ $\quad\quad\quad\quad\quad\quad 1'$, or C^{int}
$3'$ $(b)(Ea)p(a,b) \neq 0$ $\quad\quad\quad\quad\quad\quad\quad\quad\quad\quad\quad\quad\quad\quad\quad\quad\quad 2'$

which is the same as 8 above, so that we can obtain 10, above, i.e. $p(a,a)=1$.
We thus have
$4'$ $p(a,b(-b))=1$ $\quad\quad\quad\quad\quad\quad\quad\quad\quad\quad\quad\quad\quad\quad\quad\quad 2', 10$
$5'$ $p(a(-a),c) \leqslant p(b,c)$ $\quad\quad\quad\quad\quad\quad\quad\quad\quad\quad\quad\quad\quad\quad\quad 2'$
$6'$ $p(-(a(-a)),c) \geqslant p(b,c)$ $\quad\quad\quad\quad\quad\quad\quad\quad\quad\quad\quad\quad\quad 2', 1'$

6. See Quine [9], p. 47.

7. The formulae resulting from this somewhat clumsy method can be in many cases very considerably simplified, as shown by the example of the transition from C_{int} to C^{int} (see above, end of Section IV).

8. We may prove it as follows. (References in square brackets are to [2], appendix *v.)

(25.1) $p(b, c) = 1 \rightarrow p(a, bc) = p(ab, c)$ 18, [B2]

(25.2) $p(a, c) = p(ab, c) + p(a(- b), c) - p(- c, c)$ [70]

(25.3) $p(b, c) = 1 \rightarrow p(- b, c) = p(- c, c)$ [64]

(25.4) $p(b, c) = 1 \rightarrow p(a(- b), c) = p(- c, c)$ 25.3, 2, [B1]

(25.5) $p(b, c) = 1 \rightarrow p(a, c) = p(ab, c) = p(a, bc)$ 25.2 to 4; 25.1

The principle of redundancy allows us to derive (12), which corresponds to A 2$^+$ of [2], from 14, i.e., A 2, and the law of commutation (i.e. 40 of [2]), and thus to prove the inter-deducibility of A2 and A2$^+$, i.e. of 12 and 14.

9. Cf. L. Kalmár [11].

10. Hilbert-Bernays [10], vol. I, p. 66.

11. See Hilbert-Bernays [10] vol. II, pp. 13, 15. The two ε-theorems (pp. 18 ff.) state that the ε-formula can be eliminated (i) without reducing the set of derivable formulae which do not contain the ε-symbol (or an expression defined in terms of the ε-symbol), and (ii) from the proofs of these formulae.

12. See also Suppes [4], p. 154. The key-term 'preceding' has been omitted by Suppes' reviewer in *Mind;* see [12] p. 110.

13. Thus in view of CR (or of a corresponding *axiom schema*) the *'rule for defining relation symbols'* in [4], p. 156, fails to 'guarantee ... the ... non-creativity' (p. 155) of D=. Our trivial procedure for making definitions creative is fairly generally applicable.

BIBLIOGRAPHY

[1] J. H. Woodger, *The Axiomatic Method in Biology*, 1937.

[2] K. R. Popper, *The Logic of Scientific Discovery*, (1934), 1959, 1960.

[3] K. R. Popper, 'The Propensity Interpretation of Probability' *British Journal for the Philosophy of Science* **10** (1959) 25–42.

[4] P. Suppes, *Introduction to Logic*, 1957.

[5] J. D. Gergonne: 'Essai sur la théorie des définitions', *Ann. de Math. p. et app.* **9** (1818), 1–35.

[6] *Bibliothèque du Congrès Internationale de Philosophie* **1** (1900), **3** (1903) 279–288

[7] A. Tarski, 'Einige methodologische Untersuchungen über die Definierbarkeit der Begriffe', *Erkenntnis* **5** (1935–1936) 80–100.

[8] J. C. C. McKinsey, 'On the independence of undefined ideas', *Bulletin of the American Math. Soc.* **41** (1935) 291–297.

[9] W. v. O. Quine, *Mathematical Logic*, 1951.

[10] D. Hilbert and P. Bernays, *Grundlagen der Mathematik* I (1934); II (1939).

[11] L. Kalmár, 'On the possibility of definition by recursion', *Acta Scient. Math. Szeged* **9** (1940) 227–232.

[12] J. A. Farris, 'Introduction to Logic by Patrick Suppes' *Mind* 49 (1960), p. 109–110.

FREDERIC B. FITCH

ALGEBRAIC SIMPLIFICATION OF REDUNDANT
SEQUENTIAL CIRCUITS

The purpose of this essay [1] is to show that if a sequential circuit can be simplified by merging "equivalent states" in the sense of E. F. Moore [2] and G. H. Mealy [3], then the equations defining the original circuit can be transformed algebraically into equations defining the simplified circuit. The transformations are purely algebraic processes occurring in a kind of extended Boolean algebra here called "discrete delay algebra". Furthermore, the procedure for deciding which states, if any, can be thus merged, is shown to be an essentially algebraic procedure, and it is shown that the various states themselves can be viewed as special compound entities of the algebra. This outcome suggests that simplification of the Mealy-Moore sort is primarily algebraic in character and is perhaps a special sort of some more general kind of algebraic simplification that has wider application [4].

By a (synchronous) sequential circuit will be meant an idealized circuit with a finite number of input terminals and a finite number of output terminals which has the following properties: At each of the discrete times $0, d, 2d, 3d$, and so on, a finite number of the input terminals (possibly none) are activated. The activation of the output terminals at the discrete times $0, d, 2d, 3d$, and so on, depends, through intermediate elements, on the activation of the input terminals at earlier or simultaneous times. The intermediate elements are the familiar idealized *and*-elements, *or*-elements, *not*-elements, and delay elements. All the delay elements have the same fixed amount of delay d, so that the output of a delay element is activated or not activated at a given discrete time depending on whether or not its input was activated at the previous discrete time, d earlier. The output of a *not*-element is activated at a given time if and only if its input is not activated at that time. The output of an *or*-element is activated at a given time if and only if at least one of its two inputs is activated at that time. The output of an *and*-element is activated at a given time if and only if each of its two inputs are activated at that time.

A Boolean algebra will be employed in which $x_1, x_2, \ldots, y_1, y_2, \ldots,$

$q_1, q_2, \ldots$, will serve as variables. More specifically, $x_1, x_2, \ldots$, are "input variables" representing the inputs of a sequential circuit, while $y_1, y_2, \ldots$, are "output variables" representing outputs of the circuit, and $q_1, q_2, \ldots$, are "state variables" representing the outputs of certain delay elements whose activation at a given time suffices to specify the "state" of the circuit at that time. In this Boolean algebra in addition to the usual Boolean operations of sum, product and negate, we also employ a "delay operation" to represent a time lapse of one discrete unit of time, d. The resulting Boolean algebra will be said to be a discrete delay algebra. If a and b are entities of the algebra, then a' is the Boolean negate of a, while ab is the Boolean product of a with b, and $a + b$ is the Boolean sum of a with b, and a^* is the delay transform of a. The properties of the Boolean negate, sum, and product are well known. The properties of the delay operation are given by the following postulates:

Postulate I. $(ab)^* = a^* \, b^*$

Postulate II. $(a + b)^* = a^* + b^*$

Postulate III. $a'^* = a^{*\prime}$

These three postulates, in the case of a discrete delay algebra, must be added to the usual postulates for Boolean algebra.

We may think of each x_i as a variable which is a function of time and which takes the Boolean value 1 at every time when the ith input is activated, and the Boolean value 0 at every time when the ith input is not activated. Similarly, y_i takes the value 1 or 0 at a given time, depending on whether or not the ith output is activated at that time, and q_i takes the value 1 or 0 at a given time depending on whether or not the output of the ith delay element is activated at that time. If z is a variable representing the input of a *not*-element at some time, then z' represents the output of that *not*-element at that time. If z and w are variables representing the inputs of an *or*-element at some time, then $z + w$ represents the output of that *or*-element at that time. If z and w are variables representing the inputs of an *and*-element at some time, then zw represents the output of that *and*-element at that time. Finally, if z is a variable representing the *output* of a delay element at some time, then z^* represents the *input* of that delay element at that time. For example, if z has the value 1 at time t, indicating that the output of the delay element is activated at time t, then z^* has the value 1 at time $t - d$, indicating that the input of the delay element must have been activated at the previous discrete time. Indeed, more generally, z^* can be taken to represent the value of z at a time that is one unit of delay earlier, even if z does not actually represent the output of a delay element. Similarly, in general, $z + w$ can be taken to represent the condition that z or w (at least one of them) is activated at a given time, even though z and w do not actually represent inputs of an *or*-element. In the same way, zw can represent the condition that z and w are both activated, while z' can represent the condition that z is not activated. (Instead of speaking of the activation of z and w, it would be more accurate to speak of the activation of whatever things z and w represent.) For example, the equation $zw = 1$ can be taken to mean that the things represented by z and w are both activated at the time in question, while the equation $z' = 1$ can be taken to mean that the thing represented by z is not activated at the time in question. Thus equality to 1 indicates that the condition holds, and equality to 0 that it fails.

Each output of a sequential circuit can be viewed as a Boolean function of the inputs of the circuit and of the outputs of the delay elements of the circuit. Similarly, each input of each delay element of the circuit can also be viewed as a Boolean function of the inputs of the circuit and of the outputs of the delay elements of the circuit.

The state (that is, the internal state) of a sequential circuit is simply the state of the outputs of the delay elements. For example, if a sequential circuit contains only two delay elements, then there are at most four possible (internal) states of the circuit: (1) Neither delay output is activated, so that $q_1 = 0$ and $q_2 = 0$. (2) The first delay output is not activated but the second one is, so that $q_1 = 0$ and $q_2 = 1$. (3) The first delay output is activated, but the second one is not, so that $q_1 = 1$ and $q_2 = 0$. (4) Both delay outputs are activated, so that $q_1 = 1$ and $q_2 = 1$.

Two states might be equivalent in the sense that the external behavior of the circuit is the same no matter which of the two states it is in. In other words, external observation of how the activation of the outputs of the circuit depends on the activation of the inputs of the circuit would not, in such a case, suffice to determine which of the two states the circuit is in. For example, states (2) and (3) mentioned above might be such that whenever the circuit is in either of these states all the outputs of the circuit are activated regardless of whether the inputs are activated or not, and the next state of the circuit is always state (1). In such a case, no observations of the relation between activation of inputs of the total circuit and activation of outputs of the total circuit would make it possible to decide which of the states (2) or (3) the circuit is in. A method has been provided by E. F. Moore [5] and G. H. Mealy [6] for determining which pairs of states of a sequential circuit are equivalent in this sense. Every group of states that are all equivalent to each other are then classified as belonging to the same "equivalence class" of states, so that every state of the circuit belongs to one and only one equivalence class. A state that is equivalent to no other state falls into an equivalence class of which it is the only member. Thus the class of all states of the circuit is the union of a class of mutually exclusive equivalence classes of states. Every state in an equivalence class is equivalent to every other state in that equivalence class and to no state in any other equivalence class.

Corresponding to every sequential circuit, a "state diagram" can be constructed. This diagram specifies how the activation of the outputs of the circuit depends on the activation of the inputs for each of the several states of the circuit, and it also specifies the conditions under which the circuit changes from one state to another. The method of Moore and Mealy provides a way of taking such a state diagram and modifying it by "merging" all the states of each equivalence class into a single state. We can thus form a state diagram for a circuit which has all the input-output properties of the original circuit but which has fewer states (at least if any of the original equivalence classes had more than one member), since each equivalence class of states is converted into an equivalence class that has only one state as a member. The simpler circuit itself can easily be constructed on the basis of the new, simplified state diagram. Circuits which can be thus simplified will be called "state-redundant" circuits. It will now be shown that this simplification process can be carried out in a purely algebraic way.

Suppose that q_1, q_2, ..., q_t are the state variables of the original state-redundant circuit C_0, and that x_1, x_2, ..., x_s are its input variables, and that y_1, y_2, ..., y_u are its output variables. All these variables take the Boolean values 1 and 0 only. The dependence of the nature of the next state of the circuit on the nature of its present state and on the nature of the present input is expressed by the following t equations which use t Boolean functions f_i (for $i = 1, 2, ..., t$)

194

$$q_i^* = f_i(q_1, \ldots, q_t, x_1, \ldots, x_s): \tag{1}$$

The dependence of the present output of the circuit on the present state and on the present input is expressed by the following u equations which use u Boolean functions g_i (for $i = 1, 2, \ldots, u$):

$$y_i = g_i(q_1, \ldots, q_t, x_1, \ldots, x_s) \tag{2}$$

There are 2^t states altogether. Let $Q_{00\ldots00}$ stand for $q_1' q_2' \ldots q_{t-1}' q_t'$ and let $Q_{00\ldots01}$ stand for $q_1' q_2' \ldots q_{t-1}' q_t$, and let $Q_{00\ldots10}$ stand for $q_1' q_2' \ldots q_{t-1} q_t'$, and so on. For example, in the case $t = 3$, Q_{101} would stand for $q_1 q_2' q_3$. Furthermore, we may abbreviate $Q_{000}, Q_{001}, Q_{010}, Q_{011}, Q_{100}, Q_{101}, Q_{110}, Q_{111}$ respectively as $Q_0, Q_1, Q_2, Q_3, Q_4, Q_5, Q_6, Q_7$. This amounts to changing the subscripts from binary notation to decimal notation. The 2^t states correspond to the 2^t conditions expressed by the equations $Q_0 = 1, Q_1 = 1, \ldots, Q_T = 1$, where $T = 2^t - 1$. For example, if $t = 3$, then the fifth state is the state expressed by the equation $Q_4 = 1$, that is, by the equation $Q_{100} = 1$ or the equation $q_1 q_2' q_3' = 1$. This latter equation holds exactly when $q_1 = 1$, $q_2 = 0$, and $q_3 = 0$. So in the fifth state the output of the first delay element would be activated, but not the outputs of the second and third delay elements.

In a similar way we let $X_{00\ldots00}$ stand for $x_1' x_2' \ldots x_{s-1}' x_s'$, and we let $X_{00\ldots01}$ stand for $x_1' x_2' \ldots x_{s-1}' x_s$, and so on. The subscripts can be changed from binary to decimal notation, as before. There are 2^s possible input signals, corresponding to the 2^s conditions expressed by the equations $X_0 = 1, X_1 = 1, \ldots, X_S = 1$, where $S = 2^s - 1$.

Similarly we let $Y_{00\ldots00}$ stand for $y_1' y_2' \ldots y_{u-1}' y_u'$, and we let $Y_{00\ldots01}$ stand for $y_1' y_2' \ldots y_{u-1}' y_u$, and so on. The subscripts can be changed from binary to decimal notation. There are 2^u possible output signals, corresponding to the 2^u conditions expressed by the equations $Y_0 = 1$, $Y_1 = 1, \ldots, Y_U = 1$, where $U = 2^u - 1$.

Suppose that whenever the circuit is in the state Q_j the effect of the input signal X_i is to produce the output signal Y_k. This situation may be expressed by either of the following two equivalent Boolean equations

$$Y_k' Q_j X_i = 0 \tag{3}$$

$$Y_k + Q_j' + X_i' = 1 \tag{3'}$$

Each of these equations says, in effect, that the occurrence of the $i + 1$st

195

input signal X_i, while the circuit is in the $j + 1$st state Q_j, implies the occurrence of the $k + 1$st output signal Y_k.

Suppose, furthermore, that whenever the circuit is in the state Q_j the effect of the input signal X_i is to cause the next state of the circuit to be Q_h. This situation may be expressed by either of the following two equivalent Boolean equations:

$$Q_h^{*'} Q_j' X_i = 0 \qquad (4)$$

$$Q_h^* + Q'_j + X'_i = 1 \qquad (4')$$

More specifically, if Equation (4) or Equation (4') holds, and if $Q_j = 1$ and $X_i = 1$, then by principles of Boolean algebra we must have the result $Q_h^* = 1$, and this last equation asserts, in effect, that the next state of the circuit will be the state Q_h. Suppose, for example, that $h = 2$ and that and that there are three delay elements in the circuit, so that $t = 3$. Then consider the significance of the equation $Q_h^* = 1$. We have $Q_h = Q_2 = Q_{010} = q_1' q_2 q_3' = 1$, so that the state Q_h would be the state the circuit is in when the output of the second delay element is activated but not the outputs of the first and third delay elements. The equation $Q_h^* = 1$ becomes $(q_1' \, q_2 \, q_3')^* = 1$. By Postulates I and III this may be converted into the equation $q_1^{*'} \, q_2^* \, q_3^{*'} = 1$, so that $q_1^* = 0$, $q_2^* = 1$, and $q_3 = 0$. Hence, after one unit of delay, we will have $q_1 = 0$, $q_2 = 1$, and $q_3 = 0$, and the circuit will be in the state Q_h. This shows that the equation $Q_h^* = 1$ guarantees that the next state of the circuit is the state Q_h. Postulate II, incidentally, can be derived by Boolean algebra from Postulates I and III, and Postulate I can be derived by Boolean algebra from Postulates II and III.

Instead of describing the circuit C_0 by Equations (1) and (2), we could equally well describe it by sufficiently many equations of the form (3) and (4), or of the form (3') and (4'). In fact, regarding (3) and (4) as *sets* of equations by choosing h, i, j, k in various suitable ways, it can be shown that (1) and (2) are derivable from (3) and (4), and conversely, using ordinary Boolean procedures and Postulates I–III. It should also be mentioned that the two sets of Equations (3) and (4) can be combined into a single set of equations of the form:

$$(Q_h^{*'} + Y_k') Q_j X_i = 0 \qquad (5)$$

A state Q_i will be said to have an output signal Y_j relatively to a finite

196

sequence of input signals, $X_{k_1}, X_{k_2}, \ldots, X_{k_n}$, if there exist states $Q_{h_1}, Q_{h_2},$
$\ldots, Q_{h_{n-1}}$ such that

$$Q_{h_1}^{*\prime} Q_i X_{k_1} = 0$$
$$Q_{h_2}^{*\prime} Q_{h_1} X_{k_2} = 0$$
$$Q_{h_3}^{*\prime} Q_{h_2} X_{k_3} = 0$$
$$\vdots$$
$$Q_{h_{n-1}}^{*\prime} Q_{h_{n-2}} X_{k_{n-1}} = 0$$
$$Y_j' Q_{h_{n-1}} X_{k_n} = 0 \tag{6}$$

Equivalence between states may now be defined as follows: Two states will be said to be equivalent if they have the same output signal relatively to every finite sequence of input signals. It is easy to see that the relation of equivalence among states is reflexive, symmetrical and transitive, and that the class of all the states of the circuit is decomposable into a mutually exclusive class of equivalence classes such that every member of each equivalence is equivalent to every member of that equivalence class and to no member of any other equivalence class. Convenient procedures for the actual classification of states into such equivalence classes have been given by Moore and Mealy, and need not be repeated here.

Two equivalent states clearly must give rise to the same output signal for any given input signal. Furthermore, for any given input signal, they must give rise to equivalent next states, for otherwise they would not themselves be equivalent.

Let the various equivalence classes of the circuit C_0 be the m equivalence classes $E_0, E_1, \ldots, E_m$, where m can be no larger than 2^t, the total number of states of the circuit C_0. If $Q_{i_1}, Q_{i_2}, \ldots, Q_{i_n}$ are all the members of an equivalence class E_j, then let $R_j = Q_{i_1} + Q_{i_2} + \ldots + Q_{i_n}$. In other words, let R_j be the Boolean sum of the members of E_j. We will call $R_0, R_1, \ldots, R_{m-1}$ "pseudo-states" of the circuit C_0 because they act very much like states and because they can be viewed as being the states of a new circuit C_1 whose states are obtained by merging the equivalent states of the circuit C_0. We have already noted that circuit C_0 can be defined by means of the sets of Equations (3) and (4). The corresponding sets of equations for defining the new circuit C_1 are given by (7) and (8) below. It is seen that the R's play the role of states in (7) and (8), just as the Q's play the role of states in (3) and (4). Furthermore, (7) and (8) are derivable algebraically from (3) and (4), so that the defining equations for the new simplified circuit C_1 are derivable algebraically from the defining equations for the original state-redundant circuit C_0. Letting Q_j and Q_h of (3) and (4) respectively belong to equivalence classes E_a and E_b, we can show:

$$Y'_k \, R_a \, X_i = 0 \tag{7}$$

$$R_b^{*\prime} \, R_a \, X_i = s \tag{8}$$

For example, suppose that E_1 has the states Q_0 and Q_1 as its members, and that E_2 has the states Q_2 and Q_3 as its members, so that,

$$R_1 = Q_0 + Q_1 \tag{9}$$

$$R_2 = Q_2 + Q_3 \tag{10}$$

Suppose also that we have as special cases of (3) and (4):

$$Y'_6 \, Q_1 \, X_5 = 0 \tag{11}$$

$$Q_3^{*\prime} \, Q_0 \, X_7 = 0 \tag{12}$$

Since Q_0 is equivalent to Q_1, we get, using (11):

$$Y'_6 \, Q_0 \, X_5 = 0 \tag{13}$$

Furthermore, since Q_0 is equivalent to Q_1, and since Q_2 is equivalent to Q_3, we get, using (12):

$$Q_3^{*\prime} \, Q_1 \, X_7 = 0 \tag{14}$$

or else,

$$Q_2^{*\prime} \, Q_1 \, X_7 = 0 \tag{15}$$

From (9), (10), (12), and (14) we get:

$$
\begin{aligned}
R_2^{*\prime} \, R_1 \, X_7 &= (Q_2 + Q_3)^{*\prime} \, (Q_0 + Q_1) \, X_7 \\
&= Q_2^{*\prime} \, Q_3^{*\prime} \, (Q_0 + Q_1) \, X_7 \\
&= 0
\end{aligned}
\tag{17}
$$

Similarly from (9), (10), (12), and (15) we also get (17). So (17) must be true, because either (14) or (15) is true. We have, therefore, derived (16) and (17). It is seen that (16) and (17) are simply special cases of (7) and (8) respectively. The proofs of the general cases of (7) and (8) can be carried out in an exactly parallel way. Thus it is seen that (7) and (8) themselves are derivable.

198

It has therefore been shown that R_0, R_1, ..., R_{m-1} act like states of the simplified circuit C_1 which is obtained by merging equivalent states of C_0. As yet, however, we have not defined state variables for the circuit C_1. If v is such that $2^{v-1} < m \leqslant 2^v$, we may define state variables r_1, r_2, ..., r_v for C_1 in the following way: We add further states to the sequence R_0, R_1, ..., R_{m-1}, so that there are 2^v states altogether for the circuit C_1, these states being R_0, R_1, ..., R_V, where $T = 2^v - 1$. If $m = 2^v$, no states would need to be added. The additional states, if any, are R_m, R_{m+1}, ..., R_V. The circuit C_1 will be constructed in such a way that these additional states will in fact never occur, and the equations of the form (7) and (8) involving these states can be chosen in whatever way will make the circuit C_1 as simple as possible. Such states which in fact never occur are often called "don't care" states. The subscripts of the R's are next transformed into binary notation. We then define the ith state variable r_i as the Boolean sum of all those R's each of which has a 1 in the ith digit of the v-digit binary number which is the subscript of that R. For example, if there are just four R's, R_{00}, R_{01}, R_{10}, R_{11}, then there would be just two state variables r_1 and r_2, where by definition we have $r_1 = R_{10} + R_{11}$ and $r_2 = R_{01} + R_{11}$. The following equations could then be derived: $R_{00} = r'_1 r'_2$, $R_{01} = r'_1 r_2$, $R_{10} = r_1 r'_2$, $R_{11} = r_1 r_2$.

From the equations (7) and (8) it is clearly possible to obtain equations for the simplified circuit C_1 in the state variables themselves, like the equations (1) and (2) in the case of the original circuit C_0. In order to make these equations as simple as possible (so that C_1 would be as simple as possible), such methods as Quine's method of prime implicants (and improved versions of it) would have to be used as far as possible, and the presence of any "don't care" states would have to be exploited as far as possible. The resulting equations would define in detail the structure of the simplified circuit C_1. Thus the equations for a state-redundant sequential circuit may be transformed by purely algebraic procedures into the equations for a reduced circuit C_1 from which the state redundancies have been removed.

It is even possible to view from an essentially algebraic standpoint the process of finding the equivalence classes, E_1, E_1, ..., E_{m-1} of the original circuit C_0. This is done as follows:

The set of Equations (5) can be combined into a single equation which characterizes the operation of the circuit C_0:

$$\sum_{j=0}^{T} \sum_{i=0}^{S} (Q^*{}_{F(i,\,j)} \, Y_{G(i,\,j)})' \, Q_j \, X_i = 0 \qquad (18)$$

Here $F(i,j)$ corresponds to h of (5) and $G(i,j)$ corresponds to k of (5). The numerical functions F and G can be determined from the Boolean functions f and g of (1) and (2), and conversely, though not in any simple way. The equation (18), incidentally, can be written in more compact form by writing $(QX)_{a_1 \ldots a_t b_1 \cdots \ldots b_s}$ for $Q_{a_1 \ldots a_t} X_{b_1 \ldots b_s}$ and by writing $(Q^* \, Y)_{a_1 \ldots a_t b_1 \ldots b_u}$ for $Q^*{}_{a_1 \ldots a_t} Y_{b_1 \ldots b_u}$. Here the a's and b's are digits of binary subscript numbers and are therefore 0's or 1's. The equation (18) then becomes:

$$\sum_{e=0}^{TS} (Q^* \, Y)'_{H(e)} \, (QX)_e = 0 \qquad (19)$$

Here the numerical function H can be determined from F and G. The equation (19) would give the same information about the circuit C_0 as table of the following form:

State			Input				Next state			Output		
q_1	...	q_t	x_1	...	x_{s-1}	x_s	$q_1{}^*$	...	$q_t{}^*$	y_1	...	y_u
0	...	0	0	...	0	0	−	...	−	−	...	−
0	...	0	0	...	0	1	−	...	−	−	...	−
0	...	0	0	...	1	0	−	...	−	−	...	−
0	...	0	0	...	1	1	−	...	−	−	...	−
0	...	0	0	...	0	0	−	...	−	−	...	−
.		.	.		.	.	.		.	.		.
.		.	.		.	.	.		.	.		.
.		.	.		.	.	.		.	.		.
1	...	1	1	...	1	1	−	...	−	−	...	−

The rows to the left of the heavy vertical line would be successive values of e from 0 to TS in binary notation, while the corresponding rows to the right of the heavy vertical line would be the corresponding values of $H(e)$ in binary notation. The equation (19), or the table itself, can be viewed as an implicative statement that asserts that if the state-input condition of the circuit has value e, then the next-state-output condition of the circuit has value $H(e)$.

200

Just as the sets of Equations (3) and (4) can be combined to form the set of Equations (5), and this latter set equations combined into the single Equation (18), rewritten as (19), so also the sets of Equations (7) and (8) can be combined into the following single equation:

$$\sum_{j=0}^{m} \sum_{i=0}^{s} (R^{*}_{H(i,\,j)} \, Y_{I(i,\,j)})' \, R_j \, X_i = 0 \tag{20}$$

This equation is derivable from (18) merely by rearranging the left side of (18). This rearrangement consists in grouping the Q's into Boolean sums by use of Boolean distributive principles and the distributive principles embodied in Postulates I–III. The proper grouping of the Q's to form the R's amounts to finding the requisite equivalence classes of the Q's, but the problem can also be viewed in the guise of the algebraic problem of transforming left side of (18) into the expression of the general form of the left side of (20) by use of discrete delay algebra. Thus even the problem of finding equivalence classes can be viewed as an algebraic problem. This is true, furthermore, in more general situations that those covered by the Mealy-Moore method for finding equivalence classes. These more general situations allow for states that are *partially* "don't care", that is, they are "don't care" with respect to some input signals but not with respect to others.[7] In such cases a Mealy-Moore decomposition into equivalence classes may not exist at all because of the presence of the peculiar partially "don't care" states. Nevertheless, insofar as simplification can be achieved at all, it can be achieved by properly grouping the original Q's, and by constructing the R's as Boolean sums of such groups of Q's, even though the groups of Q's will not constitute equivalence classes in the sense of Moore and Mealy. In practice the optimal grouping for achieving simplicity can usually be found without much difficulty, but no method as straightforward as that of Mealy and Moore seems to be known when these partially "don't care" states are involved. In any case the problem can always be viewed as the algebraic one of making the transition from Equation (18) to Equation (20), whether partially "don't care" states are involved or not.

After state variables for the simplified circuit have been defined, the final equations for the simplified circuit may be expressed in either of the following two forms

$$\sum_{j=0}^{S} \sum_{i=0}^{V} (R^{*}_{J(i,j)} Y_{L(i,j)})' R_j X_i = 0 \qquad (21)$$

$$\sum_{e=0}^{VS} (R^{*} Y)'_{M(e)} (RX)_e = 0 \qquad (22)$$

Yale University, New Haven, Connecticut, U.S.A.

REFERENCES

1. This essay embodies research done by the author for the Bell Telephone Laboratories Murray Hill, New Jersey, during the summer of 1958.
2. E. F. Moore, *Gedanken-Experiments on Sequential Machines*, Automata Studies, Princeton University Press, 1956, pp. 129–153.
3. G. H. Mealy, A Method for Synthesizing Sequential Circuits, *Bell System Technical Journal* **34** (1955) 1045–1079. The term "merging" is being used below in the sense of Mealy, and not in quite the same sense as is used by such writers as Huffman and Caldwell.
4. For a somewhat different kind of simplification see paragraphs 8.10 and 8.11 of my paper, 'Representation of Sequential Circuits in Combinatory Logic', *Philosophy of Science* **25** (1958) 263–279.
5. *Loc. cit.*
6. *Loc. cit.*
7. The existence of these more general situations, and the problems they raise, were first pointed out to me by Chester Y. Lee of the Bell Telephone Laboratories in a letter dated January 10, 1959.

CZESŁAW LEJEWSKI

ARISTOTLE'S SYLLOGISTIC AND ITS EXTENSIONS

The task I have set myself in this paper can be described as bridging the gap between Aristotle's syllogistic and Leśniewski's ontology. I propose to suggest a number of successive extensions of syllogistic culminating in a system of what may be regarded as basic ontology. In this way I hope to throw new light on the significance of the Aristotelian logic. At the same time I hope to add a little to the understanding of Leśniewski's ontology, which interestingly enough was conceived by its originator as a modernised continuation of the ancient and medieval tradition.[1]

I. ŁUKASIEWICZ'S VERSION OF THE ARISTOTELIAN SYLLOGISTIC

Throughout the paper I shall be using Leśniewski's variation of the Peano-Russellian symbolism. Instead of 'every a is b' I shall write '$a \subset b$', and 'some a is b' will be rendered by '$a \triangle b$'. In accordance with this convention Łukasiewicz's axiom system for syllogistic assumes the following form

A1.1	$a \subset a$
A1.2	$a \triangle a$
A1.3	$b \subset c . a \subset b . \supset . a \subset c$
A1.4	$b \subset c . b \triangle a . \supset . a \triangle c$ [2]

I am making Łukasiewicz's version of Aristotle's syllogistic the point of departure in my inquiry because to my mind this version approximates Aristotle's own treatment of the subject most closely both in letter and in spirit. A1.1 and A1.2 are the two laws of identity for the affirmative categorical propositions; A1.3 and A1.4 are the syllogistic laws *Barbara* and *Datisi* respectively. The negative categorical propositions 'no a is b' (in symbols: $a \neq b$) and 'some a is not b' (in symbols: $a \triangledown b$) can be introduced, after the manner of Łukasiewicz, by means of the following definitions

D1.1 $\qquad a \not\equiv b . \equiv . \sim (a \triangle b)$

D1.2 $\qquad a \triangledown b . \equiv . \sim (a \subset b)$ [3]

The deductive system characterised by the above axioms will be referred to as S1. Like any other system of syllogistic it presupposes the logic of propositions. Further theses are added to it in accordance with the rule for introducing propositional definitions, of which D1.1 and D1.2 are examples, the rule of substitution, and the rule of detachment. Since there is no provision for introducing into the system constant names or name-forming functors, all that we can substitute for the nominal variables are other nominal variables.

The meaning of the four syllogistic functors, ' $\subset$ ', ' $\triangle$ ', ' $\not\equiv$ ', and ' $\triangledown$ ' seems to coincide with the meaning in which the expressions 'every ... is ...', 'some ... is ...', 'no ... is ...' and 'some ... is not ...' are normally used in ordinary language. Thus, for instance, each of the four categorical propositions has existential import. In other words from the truth of the proposition which says that 'every a is b', to take the universal affirmative proposition as an example, we can validly infer the truth of the proposition which says that a's exist and b's exist (in symbols: $ex(a) . ex(b)$). And the same holds for the remaining categorical propositions.

In addition to existential import the Aristotelian categorical propositions seem to have what one might call the import of generality.[4] This means that from the truth of the proposition which says that every a is b or some a is b or no a is b or some a is not b, one could safely infer that there are several a's and that there are several b's. In accordance with this singular propositions of the form a is b cannot be regarded as special cases of the corresponding universal or particular affirmative propositions. In tacitly assuming that the categorical propositions should have the import of generality, Aristotle appears to have followed ordinary usage. It is interesting to note that his successors were more liberal. In Sextus Empiricus we find the following inference, which is given as an example of a Peripatetic syllogism

> Socrates (is a) man;
> every man (is an) animal;
> therefore, Socrates (is an) animal.[5]

204

This inference does not suggest that the original syllogistic was subsequently extended, and that singular propositions were introduced into syllogistic as a new type of categorical proposition. As far as our information goes no logical law which holds for singular propositions but fails to hold for universal or particular, was known to the logicians of the Peripatetic school. The inference, which undoubtedly was meant to illustrate the mood *Barbara*, shows only that singular affirmative propositions of ordinary language had begun to be regarded as special cases of universal affirmative propositions, whose meaning was thus generalised. A similar generalisation can be obtained for the remaining categorical propositions. In what follows I shall no longer presuppose the import of generality for categorical propositions but I shall continue to assume that the existential import constitutes one of their distinguishing characteristics.

II. SYLLOGISTIC WITH QUANTIFIERS

We now proceed to the discussion of the first stage in extending Aristotelian syllogistic. It consists in introducing the quantifiers 'for all a, b, c, ...' (in symbols: [$a\,b\,c\,...$]) and 'for some a, b, c, ...' (in symbols: [$\exists\,a\,b\,c\,...$]) together with their usual rules. The theory of quantification in its modern form was not known to Aristotle but there is sufficient evidence that he felt the need for expressing certain ideas which logicians have since learnt to express with the aid of the quantifiers. Thus, for instance, the universal quantifier at the head of an asserted expression indicates that the formula which follows it is true for all values of the variables which are bound by it. In Aristotle this idea is expressed by the appeal to what is known as syllogistic necessity.[6] And again, Aristotle's proofs by *ecthesis* are based on an argument which in modern logic involves the use of the particular quantifier.[7] One can, therefore, maintain that adding the quantifiers to syllogistic amounts to taking into account and elaborating certain conceptions which, admittedly, in a very vague and rudimentary form had already begun to dawn in Aristotle's mind.

By allowing for the use of quantifiers we turn S1 into a new system which will be referred to as S2. Its axioms and definitions take the form of the following expressions:

A2.1	$[a] . a \mathrel{\mathsf{C}} a$
A2.2	$[a] . a \mathbin{\triangle} a$
A2.3	$[a\,b\,c]: b \mathrel{\mathsf{C}} c . a \mathrel{\mathsf{C}} b . \supset . a \mathrel{\mathsf{C}} c$
A2.4	$[a\,b\,c]: b \mathrel{\mathsf{C}} c . b \mathbin{\triangle} a . \supset . a \mathbin{\triangle} c$
D2.1	$[a\,b]: a \neq b . \equiv . \sim (a \mathbin{\triangle} b)$
D2.2	$[a\,b]: a \mathbin{\triangledown} b . \equiv . \sim (a \mathrel{\mathsf{C}} b)$

The construction of the system can be carried on by applying the rule of propositional definitions, the rule of substitution, the rules for the use of the quantifiers, and the rule of detachment. It goes without saying that S2 presupposes the logic of propositions.

III. INTERPRETATION OF THE QUANTIFIERS

As long as the only values we can substitute for the variables of S2 are other variables, no problem arises. But let us suppose that we have subjoined to S2 a theory in which expressions like 'man' and 'centaur' occur among other constant names. It would appear that propositions

(1)   man $\triangle$ man

i.e., some man is a man, and

(2)   centaur $\triangle$ centaur

i.e., some centaur is a centaur, should follow from A2.2. Now proposition (1) happens to be true but proposition (2) is false as it implies that centaurs exist. This clearly shows that the range of values of the variables in S2 cannot include just any constant names. It has to be limited to variables and to referential names, i.e., to names that designate something. Non-referential names, which like 'centaur' do not designate anything, cannot be regarded as suitable values for the purpose of substitution. This leads to the conclusion that in A2.2, and in any other quantified expression of S2, the universal quantifier has a restricted interpretation.[8] If the usual meaning of the particular affirmative proposition and the truth of A2.2 is to be preserved then the meaning of A2.2 has to be equated with the meaning of the proposition which says that

(3)   for all a such that a's exist, – some a is a

206

And to avoid confusion with expressions in which the universal quantifier allows for unrestricted interpretation, A2.2 should be replaced by

(4) $$[\ a\ ].\, a \bigtriangleup a$$
$$\text{ex}(a)$$

which shows explicitly that the interpretation of the quantifier has been restricted in a certain specific way. Similar rephrasing ought to be applied to all theses of S2.

Before we go any further it may be worth remarking that our restricted interpretation of the universal quantifier, which results in barring non-referential names as possible values of the syllogistic variables, appears to be quite in harmony with Aristotle's practice. Although he knew that there were names that designated nothing, he hardly made any use of them in his exposition of the syllogisms.

Quantifiers with restricted interpretation raise several problems concerning the range of values substitutable for the variables. We have seen that the suitability of a constant name as a value to be substituted for a syllogistic variable can be decided provided we have sufficient empirical information as to whether it designates anything or not. The position becomes more complicated if we contemplate admitting name-forming functors as meaningful expressions of S2. Let us suppose that the nominal functions 'a or b' (in symbols: $a \cup b$) and 'a and b' (in symbols: $a \cap b$) have been introduced into S2 by means of additional axioms or by means of suitable definitions. Can these functions be regarded as possible values of the variables? If in (4) we substitute '$a \cup b$' for 'a' then the result of this substitution, which is

(5) $$[\ a\quad b\ ].\, a \cup b \bigtriangleup a \cup b$$
$$\text{ex}(a),\ \text{ex}(b)$$

turns out to be a true proposition. On the other hand the proposition

(6) $$[\ a\quad b\ ].\, a \cap b \bigtriangleup a \cap b$$
$$\text{ex}(a),\ \text{ex}(b)$$

which is the result of substituting '$a \cap b$' for 'a' in (4) does not always hold. On purely intuitive grounds it implies that

(7) $$[\ a\quad b\ ].\, a \bigtriangleup b$$
$$\text{ex}(a),\ \text{ex}(b)$$

which is patently false except on the assumption that there exists exactly one object.[9]

207

If we want to use nominal functions within the framework of S2 then a further limitation of the range of values of the variables is necessary. We can obtain it by demanding on the metalogical level that only those nominal functions of the type $\varphi(a\ b\ c\ \ldots)$ should be substituted for the variables of S2 which satisfy the following condition:

(8) $\qquad\qquad \text{ex}\big(\varphi(a\ b\ c\ \ldots)\big) . \text{ex}(a) . \text{ex}(b) . \text{ex}(c) . \ldots .$[10]

IV. GENERALISED SYLLOGISTIC

The foregoing considerations show quite clearly how unwieldy a system becomes if it allows for quantifiers with restricted interpretation. The richer the syntax of the system the more elaborate has to be the rule of substitution, which determines the range of values substitutable for the variables of the system. This is why it appears to be preferable to have all the limitations stated 'within the system' leaving the quantifiers open to unrestricted interpretation.[11] The required readjustment is secured by preceding the subquantificate in each of the axioms or definitions of S2 with an antecedent which for each variable, 'a', 'b', and 'c', occurring in the quantifier states that a's exist, b's exist, and c's exist. In this way we reach System S3, among the axioms of which we have the following theses:

A3.1 $\qquad [a] : \text{ex}(a) . \supset . a \subset a$

A3.2 $\qquad [a] : \text{ex}(a) . \supset . a \triangle a$

A3.3 $\qquad [a\ b\ c] : \text{ex}(a) . \text{ex}(b) . \text{ex}(c) . b \subset c . a \subset b . \supset . a \subset c$

A3.4 $\qquad [a\ b\ c] : \text{ex}(a) . \text{ex}(b) . \text{ex}(c) . b \subset c . b \triangle a . \supset . a \triangle c$

A3.5 $\qquad [a\ b] : . \text{ex}(a) . \text{ex}(b) . \supset : a \neq b . \equiv . \sim(a \triangle b)$

A3.6 $\qquad [a\ b] : . \text{ex}(a) . \text{ex}(b) . \supset : a \triangledown b . \equiv . \sim(a \subset b)$

S3 has the same rules of inference as S2 except that the rule of substitution presupposes no longer any special restrictions concerning the range of values substitutable for the variables of the system. D2.1 and D2.2 become additional axioms because on reformulation neither of them satisfies the conditions of a well-formed definition.

The present stage of S3 has been arrived at in two moves. First the quantifiers and the rules for their use were added to S1. Secondly, the system obtained in this way was reformulated for the purpose of the unrestricted interpretation of the quantifiers. Thus, perhaps, it would be more correct to say that so far not syllogistic itself but the presuppositions of syllogistic have been extended. There is no difference between the axiom system of S1 and the axiom system of S2 as regards their contents. And the same applies to the definitions of S1 and S2. Similarly, the first six axioms of S3 say exactly the same as the axioms and the two definitions of S1 or S2, but they do so in a more explicit manner, which has the advantage of dispensing with any need for limiting the range of values substitutable for the variables.

As a whole, however, S3 does constitute an extension of syllogistic proper since the theses given below are also included in the list of its axioms.

A3.7	$[a\,b]: a \subset b . \supset . \operatorname{ex}(a)$
A3.8	$[a\,b]: a \subset b . \supset . \operatorname{ex}(b)$
A3.9	$[a\,b]: a \neq b . \supset . \operatorname{ex}(a)$
A3.10	$[a\,b]: a \neq b . \supset . \operatorname{ex}(b)$
A3.11	$[a\,b]: a \triangledown b . \supset . \operatorname{ex}(a)$
A3.12	$[a\,b]: a \triangledown b . \supset . \operatorname{ex}(b)$
A3.13	$[a\,b]: a \triangle b . \supset . [\exists c] . c \subset a . c \subset b$

Axioms A3.7 — A3.12 formally recognize the fact that the universal affirmative, the universal negative, and the particular negative propositions have existential import. The corresponding theses concerning the existential import of the particular affirmative proposition follow at once from A3.8 and A3.13.

Axiom A3.13, which raises no objections as regards intuitiveness, seems to be presupposed in Aristotle's proofs by *ecthesis*.[12]

Since the axioms of S3 give expressions to conceptions that can be regarded as genuinely Aristotelian, I propose to describe the totality of theses derivable within the framework of S3 as *generalised syllogistic*. I will now show that S3 is inferentially equivalent to a system with much simpler axiomatic foundations. This new system, to be called S4, presupposes the same rules of inference as S3 but is based on a single axiom of the following form:

| A4.1 | $[a\,b] :: a \subset b . \equiv :. [\exists c] . c \subset a :. [c] : c \subset a . \supset . c \subset b$ |

The proof that S3 and S4 are inferentially equivalent involves the following deductions within S3.

T3.1 $[a\,b]: a \subset b . \supset .[\exists\, c]. c \subset a$

Proof:

$[a\,b]:.$

(1) $a \subset b . \supset :$

(2) $\mathrm{ex}(a).$ (follows from A3.7, 1)

(3) $a \subset a :$ (from A3.1, 2)

        $[\exists\, c]. c \subset a$ (3)

T3.2 $[a\,b\,c]: a \subset b . c \subset a . \supset . c \subset b$

Proof:

$[a\,b\,c]:$

(1) $a \subset b.$

(2) $c \subset a . \supset .$

(3) $\mathrm{ex}(a).$ (A3.7, 1)

(4) $\mathrm{ex}(b).$ (A3.8, 1)

(5) $\mathrm{ex}(c).$ (A3.7, 2)

        $c \subset b$ (A3.3, 5, 3, 4, 1, 2)

T3.3 $[a\,b\,c]:: c \subset a :. [d]: d \subset a . \supset . d \subset b :. \supset . a \subset b$

Proof:

$[a\,b\,c]::$

(1) $c \subset a :.$

(2) $[d]: d \subset a . \supset . d \subset b :. \supset .$

(3) $\mathrm{ex}(a).$ (A3.8, 1)

(4) $a \subset a.$ (A3.1, 3)

        $a \subset b$ (2,4)

T3.4 = A4.1 $[a\,b]:: a \subset b . \equiv :. [\exists\, c]. c \subset a :. [c]: c \subset a . \supset . c$

           $\subset b$ (T3.1, T3.2, T3.3)

T3.5 = D4.1 $[a]: \mathrm{ex}(a). \equiv . a \subset a$ (A3.1, A3.7)

T3.6 $[a\,b\,c]: c \subset a . c \subset b . \supset . a \triangle b$

Proof:

$[a\,b\,c]:$

(1) $c \subset a.$

(2) $c \subset b . \supset .$

(3)	ex(c).	(A3.7, 1)
(4)	ex(a).	(A3.8, 1)
(5)	ex(b).	(A3.8, 2)
(6)	$c \triangle c$.	(A3.2, 3)
(7)	$c \triangle a$.	(A3.4, 3, 5, 4, 1, 6)
	$a \triangle b$	(A3.4, 4, 3, 5, 2, 7)

T3.7 = D4.2 $[a\,b] : a \triangle b . \equiv . [\exists c] . c \mathrel{\mathsf C} a . c \mathrel{\mathsf C} b$

(A3.13, T3.6)

T3.8 = D4.3 $[a\,b] : a \mathrel{\#} b . \equiv . \text{ex}(a) . \text{ex}(b) . \sim (a \triangle b)$

(A3.9, A3.10, A3.5)

T3.9 = D4.4 $[a\,b] : a \triangledown b . \equiv . \text{ex}(a) . \text{ex}(b) . \sim (a \mathrel{\mathsf C} b)$

(A3.11, A3.12, T3.6)

Theses T3.5, T3.7, T3.8, and T3.9 show that the functors 'ex', '$\triangle$', '$\#$', and '$\triangledown$' can eventually be defined in terms of '$\mathsf C$'. By deriving these theses within the framework of S3 and by deriving T3.4 = A4.1 we have shown that any theses obtainable in S4 can also be obtained in S3. It remains to prove the converse. This will be done by deriving A3.1 − A3.12 from A4.1 and D4.1 − D4.4. The required deductions proceed as follows

T4.1 = A3.1 $[a] : \text{ex}(a) . \supset . a \mathrel{\mathsf C} a$ (D4.1)

T4.2 = A3.2 $[a] : \text{ex}(a) . \supset . a \triangle a$

Proof:

$[a]$:

(1)	ex(a). $\supset$.	
(2)	$a \mathrel{\mathsf C} a$.	(D4.1, 1)
	$a \triangle a$	(D4.2, 2)

T4.3 = A3.3 $[a\,b\,c] : \text{ex}(a) . \text{ex}(b) . \text{ex}(c) . b \mathrel{\mathsf C} c . a \mathrel{\mathsf C} b . \supset . a \mathrel{\mathsf C} c$

(A4.1)

T4.4 $[a\,b\,c] : b \mathrel{\mathsf C} c . b \triangle a . \supset . a \triangle c$

Proof:

$[a\,b\,c]$:.

(1)	$b \mathrel{\mathsf C} c$.
(2)	$b \triangle a . \supset :$

$$[\exists\, d]\,.$$

(3) $d \subset b\,.$ $\Big\}$ (D4.2, 2)

(4) $d \subset a\,.$

(5) $d \subset c\!:$ (A4.1, 1, 3)

$a \bigtriangleup c$ (D4.2, 4, 5)

T4.5 = A3.4 $[a\,b\,c]\!: \mathrm{ex}(a)\,.\,\mathrm{ex}(b)\,.\,\mathrm{ex}(c)\,.\,b \subset c\,.b \bigtriangleup a\,.\,\supset\,.\,a \bigtriangleup c$

(T4.4)

T4.6 = A3.5 $[a\,b]\!:.\,\mathrm{ex}(a)\,.\,\mathrm{ex}(b)\,.\,\supset\,:\,a \mathbin{\#} b\,.\equiv\,.\sim (a \bigtriangleup b)$

(D4.3)

T4.7 = A3.6 $[a\,b]\!:.\,\mathrm{ex}(a)\,.\,\mathrm{ex}(b)\,.\,\supset\,:\,a \bigtriangledown b\,.\equiv\,.\sim (a \subset b)$

(D4.4)

T4.8 = A3.7 $[a\,b]\!: a \subset b\,.\,\supset\,.\,\mathrm{ex}(a)$
Proof:
$[a\,b]\!:.$
(1) $a \subset b\,.\,\supset\!:$
(2) $[\exists\, c]\,.\,c \subset a\!:$ (A4.1, 1)
(3) $a \subset a\,.$ (A4.1, 2)
$\mathrm{ex}(a)$ (D4.1, 3)

T4.9 = A3.8 $[a\,b]\!: a \subset b\,.\,\supset\,.\,\mathrm{ex}(b)$
Proof:
$[a\,b]\!:$
(1) $a \subset b\,.\,\supset\,.$
(2) $b \subset b\,.$ (A4.1, 1)
$\mathrm{ex}(b)$ (D4.1, 2)

T4.10 = A3.9 $[a\,b]\!: a \mathbin{\#} b\,.\,\supset\,.\,\mathrm{ex}(a)$ (D4.3)

T4.11 = A3.10 $[a\,b]\!: a \mathbin{\#} b\,.\,\supset\,.\,\mathrm{ex}(b)$ (D4.3)

T4.12 = A3.11 $[a\,b]\!: a \bigtriangledown b\,.\,\supset\!:\,\mathrm{ex}(a)$ (D4.4)

T4.13 = A3.12 $[a\,b]\!: a \bigtriangledown b\,.\,\supset\,.\,\mathrm{ex}(b)$ (D4.4)

T4.14 = A3.13 $[a\,b]\!: a \bigtriangleup b\,.\,\supset\,.\,[\exists\, c]\,.\,c \subset a\,.\,c \subset b$ (D4.2)

From theses T4.1–T4.3, T4.5–T4.14 we can see that any thesis derivable in S3 is also derivable in S4, which completes the proof that the two systems are inferentially equivalent.

In a system of generalised syllogistic we can derive, among other theses, the 24 syllogistic laws, the laws of conversion, the laws of contrariety, and the laws of subalternation. We can also prove that

(9) $[a\,b]: . \, \mathrm{ex}(a) . \, \mathrm{ex}(b) . \supset : a \neq b . \equiv . \sim (a \,\triangle\, b)$

(10) $[a\,b]: . \, \mathrm{ex}(a) . \, ex(b) . \supset : a \,\triangledown\, b . \equiv . \sim (a \,\mathsf{C}\, b)$

(11) $[a\,b]: . \, \mathrm{ex}(a) . \, \mathrm{ex}(b) . \supset : a \,\triangle\, b . \lor . a \,\triangledown\, b$

but propositions which say that

(12) $[a\,b]: a \neq b . \equiv . \sim (a \,\triangle\, b)$

(13) $[a\,b]: a \,\triangle\, b . \equiv . \sim (a \,\mathsf{C}\, b)$

(14) $[a\,b]: a \,\triangle\, b . \lor . a \,\triangledown\, b$

cannot be proved in a system of generalised syllogistic. This means that on the interpretation we have accorded to the categorical propositions the traditional laws of contradiction and subcontrariety hold only for referential names. They do not hold for non-referential ones.

V. EXTENDED SYLLOGISTIC

We have already said that in our view the four categorical propositions have existential import. While this is generally accepted as regards the particular affirmative proposition, alternative interpretations of the remaining types of categorical proposition have been considered by various logicians. In particular the possibility of depriving the universal propositions of their existential import has been explored in some detail.[13] In this connection it may not be out of place to note the following.

Ordinary usage makes hardly any distinction between propositions with existential import on the one hand and the corresponding propositions without existential import on the other. This is not surprising in view of the fact that ordinary usage has little room for the use of non-referential names in contexts other than fiction. To put it in different terms, ordinary usage seems to presuppose a restricted interpretation of the quantifiers, which is reflected in the sense we usually attach to such pronouns as 'whoever', 'whatever', 'whichever', 'somebody', 'something', etc.[14] The same holds for Aristotle's syllogistic. Since Aristotle tacitly assumed what amounted to the restricted interpretation of the quantifiers on the line of the restrictions inherent in S2, the problem of existential import of propositions did not present itself to him. It occurred to logicians after they have recognized the legitimacy and usefulness of reasonings involving non-referential names. In Aristotle's syllogistic, just as in S2, any type of proposition has existential import. If we want to have propositions without existential import, we have to drop the restrictions concerning the interpretation of the quantifiers. S4 has no such restrictions, and we may now turn to the problem of introducing into S4 the universal affirmative proposition without existential import.

By the universal affirmative proposition without existential import we shall understand the proposition of the type 'all a is b' (in symbols: $a \subset b$). Its meaning will be determined by the following definition

D4.5 $\qquad [a\,b]:.\,a \subset b\,.\equiv:[c]:c \; \mathsf{C}\,a\,.\supset.\,c \; \mathsf{C}\,b$

It is evident from the definiens that a proposition of the type '$a \subset b$' is true even if the corresponding proposition of the type 'ex(a)' is false. For if a proposition of the type 'ex(a)' is false then the corresponding proposition of the type '$[c]\,.\sim(c \; \mathsf{C}\,a)$' is true, and consequently the corresponding proposition of the type '$[c]:c \; \mathsf{C}\,a\,.\supset.\,c \; \mathsf{C}\,b$' is also true. It is further evident that a proposition of the type '$a \subset b$' is implied by the corresponding proposition of the type '$a \; \mathsf{C}\,b$'; but the converse does not hold. This is why in Leśniewski's ontology the universal affirmative proposition with existential import has been called *strong inclusion* whereas the name of *weak inclusion* has been given to the universal affirmative proposition without existential import. Although strong inclusion is not implied by the corresponding weak inclusion the latter suffices for the purpose of defining the former, and the following equivalence offers itself as a possible definition

(15) $\qquad [a\,b]:.\,a \; \mathsf{C}\,b\,.\equiv:a \subset b:[\exists c]\,.\sim(a \subset c)$

Now, the interesting thing about (15) is this. It does define 'C' in terms of '$\subset$' but is not deducible from A4.1 supplemented by D4.5, which means that there can be no system which has '$\subset$' as the only primitive term and is weak enough to be inferentially equivalent to S4. If we wish to prove (15), we have to extend S4, and the weakest extension that can still serve our purpose consists in assuming

(16) $\qquad [a\,b]:a \; \mathsf{C}\,b\,.\supset.\,[\exists c]\,.\sim(a \; \mathsf{C}\,c)$

as an additional axiom.

214

If we interpret the variables of A4.1 and (16) as propositional variables and if we interpret the constant term ' C ' as 'vr', i.e. as the functor which forms a true proposition irrespective of the truth value of either of its two arguments, then, on this interpretation, A4.1, turns out to be true whereas (16) becomes false. This shows that (16) is not derivable from A4.1. Nor is A4.1 derivable from (16). For if again we interpret the variables of A4.1 and (16) as propositional variables while interpreting ' C ' as the functor of exclusive disjunction, which forms propositions equivalent to the corresponding propositions of the type ' $\sim (p \equiv q)$ ', then, on this interpretion, (16) is true but A4.1 is false. Thus A4.1 and (16) are found to be mutually independent.

The totality of theses derivable from A4.1 and (16) by means of the same rules of inference as those available in S4 may, for terminological convenience, be described as *extended syllogistic*. In what follows two systems of extended syllogistic, System S5 and System S6, will be outlined and shown to be inferentially equivalent.

Both S5 and S6 presuppose the same rules of inference as those in S4. S5 is based on a single axiom, which takes the form of the following proposition

A5.1 $\qquad [a\,b]::a\ \mathsf{C}\,b\,.\equiv:.\,[\exists\,c\,d]\,.\,c\ \mathsf{C}\,a\,.\,\sim(c\ \mathsf{C}\,d):.\,[c]:$
$\qquad\qquad c\ \mathsf{C}\,a\,.\supset.\,c\ \mathsf{C}\,b$

The proposition which says that

A6.1 $\qquad [a\,b]:.\,a\subset b\,.\equiv:[c]:c\subset a\,.\supset.\,c\subset b$

serves as the only axiom of S6.

Before we proceed to establishing inferential equivalence between S5 and S6, it may be worth our while to convince ourselves that S5 is in fact a system of extended syllogistic in our sense of the term, or, to put it in other words, that A5.1 is inferentially equivalent to A4.1 and (16) taken together. Let us, therefore, assume

M1 = A4.1 $\qquad [a\,b]::a\ \mathsf{C}\,b\,.\equiv:.\,[\exists\,c]\,.\,c\ \mathsf{C}\,a:.\,[c]:c\ \mathsf{C}\,a\,.\supset.\,c\ \mathsf{C}\,b$

M2 = (16) $\qquad [a\,b]:a\ \mathsf{C}\,b\,.\supset.\,[\exists\,c]\,.\sim(a\ \mathsf{C}\,c)$

and let us proceed with the deductions as follows

M3 $\qquad [a\,b]:a\ \mathsf{C}\,b\,.\supset.\,[\exists\,c\,d]\,.\,c\ \mathsf{C}\,a\,.\,\sim(c\ \mathsf{C}\,d)$

215

CZESŁAW LEJEWSKI

Proof:

$[a\,b]:.$

(1) $a \subset b . \supset :$

(2) $[\exists c] . c \subset a :$ (M1, 1)

(3) $a \subset a :$ (M1, 2)

(4) $[\exists c] . {\sim} (a \subset c) :$ (M2, 1)

     $[\exists c\,d] . c \subset a . {\sim} (c \subset d)$ (3, 4)

M4 = A5.1 $[a\,b] :: a \subset b . \equiv :. [\exists c\,d] . c \subset a . {\sim} (c \subset d) :. [c] : c \subset a . \supset . c \subset b$ (M1, M3)

Now, let us assume

N1 = A5.1 $[a\,b] :: a \subset b . \equiv :. [\exists c\,d] . c \subset a . {\sim} (c \subset d) :. [c] : c \subset a . \supset . c \subset b$

and let our deductions develop as follows

N2 $[a\,b\,c] : a \subset b . c \subset a . \supset . c \subset b$ (N1)

N3 $[a\,b\,c] :: c \subset a :. [d] : d \subset a . \supset . d \subset b :. \supset . a \subset b$

Proof:

$[a\,b\,c] ::$

(1) $c \subset a :.$

(2) $[d] : d \subset a . \supset . d \subset b :. \supset :$

     $[\exists d\,e] .$

(3) $d \subset c .$ } (N1, 1)

(4) ${\sim} (d \subset e) .$

(5) $d \subset a :$ (N2, 1, 3)

(6) $a \subset a :$ (N1, 5, 4)

     $a \subset b$ (2, 6)

N4 = A4.1   (N1, N3)

N5. = (16)

Proof:

$[a\,b]:.$

(1) $a \subset b . \supset :$

     $[\exists c\,d] .$

(2) $c \subset a .$ } (N1, 1)

(3) ${\sim} (c \subset d) .$

216

$$(4) \qquad \sim (a \; \mathbf{C} \; d): \qquad\qquad (N2, 2, 3)$$
$$[\exists \; c]: \sim (a \; \mathbf{C} \; c) \qquad\qquad (4)$$

It is clear from the above considerations that the totality of theses derivable from A4.1 and (16) coincides with the totality of theses derivable from A5.1, which means that S5 is a system of what we have suggested to call extended syllogistic.

The proof that S5 and S6 are inferentially equivalent involves the following deductions within S5.

D5.1 $[a \, b]: . \, a \subset b . \equiv : [c]: c \; \mathbf{C} \; a . \supset . c \; \mathbf{C} \; b$

T5.1 $[a \, b]: a \; \mathbf{C} \; b . \supset . a \subset b$ (A5.1, D5.1)

T5.2 $[a \, b]: a \; \mathbf{C} \; b . \supset . [\exists \; c] . \sim (a \subset c)$
Proof:
$[a \, b]: .$
(1) $a \; \mathbf{C} \; b . \supset :$
       $[\exists \; c \, d] .$
(2) $c \; \mathbf{C} \; a .$ $\Big\}$ (A5.1, 1)
(3) $\sim (c \; \mathbf{C} \; d) .$
(4) $\sim (a \subset d) :$ (D5.1, 2, 3)
       $[\exists \; c] . \sim (a \subset c)$ (4)

T5.3 $[a \, b \, c]: a \subset b . \sim (a \subset c) . \supset . a \; \mathbf{C} \; b$
Proof:
$[a \, b \, c] ::$
(1) $a \subset b .$
(2) $\sim (a \subset c) . \supset : .$
       $[\exists \; d] .$
(3) $d \; \mathbf{C} \; a .$ $\Big\}$ (D5.1, 2)
(4) $\sim (d \; \mathbf{C} \; c) : .$
(5) $[d]: d \; \mathbf{C} \; a . \supset . d \; \mathbf{C} \; b : .$ (D5.1, 1)
       $a \; \mathbf{C} \; b$ (A5.1, 3, 4, 5)

T5.4 = D6.1 $[a \, b]: . \, a \; \mathbf{C} \; b . \equiv : a \subset b : [\exists \; c] . \sim (a \subset c)$
                                        (T5.1, T5.2, T5.3)

T5.5 $[a]: a \subset a$ (D5.1)

T5.6 $[a \, b \, c]: a \subset b . c \subset a . \supset . c \subset b$

Proof:

$[a\,b\,c]::$

(1) $\quad a \subset b.$

(2) $\quad c \subset a . \supset :.$

(3) $\quad [d]: d \,\mathsf{C}\, a . \supset . d \,\mathsf{C}\, b:$ (D5.1, 1)

(4) $\quad [d]: d \,\mathsf{C}\, c . \supset . d \,\mathsf{C}\, a:.$ (D5.1, 2)

(5) $\quad [d]: d \,\mathsf{C}\, c . \supset . d \,\mathsf{C}\, b:.$ (4, 3)

$\quad\quad c \subset b$ (D5.1, 5)

T5.7 $\quad [a\,b]:. [c]: c \subset a . \supset . c \subset b: \supset . a \subset b$ (T5.5)

T5.8 = A6.1 $\quad [a\,b]:. a \subset b . \equiv : [c]: c \subset a . \supset . c \subset b$ (T5,6. T5.7)

By deriving T5.8 = A6.1 and T5.4 = D6.1 we have satisfied ourselves that any thesis obtainable in S6 can also be obtained in S5. It now remains to show that conversely any thesis obtainable in S5 can also be obtained in S6. This will be done by deriving A5.1 and D5.1 within the framework of S6. Our deductions proceed as follows.

D6.1 $\quad [a\,b]:. a \,\mathsf{C}\, b . \equiv : a \subset b : [\exists c]. \sim (a \subset c)$

T6.1 $\quad [a] . a \subset a$ (A6.1)

T6.2 $\quad [a\,b\,c]: a \subset b . c \,\mathsf{C}\, a . \supset . c \,\mathsf{C}\, b$

Proof:

$[a\,b\,c]:.$

(1) $\quad a \subset b.$

(2) $\quad c \,\mathsf{C}\, a . \supset :$

(3) $\quad c \subset a:$

(4) $\quad [\exists d] . \sim (c \subset d):$ (D6.1, 2)

(5) $\quad c \subset b:$ (A6.1, 1, 3)

$\quad\quad c \,\mathsf{C}\, b$ (D6.1, 5, 4)

T6.3 $\quad [a\,b]:: [c]: c \,\mathsf{C}\, a . \supset . c \,\mathsf{C}\, b:. \sim (a \subset b):. \supset . a \supset b$

Proof:

$[a\,b]::$

(1) $\quad [c]: c \,\mathsf{C}\, a . \supset . c \,\mathsf{C}\, b:.$

(2) $\quad \sim (a \subset b):. \supset .$

(3) $\quad a \,\mathsf{C}\, a.$ (D6.1, T6.1, 2)

(4) $\quad a \,\mathsf{C}\, b.$ (1, 3)

$\quad\quad a \subset b$ (D6.1, 4)

218

T6.4 $\qquad$ $[a\,b]:.\,[c]:c\,\subset a\,.\supset.\,c\,\subset b:\supset.\,a\subset b$ $\qquad$ (T6.3)

T6.5 = D5.1 $\qquad\qquad\qquad\qquad\qquad\qquad\qquad$ (T6.2, T6.4)

T6.6 $\qquad$ $[a\,b]:a\,\subset b\,.\supset.\,[\exists\,c\,d].\,c\,\subset a\,.\sim(c\,\subset d)$

Proof:

$[a\,b]:.$

(1) $\qquad a\,\subset b\,.\supset:$
$\qquad\qquad [\exists\,c].$

(2) $\qquad\qquad\qquad \sim(a\subset c).$ $\qquad\qquad\qquad$ (D6.1, 1)

(3) $\qquad\qquad\qquad \sim(a\,\subset c):$ $\qquad\qquad\qquad$ (D6.1, 2)

(4) $\qquad\qquad a\,\subset a:$ $\qquad\qquad\qquad$ (D6.1, T6.1, 2)
$\qquad\qquad [\exists\,c\,d].\,c\,\subset a\,.\sim(c\,\subset d)$ $\qquad\qquad\qquad$ (4, 3)

T6.7 $\qquad$ $[a\,b\,c]:a\,\subset b\,.\,c\,\subset a\,.\supset.\,c\,\subset b$

Proof:

$[a\,b\,c]:.$

(1) $\qquad a\,\subset b.$

(2) $\qquad c\,\subset a\,.\supset:$

(3) $\qquad a\,\subset b.$ $\qquad\qquad\qquad$ (D6.1, 1)
$\qquad\qquad c\,\subset b$ $\qquad\qquad\qquad$ (D6.2, 3, 2)

T6.8 $\qquad$ $[a\,b\,c]::c\,\subset a:.\,[d]:d\,\subset a\,.\supset.\,d\,\subset b:.\supset.\,a\,\subset b$

Proof:

$[a\,b\,c]:.$

(1) $\qquad c\,\subset a:.$

(2) $\qquad [d]:d\,\subset a\,.\supset.\,d\,\subset b:.\supset:$
$\qquad\qquad [\exists\,d].$

(3) $\qquad\qquad\qquad \sim(c\subset d).$ $\qquad\qquad\qquad$ (D6.1, 1)

(4) $\qquad\qquad\qquad \sim(c\,\subset d).$ $\qquad\qquad\qquad$ (D6.1, 3)

(5) $\qquad\qquad\qquad \sim(a\subset d):$ $\qquad\qquad\qquad$ (T6.2, 1, 4)

(6) $\qquad\qquad a\,\subset a:$ $\qquad\qquad\qquad$ (D6.1, T6.1, 5)
$\qquad\qquad a\,\subset b$ $\qquad\qquad\qquad$ (2, 6)

T6.9 = A5.1 $\qquad\qquad\qquad\qquad\qquad\qquad\qquad$ (T6.6, T6.7, T6.8)

From T6.9 = A5.1 and T6.5 = D5.1 we can see that any thesis derivable in S5 is also derivable in S6, which completes the proof that the two systems are inferentially equivalent.

VI. BASIC BOOLEAN ALGEBRA

Our next task is to consider a further extension of the systems of extended syllogistic. Since S4 is part of S5, it is obvious that D4.2, which defines the particular affirmative proposition or partial inclusion in terms of strong conclusion, can be introduced into S5 as

D5.2 $\qquad [a\,b]: a \wedge b . \equiv . [\exists c] . c \subset a . c \subset b$

On purely intuitive grounds it would appear that, conversely, strong inclusion could be defined in terms of partial inclusion, and the proposition which says that

(17) $\qquad [a\,b]:: a \subset b . \equiv : . a \wedge a : . [c] : c \wedge a . \supset . c \wedge b$

seems to qualify as a possible definition in a system in which the functor of partial inclusion is used as a primitive term. It turns out, however, that we cannot have a system which is based on '$\wedge$' as the only primitive term and which is weak enough to be inferentially equivalent to S5. For A5.1 and D5.2 do not entail (17). If we want to prove (17), we have to strengthen A5.1 by subjoining to it a thesis which is not weaker than the following

(18) $\qquad [a\,b\,c]:: c \subset a : . [d\,e] : d \subset e . d \subset a . \supset . [\exists f] . f \subset e .$
$f \subset b : . \supset . a \subset b$

A5.1 and (18) are, of course, mutually independent. If we interpret the variables of A5.1 and (18) as representing integers and if we interpret '$\subset$' as '$\leqslant$' in its arithmetical sense then, on this interpretation, A5.1 becomes true whereas for some values of 'a' and 'b' (18) shows itself to be false. Again if we interpret the variables of A5.1 and (18) as propositional variables and if we interpret '$\subset$' as 'vr' then, on this interpretation, (18) becomes true whereas A5.1 turns out to be false.

The totality of theses derivable from A5.1 and (18) by means of the rule of propositional definitions, the rule of substitution, the rules for the use of the quantifiers, and the rule of detachment, will be described as *basic Boolean algebra*. I propose to describe it in this way because given appropriate rules of nominal definition it extends to coincide with a totality containing systems of ordinary Boolean Algebra as has been shown in Lejewski [5].

The functors 'C', '$\subset$', and '$\triangle$' are all suitable as single primitive terms to be used in systems of basic Boolean algebra. We may refer to these systems as S7, S8, and S9 respectively.

S7 has the following single axiom

A7.1 $[a\,b]::a\ C\,b.\equiv:.[\exists\,c\,d].c\ C\,a.\sim(c\ C\,d):.[c\,d]:$
         $c\ C\,d.c\ C\,a.\supset.[\exists f].f\ C\,d.f\ C\,b$

The single axiom of S8 says that

A8.1 $[a\,b]:.a\subset b.\equiv:[c\,d\,e]:c\subset d.c\subset a.\sim(c\subset e).$
         $\supset.[\exists f\,g].f\subset d.f\subset b.\sim(f\subset g)$

And, finally, the following proposition is used as the single axiom of S9

A9.1 $[a\,b]::a\triangle b.\equiv:.[\exists\,c\,d]:.c\triangle a.\sim(c\triangle d):.[e]:$
         $e\triangle c.\supset.e\triangle a.e\triangle b$ [15]

The proof that S7, S8, and S9 are inferentially equivalent will not be given here because it can easily be reconstructed on the basis of the proofs contained in my paper mentioned above. I will, however, offer a proof that the totality of theses derivable from A7.1 coincides with the totality of theses derivable from A5.1 and (18). The proof, which is very simple, involves the following steps. First we assume

P1 = A5.1 $[a\,b]::a\ C\,b.\equiv:.[\exists\,c\,d].c\ C\,a.\sim(c\ C\,d):.[c]:c$
         $C\,a.\supset.c\ C\,b$

P2 = (18) $[a\,b\,c]::c\ C\,a:.[d\,e]:d\ C\,e.d\ C\,a.\supset.[\exists f].f\ C\,e$
         $.f\ C\,b:.\supset.a\ C\,b$

and proceed by deriving

P3 $[a\,b\,c\,d]:a\ C\,b.c\ C\,d.c\ C\,a.\supset.[\exists f].f\ C\,d.f\ C\,b$
     Proof:
     $[a\,b\,c\,d]:.$
     (1) $a\ C\,b.$
     (2) $c\ C\,d.$
     (3) $c\ C\,a.\supset:$
     (4) $c\ C\,b:$ (P1, 1, 3)
         $[\exists f].f\ C\,d.f\ C\,b$ (2, 4)

P4 = A7.1 $[a\,b]::a \subset b . \equiv :. [\exists\,c\,d].c \subset a. \sim (a \subset d):.[c\,d]:$
$c \subset d.c \subset a. \supset .[\exists f].f \subset d.f \subset b$ (P1, P3, P2)

Secondly we assume

Q1 = A7.1

from which we derive

Q2 $[a\,b\,c]:a \subset b.c \subset a. \supset .c \subset b$
Proof:
$[a\,b\,c]::$
(1) $a \subset b.$
(2) $c \subset a. \supset :.$
(3) $[\exists\,d\,e].d \subset c. \sim (d \subset e):.$
(4) $[d\,e]:d \subset e.d \subset c. \supset .[\exists f].f \subset e.$ }(Q1, 2)
       $f \subset a:.$
(5) $[d\,e]:d \subset e.d \subset a. \supset .[\exists f].f \subset e.f \subset b:.$
                             (Q1, 1)
(6) $[d\,e]:d \subset e.d \subset c. \supset .[\exists f].f \subset e.f \subset b:.$
                                 (4, 5)
       $c \subset b$ (Q1, 3, 6)

Q3 $[a\,b]:a \subset b. \supset .[\exists c]. \sim (a \subset c)$
Proof:
$[a\,b]:.$
(1) $a \subset b. \supset :$
       $[\exists\,c\,d].$
(2) $c \subset a.$ }(Q1, 1)
(3) $\sim (c \subset d).$
(4) $\sim (a \subset d):$ (Q2, 2, 3)
       $[\exists c]. \sim (a \subset c)$ (4)

Q4 $[a\,b\,c]::c \subset a:.[d]:d \subset a. \supset .d \subset b:. \supset .a \subset b$
Proof:
$[a\,b\,c]::$
(1) $c \subset a:.$
(2) $[d]:d \subset a. \supset .d \subset b:. \supset :$
(3) $[\exists d]. \sim (c \subset d):$ (Q3, 1)
(4) $a \subset a:$ (Q1, 1, 3)
       $a \subset b$ (2, 4)

Q5 = A5.1   (Q1, Q2, Q4)

Q6 = (18)   (Q1)

It is clear from P1–P4 and Q1–Q6 that A7.1 is inferentially equivalent to A5.1 and (18) taken together.

VII. BASIC ONTOLOGY

The clue to the final extension of syllogistic to be considered in this paper is derived from the examination of the interdefinability of strong and singular inclusions.

In S7 the functor of singular inclusion can be defined as follows

D7.1 $[a\,b]::a\,\varepsilon\,b\,.\equiv:.a\,\mathsf{C}\,b:.\,[c]:c\,\mathsf{C}\,a\,.\supset.\,a\,\mathsf{C}\,c$

Now, the definition of strong inclusion in terms of singular inclusion takes the form of the following familiar equivalence

(19) $[a\,b]::a\,\mathsf{C}\,b\,.\equiv:.\,[\exists\,c]\,.\,c\,\varepsilon\,a:.\,[c]:c\,\varepsilon\,a\,.\supset.\,c\,\varepsilon\,b$

This equivalence, however, cannot be derived from A7.1 and D7.1, which means that we cannot have a system of basic Boolean algebra with 'ε' as the only undefined term. If we want to prove (19), we have to strengthen A7.1 by subjoining to it an additional axiom not weaker than the proposition which says that

(20) $[a\,b]::a\,\mathsf{C}\,b\,.\supset:.\,[\exists\,c]:.\,c\,\mathsf{C}\,a:\,[d]:d\,\mathsf{C}\,c\,.\supset.$
                  $c\,\mathsf{C}\,d$

The independence of (20) from A7.1 can be established with the aid of the following interpretation. Let the variables of A7.1 and (20) represent ordered couples of rational numbers $\langle x, y \rangle$, $\langle v, w \rangle$, ..., and let the functor 'C' be interpreted as the relation that holds between any two such couples, say $\langle x, y \rangle$ and $\langle v, w \rangle$, if and only if $v \leqslant x < y \leqslant w$. On this interpretation A7.1 would be equivalent to the proposition which says that

(21) $[q\,r\,s\,t]::q \leqslant r < s \leqslant t\,.\equiv:.\,[\exists\,u\,v\,w\,x]\,.\,r \leqslant u < v$
                $\leqslant s\,.\sim(w \leqslant u < v \leqslant x):.\,[u\,v\,w\,x]:u \leqslant v < w \leqslant x\,.$
                $r \leqslant v < w \leqslant s\,.\supset.\,[\exists\,y\,z]\,.\,u \leqslant y < z \leqslant x\,.\,q \leqslant y$
                $< z \leqslant t$

and (20) would be equivalent to the proposition which says that

(22) $\qquad [q\,r\,s\,t] :: q \leqslant r < s \leqslant t . \supset :. [\exists\, u\, v] :. r \leqslant u < v \leqslant s :. [w\, x] : u \leqslant w < x \leqslant v . \supset . w \leqslant u < v \leqslant x$

Now, (21) is true whereas (22) is false because for some values of the variables its antecedent turns out to be true whereas its consequent is always false in view of the fact that

(23) $\qquad [r\,u\,v\,s] : r \leqslant u < v \leqslant s . \supset . [\exists\, w\, x]. u \leqslant w < x \leqslant v . \sim (w \leqslant u < v \leqslant x)$

The independence of A7.1 from (20) can easily be established by inter-preting the variables as propositional variables and by interpreting 'C' as 'vr'. For on this interpretation (20) remains true whereas A7.1 becomes false.

The totality of theses derivable from A7.1 and (20) by means of the same rules of inference as those provided for the systems S3–S9 will be referred to as *basic ontology*. The transition to a system of the original Leśniewskian ontology involves introducing a rule for writing nominal definitions, and a rule of extensionality. This form of extension, however, cannot be explored within the boundaries of the present essay.

Any of the four functors, 'C', '⊂', '△', or 'ε' can be used as a single primitive term of a system of basic ontology. We have S10 with the following single axiom

A10.1 $\qquad [a\,b] ::: a \; \mathsf{C} \; b . \equiv ::. [\exists\, c\, d]. c \; \mathsf{C} \; a . \sim (c \; \mathsf{C} \; d) ::. [c] :: c \; \mathsf{C} \; a :. [d] : d \; \mathsf{C} \; c . \supset . c \; \mathsf{C} \; d :. \supset . c \; \mathsf{C} \; b$

The single axiom of S11 takes the form of the following proposition

A11.1 $\qquad [a\,b] ::: a \subset b . \equiv ::. [c] ::. c \subset a :: [d\,e] :. d \subset c . \supset : c \subset d . \lor . d \subset e :: \supset . c \subset b$

S12 is based on an axiom which says that

A12.1 $\qquad [a\,b] :: a \triangle b . \equiv :. [\exists\, c\, d] :. c \triangle a . c \triangle b . \sim (c \triangle d) :. [d\,e] : d \triangle c . e \triangle c . \supset . d \triangle e$

Finally, the proposition which says that

A13.1 $\quad [a\,b]::a\,\varepsilon\,b.\equiv:.[\exists\,c\,d].c\,\varepsilon\,a.c\varepsilon b.\sim(c\varepsilon d):.[c\,d]$
$:c\,\varepsilon\,a.d\varepsilon a.\supset.c\,\varepsilon\,d$

serves as the axiom of S13.[16]

I now propose to give the proof that the totality of theses derivable from A10.1 coincides with the totality of theses derivable from A7.1 and (20). First, we assume

U1 = A7.1 $\quad [a\,b]::a\,\mathsf{C}\,b.\equiv:.[\exists\,c\,d].c\,\mathsf{C}\,a.\sim(c\,\mathsf{C}\,d):.$
$[c\,d]:c\,\mathsf{C}\,d.c\,\mathsf{C}\,a.\supset.[\exists f].f\,\mathsf{C}\,d.f\,\mathsf{C}\,b$

U2 = (20) $\quad [a\,b]::a\,\mathsf{C}\,b.\supset:.[\exists\,c]:.c\,\mathsf{C}\,a:.[d]:d\,\mathsf{C}\,c.\supset.$
$c\,\mathsf{C}\,d$

and derive the following theses

U3 $\quad [a\,b\,c]:a\,\mathsf{C}\,b.c\,\mathsf{C}\,a.\supset.c\,\mathsf{C}\,b \qquad\qquad (\text{U1})^{11}$

U4 $\quad [a\,b\,e\,f]:::[c]::c\,\mathsf{C}\,a:.[d]:d\,\mathsf{C}\,c.\supset.c\,\mathsf{C}\,d:.$
$\supset.c\,\mathsf{C}\,b::.e\,\mathsf{C}\,f.e\,\mathsf{C}\,a::.\supset.[\exists\,g].g\,\mathsf{C}\,f.$
$g\,\mathsf{C}\,b$
Proof:
$[a\,b\,e\,f]:::$

(1) $\quad [c]::c\,\mathsf{C}\,a:.[d]:d\,\mathsf{C}\,c.\supset.c\,\mathsf{C}\,d:.\supset.$
$\quad c\,\mathsf{C}\,b::.$

(2) $\quad e\,\mathsf{C}\,f.$

(3) $\quad e\,\mathsf{C}\,a::.\supset::.$
$\quad [\exists\,g]::$

(4) $\quad\quad g\,\mathsf{C}\,e:.$ $\qquad\qquad$ (U2, 3)

(5) $\quad\quad [h]:h\,\mathsf{C}\,g.\supset.g\,\mathsf{C}\,h:.$

(6) $\quad\quad g\,\mathsf{C}\,f.$ $\qquad\qquad$ (U3, 2, 4)

(7) $\quad\quad g\,\mathsf{C}\,a.$ $\qquad\qquad$ (U3, 3, 4)

(8) $\quad\quad g\,\mathsf{C}\,b::.$ $\qquad\qquad$ (1, 7, 5)
$\quad [\exists\,g].g\,\mathsf{C}\,f.g\,\mathsf{C}\,b$ $\qquad\qquad$ (6, 8)

U5 $\quad [a\,b\,e\,f]:::e\,\mathsf{C}\,a.\sim(e\,\mathsf{C}\,f)::.[c]::c\,\mathsf{C}\,a:.[d]:$
$d\,\mathsf{C}\,c.\supset.c\,\mathsf{C}\,d:.\supset.c\,\mathsf{C}\,b::.\supset.a\,\mathsf{C}\,b$
Proof:
$[a\,b\,e\,f]:::$

(1) $e \mathbin{C} a$.

(2) $\sim (e \mathbin{C} f) :: $.

(3) $[c] :: c \mathbin{C} a :. [d] : d \mathbin{C} c . \supset . c \mathbin{C} d :. \supset .$
$c \mathbin{C} b :: . \supset :$.

(4) $[c\,d] : c \mathbin{C} d . c \mathbin{C} a . \supset . [\exists g] . g \mathbin{C} d . g \mathbin{C} b :.$
(U4, 3)

$a \mathbin{C} b$ (U1, 1, 2, 4)

U6 = A10.1 $[a\,b] ::: a \mathbin{C} b . \equiv :: . [\exists c\,d] . c \mathbin{C} a . \sim (c \mathbin{C} d) :: .$
$[c] :: c \mathbin{C} a :. [d] : d \mathbin{C} c . \supset . c \mathbin{C} d :. \supset . c \mathbin{C} b$
(U1, U3, U5)

As a next step we assume

W1 = A10.1

from which we derive

W2 $[a\,b\,c] : a \mathbin{C} b . c \mathbin{C} a . \supset . c \mathbin{C} b$
Proof:
$[a\,b\,c] ::: $

(1) $a \mathbin{C} b$.

(2) $c \mathbin{C} a . \supset :: $.

(3) $[\exists d\,e] . d \mathbin{C} c . \sim (d \mathbin{C} e) :: $.

(4) $[d] :: d \mathbin{C} c :. [e] : e \mathbin{C} d . \supset . d \mathbin{C}$ $\left.\right\}$ (W1, 2)
$e :. \supset . d \mathbin{C} a :: $.

(5) $[d] :: d \mathbin{C} a :. [e] : e \mathbin{C} d . \supset . d \mathbin{C} e :. \supset .$
$d \mathbin{C} b :: $.   (W1, 1)

(6) $[d] :: d \mathbin{C} c :. [e] : e \mathbin{C} d . \supset . d \mathbin{C} e :. \supset .$
$d \mathbin{C} b :: $.   (4, 5)

$c \mathbin{C} b$ (W1, 3, 5)

W3 $[a\,b\,c\,d] : a \mathbin{C} b . c \mathbin{C} d . c \mathbin{C} a . \supset . [\exists f] . f \mathbin{C} d . f \mathbin{C} b$
Proof:
$[a\,b\,c\,d] :. $

(1) $a \mathbin{C} b$.

(2) $c \mathbin{C} d$.

(3) $c \mathbin{C} a . \supset :$

(4) $c \mathbin{C} b :$ (W2, 1, 3)

$[\exists f] . f \mathbin{C} d . f \mathbin{C} b$ (2, 4)

W4 $[a\,b\,c]::[d\,e]:d\subset e.d\subset a.\supset.[\exists f].f\subset e.$
$.f\subset b:.c\subset a:.[d]:d\subset c.\supset.c\subset d:.\supset.c\subset b$

Proof:
$[a\,b\,c]::$
(1) $[d\,e]:d\subset e.d\subset a.\supset.[\exists f].f\subset e.f\subset b:.$
(2) $c\subset a:.$
(3) $[d]:d\subset c.\supset.c\subset d:.\supset:.$

  $[\exists e]:$ $\qquad\qquad\qquad\qquad\qquad$ $\left.\right\}$ (W1, 2)
(4) $\qquad\qquad\qquad e\subset c.$
(5) $\qquad\qquad\qquad e\subset a:$ $\qquad$ (W2,2,4)

  $[\exists f].$ $\qquad\qquad\qquad\qquad$ $\left.\right|$
(6) $\qquad\qquad\qquad f\subset c.$ $\quad$ $\left.\right\}$ (1, 4, 5)
(7) $\qquad\qquad\qquad f\subset b.$ $\qquad$ $\left.\right|$
(8) $\qquad\qquad\qquad c\subset f:.$ $\qquad$ (3, 6)
  $c\subset b$ $\qquad\qquad\qquad\qquad$ (W2, 7, 8)

W5 $[a\,b\,g\,h]::g\subset a.\sim(g\subset h):.[c\,d]:c\subset d.c\subset a.\supset.$
$[\exists f].f\subset d.f\subset b:.\supset.a\subset b$

Proof:
$[a\,b\,g\,h]:::$
(1) $g\subset a.$
(2) $\sim(g\subset h):.$
(3) $[c\,d]:c\subset d.c\subset a.\supset.[\exists f].f\subset d.f\subset b:.$
  $\supset::.$
(4) $[c]::c\subset a:.[d]:d\subset c.\supset.c\subset d:.\supset.$
  $c\subset b::.$ $\qquad\qquad\qquad\qquad$ (W4, 3)
  $a\subset b$ $\qquad\qquad\qquad\qquad\qquad$ (W1, 1, 2, 4)

W6 = A7.1 $\qquad\qquad\qquad\qquad\qquad$ (W1, W3, W5)

W7 = (20) $\quad$ *Proof:*
  $[a\,b]::.$
(1) $\quad a\subset b.\supset::$
  $[\exists\,cd].$
(2) $\qquad\qquad\qquad\qquad c\subset a.$ $\qquad$ (W1, 1)
(3) $\qquad\qquad\qquad\qquad\sim(c\subset d).$
(4) $\qquad\qquad\qquad\qquad\sim(a\subset d)::$ $\quad$ (W2, 2, 3)
  $[\exists c]:.c\subset a:.[d]:d\subset c.\supset.c\subset d$ (W1, 2, 3, 4)

227

U1–U6 and W1–W7 show that the totality of theses derivable from A10.1 coincides with the totality of theses derivable from A7.1 and (20).

The proof that S10, S11, S12, and S13 are inferentially equivalent would take too much space to be included in the present paper. I will, however, give a very condensed outline of the proof establishing inferential equivalence between S10 and S13. The proof that S11 is inferentially equivalent to S13 is in great measure analogous. The proof that S12 and S13 are inferentially equivalent can be reconstructed on the basis of a similar proof given in Lejewski (4).

To return to the outline we note that in S10 we can derive the following theses

D10.1 $\quad [a\,b] :: a\,\varepsilon\,b . \equiv :. a\,\mathsf{C}\,b :. [c] : c\,\mathsf{C}\,a . \supset . a\,\mathsf{C}\,c$

T10.1 $\quad [a\,b\,c] : a\,\mathsf{C}\,b . c\,\mathsf{C}\,a . \supset . c\,\mathsf{C}\,b \qquad\qquad$ (A10.1)[18]

T10.2 $\quad [a\,b] :: a\,\mathsf{C}\,b . \supset :. [\exists c] :. c\,\mathsf{C}\,a :. [c] : d\,\mathsf{C}\,c . \supset .$
$\qquad\qquad c\,\mathsf{C}\,d \qquad\qquad\qquad\qquad$ (A10.1, T10.1)[19]

T10.3 $\quad [a\,b] : a\,\mathsf{C}\,b . \supset . [\exists c] . c\,\varepsilon\,a \qquad$ (T10.2, D10.1)

T10.4 $\quad [a\,b\,c] : a\,\mathsf{C}\,b . c\,\varepsilon\,a . \supset . c\,\varepsilon\,b \qquad$ (D10.1, T10.1)

T10.5 $\quad [a\,b\,c] :: [c] : c\,\varepsilon\,a . \supset . c\,\varepsilon\,b :. d\,\mathsf{C}\,a :. [e] : e\,\mathsf{C}\,d .$
$\qquad\qquad \supset . d\,\mathsf{C}\,e :. \supset . d\,\mathsf{C}\,b \qquad\qquad$ (D.10.1)

T10.6 $\quad [a\,b\,c] :: c\,\varepsilon\,a :. [d] : d\,\varepsilon\,a . \supset . d\,\varepsilon\,b :. \supset . a\,\mathsf{C}\,b$
$\qquad\qquad$ (D10.1, A10.1, T10.1, T10.5)

T10.7 = D13.1 $\quad [a\,b] :: a\,\mathsf{C}\,b . \equiv :. [\exists c] . c\,\varepsilon\,a :. [c] : c\,\varepsilon\,a . \supset . c\,\varepsilon\,b$
$\qquad\qquad$ (T10.3, T10.4, T10.6)

T10.8 $\quad [a\,b] : a\,\varepsilon\,b . \supset . [\exists c\,d] . c\,\varepsilon\,a . c\,\varepsilon\,b . \sim(c\,\varepsilon\,d)$
$\qquad\qquad$ (D10.1, T10.2, T10.1, A10.1)

T10.9 $\quad [a\,b\,c\,d] : a\,\varepsilon\,b . c\,\varepsilon\,a . d\,\varepsilon\,a . \supset . c\,\varepsilon\,d \quad$ (D10.1, T10.1)

T10.10 $\quad [a\,c\,d] :: [e\,f] : e\,\varepsilon\,a . f\,\varepsilon\,a . \supset . e\,\varepsilon\,f :. d\,\mathsf{C}\,a . c\,\varepsilon\,a :.$
$\qquad\qquad \supset . c\,\varepsilon\,d \qquad\qquad$ (T10.3, T10.4, D10.1)

T10.11 $\quad [a\,d] :: [e\,f] : e\,\varepsilon\,a . f\,\varepsilon\,a . \supset . e\,\varepsilon\,f :. d\,\mathsf{C}\,a :. \supset . a\,\mathsf{C}\,d$
$\qquad\qquad$ (T10.10, T10.3, T10.4, T10.6)

T10.12 $[a\,b\,c]::c\,\varepsilon\,a\,.\,c\,\varepsilon\,b:.\,[d\,e]:d\,\varepsilon\,a\,.\,e\,\varepsilon\,a\,.\,\supset\,.\,d\,\varepsilon\,e:.$
         $\supset\,.\,a\,\varepsilon\,b$ (T10.6, T10.11, D10.1, T10.4)

T10.13 = A13.1 $[a\,b]::a\,\varepsilon\,b\,.\,\equiv:.\,[\exists\,c\,d]\,.\,c\,\varepsilon\,a\,.\,c\,\varepsilon\,b\,.\,\sim(c\,\varepsilon\,d):.$
         $[c\,d]:c\,\varepsilon\,a\,.\,d\,\varepsilon\,a\,.\,\supset\,.\,c\,\varepsilon\,d$

                                    (T10.8, T10.9, T10.12)

We complete the proof by deriving A10.1 and D10.1 within the framework of S13. This involves the following subsidiary theses

D13.1 $[a\,b]::a\,\subset\,b\,.\,\equiv:.\,[\exists\,c]\,.\,c\,\varepsilon\,a:.\,[c]:c\,\varepsilon\,a\,.\,\supset\,.\,c\,\varepsilon\,b$

T13.1 $[a\,b]:a\,\varepsilon\,b\,.\,\supset\,.\,a\,\varepsilon\,a$ (A13.1)

T13.2 $[a\,b\,c]:a\,\varepsilon\,b\,.\,c\,\varepsilon\,a\,.\,\supset\,.\,c\,\varepsilon\,b$ (A13.1)

T13.3 $[a\,b]:a\,\varepsilon\,b\,.\,\supset\,.\,a\,\subset\,b$ (A13.1, T13.2, D13.1)

T13.4 $[a\,b\,c\,d]:a\,\varepsilon\,b\,.\,c\,\subset\,a\,.\,d\,\varepsilon\,a\,.\,\supset\,.\,d\,\varepsilon\,c$
                               (D13.1, A13.1, T13.2)

T13.5 $[a\,b\,c]:a\,\varepsilon\,b\,.\,c\,\subset\,a\,.\,\supset\,.\,a\,\subset\,c$
                               (A13.1, T13.4, D13.1)

T13.6 $[a\,d\,e]::[c]:c\,\subset\,a\,.\,\supset\,.\,a\,\subset\,c:.\,d\,\varepsilon\,a\,.\,e\,\varepsilon\,a:.\,\supset\,.$
        $d\,\varepsilon\,e$ (T13.3, D13.1)

T13.7 $[a\,b]::a\,\subset\,b:.\,[c]:c\,\subset\,a\,.\,\supset\,.\,a\,\subset\,c:.\,\supset\,.\,a\,\varepsilon\,b$
                       (D13.1, A13.1, T 13.2, T13.6)

T13.8 = D10.1   (T13.3, T13.5, T13.7)

T13.9 $[a\,b]:a\,\subset\,b\,.\,\supset\,.\,[\exists\,c\,d]\,.\,c\,\subset\,a\,.\,\sim(c\,\subset\,d]$
                          (D13.1, A13.1, T13.2)

T13.10 $[a\,b\,c]:a\,\subset\,b\,.\,c\,\subset\,a\,.\,\supset\,.\,c\,\subset\,b$ (D13.1)

T13.11 $[a\,b\,g]:::[e]::e\,\subset\,a:.\,[f]:f\,\subset\,e\,.\,\supset\,.\,e\,\subset\,f:.$
        $\supset\,.\,e\,\subset\,b:::g\,\varepsilon\,a::.\,\supset\,.\,g\,\varepsilon\,b$ (T13.8)

T13.12 $[a\,b\,c]:::c\,\subset\,a:::[d]::d\,\subset\,a:.\,[e]:e\,\subset\,d\,.\,\supset\,\cdot d\,\subset\,e:.$
        $\supset\,.\,d\,\subset\,b:::.\,\supset\,.\,a\,\subset\,b$ (13.1, T13.11)

T13.13 = A10.1   (T13.9, T13.10, T13.12)

From the above deductions in S10 and S13 it is clear that the two systems are inferentially equivalent.

VIII. CONCLUDING REMARKS

To sum up the results it may perhaps be worth our while to review the main stages in our attempt to expand Aristotelian syllogistic into a modern system of the logic of names. By introducing the quantifiers and by strengthening the axiomatic foundations of the ordinary syllogistic (A3.1–A3.6) with the *axioms of existential import* (A3.7–A3.12) and the *axiom of ecthesis* (A3.13) we arrived at a system (S3) of generalised syllogistic, which was shown to be inferentially equivalent to a system (S4) based on

A4.1 $$[a\,b]::a\,\mathsf{C}\,b.\equiv:.[\exists\,c].c\,\mathsf{C}\,a:.[c]:c\,\mathsf{C}\,a.\supset.\\c\,\mathsf{C}\,b$$

as the only axiom. By subjoining to A4.1 the thesis which said that

(16) $$[a\,b]:a\,\mathsf{C}\,b.\supset.[\exists\,c].\sim(a\,\mathsf{C}\,c)$$

we were led to systems of extended syllogistic (S5 and S6). A further extension was secured by subjoining

(18) $$[a\,b\,c]::c\,\mathsf{C}\,a:.[d\,e]:d\,\mathsf{C}\,e.d\,\mathsf{C}\,a.\supset.[\exists\,f].\\f\,\mathsf{C}\,e.f\,\mathsf{C}\,b:.\supset.a\,\mathsf{C}\,b$$

This gave rise to systems of basic Boolean algebra (S7, S8 and S9). Finally by strengthening basic Boolean algebra with

(20) $$[a\,b]::a\,\mathsf{C}\,b.\supset:.[\exists\,c]:.c\,\mathsf{C}\,a:.[d]:d\,\mathsf{C}\,c.\supset.\\c\,\mathsf{C}\,d$$

we obtained basic ontology, of which four systems (S10–S13) were outlined. While discussing basic ontology it was pointed out that the transition to systems of Leśniewskian ontology proper involved new, and rather powerful, rules of inference.

University of Manchester, Manchester, England

REFERENCES

1. For a modern treatment of Aristotle's syllogistic see Łukasiewicz [8]; a condensed but authoritative presentation of ontology is to be found in Leśniewski [6]; an elementary discussion of ontology and some of its problems is contained in Kotarbiński [2], Sobociński [12], Sobociński [13], Słupecki [11], and Lejewski [4].
2. See Łukasiewicz [7] p. 172 (or p. 87 in the 2nd edition) and Łukasiewicz [8] p. 88.
3. See Łukasiewicz [7] p. 174 (or p. 88 in the 2nd ed.) and Łukasiewicz [8] p. 88.
4. In this connection see Łukasiewicz [8] pp. 4 f.
5. See Sextus Empiricus, *Hyp. Pyrrh.* ii 164, and Łukasiewicz [8]. It has recently been pointed out by W. and M. Kneale in their monograph *The Development of Logic*, Oxford, 1962, that an example of a syllogism with a singular proposition as premiss occurs in *An. Pr.* ii 27 (70^a 27).
6. See Łukasiewicz [8] pp. 11 f.
7. See Łukasiewicz [8] pp. 61 f.
8. See Ajdukiewicz [1] pp. 220 f.
9. See Popper [10] p. 727.
10. See Ajdukiewicz [1] pp. 220 f.
11. See Lejewski [3]
12. See Łukasiewicz [8] pp. 59 f.
13. See, for instance, Słupecki [11] pp. 33 f.
14. G. E. Moore argues that '... the proposition "whoever wrote *Waverly* was Scotch, but nobody did write *Waverly*" is self-contradictory'. See Moore [9] p. 180.
15. A7.1 and A9.1 can be simplified a little if we admit nominal definitions. See Lejewski [5] pp. 91.
16. An appropriate rule for introducing nominal definitions enables us to simplify A10.1, A12.1, and A13.1. On the subject of axiomatic foundations of ontology see Leśniewski [6], Sobociński [12], Słupecki (11), and Lejewski (4).
17. See the proof of Q2 above.
18. See the proof of W2 above.
19. See the proof of W7 above.

BIBLIOGRAPHY

[1] Ajdukiewicz, K., *Główne zasady metodologji nauk i logiki formalnej* (Principles of Methodology and Formal Logic), Warszawa, 1928 (mimeographed).
[2] Kotarbiński, T., *Elementy teorji poznania, logiki formalnej i metodologji nauk* (Elements of Epistemology, Formal Logic and Methodology), Lwów, 1929.
[3] Lejewski, C., 'Logic and Existence', *The British Journal for the Philosophy of Science* 5 (1954).
[4] Lejewski, C., 'On Leśniewski's Ontology', *Ratio* 1 (1957–1958).
[5] Lejewski, C., 'Studies in the Axiomatic Foundations of Boolean Algebra', *Notre Dame Journal of Formal Logic*, 1 (1960) and 2 (1961).
[6] Leśniewski S., 'Über die Grundlagen der Ontologie', *Comptes rendus des séances de la Societé des Sciences et des Lettres de Varsovie*, Classe III, 18 (1930).
[7] Łukasiewicz, J., *Elementy logiki matematycznej* (Elements of Mathematical Logic), Warszawa, 1929 (mimeographed); 2nd edition: Warszawa, 1958.

[8] Łukasiewicz, J., *Aristotle's Syllogistic*, Oxford, 1951; 2nd edition: Oxford, 1957.
[9] Moore, G. E., 'Russell's "Theory of Descriptions"', *The Philosophy of Bertrand Russell*, The Library of Living Philosophers, vol. 5, Chicago, 1944
[10] Popper, K. R., 'The Trivialization of Mathematical Logic', *Library of the Xth International Congress of Philosophy*, vol. 1, *Proceedings of the Congress*, Amsterdam, 1948.
[11] Słupecki, J., 'S. Leśniewski's Calculus of Names', *Studia Logica* 3 (1955).
[12] Sobociński, B., 'O kolejnych uproszczeniach aksjomatyki "ontologji" prof. St. Leśniewskiego' (On Successive Simplifications of the Axiom-system of Leśniewski's 'Ontology'), *Księga Pamiątkowa – Fragmenty Filozoficzne*, Warszawa 1934.
[13] Sobociński, B., 'L'analyse de l'antinomie Russellienne par Leśniewski', *Methodos* 1 (1949) and 2 (1950).

PART III

MODELS IN SCIENCE

F. T. C. HARRIS

A REPRESENTATION OF ANIMAL GROWTH

This essay is an account of a method which can be used to describe the changing shapes of animals and plants as they develop, to record cell-lineages succinctly and to predict the spatial relationships between cells of developing organisms. It opens up possibilities both of reducing the complexities of our present catalogue of cleavage types, and also of looking for new embryological generalisations. The ideas and concepts on which the essay is based arise from work carried out by J. H. Woodger.

Woodger's early interest and work in embryology and its philosophical analysis has found little expression in his later publications, his three papers on 'The "Concept of Organism" and the Relation between Embryology and Genetics' published in 1930 and 1931 utilised the concept of hierarchical relation, a concept which originated from and was worked out by him. Section 8 of *The Axiomatic Method in Biology* [1] also deals with embryology.

The concept of hierarchy is fundamental to this essay and although Dr. Lindenmayer explains it in his essay in this volume, I hope that the mathematically sophisticated will bear with me whilst I explain it here. The commonest example of a hierarchical relation in biology, is the relation in which an asexually reproducing cell stands to either of its daughter cells. Biology gives examples of other $1 \rightarrow$ many relations, such as merogony. Such hierarchies are explored by Dr. Lindenmayer's paper. A simple example of a hierarchy would be the relation 'asexual parent of' with regard to a clone of amoebae grown under optimal conditions. The condition that the environment is optimal is added in order that every amoeba shall have two successors. If at some time the clone dies out, we will have to add the requirement that the hierarchy is finite. A hierarchy then, is a relation that is $1 \rightarrow$ many, such as the relation of binary fission between any one of our amoebae and its immediate progeny; and such that the beginner of the clone is related to every member of the clone. You notice that there are two parts to the definition of hierarchy, the first permits the relation to be multi-termed. Note that the relation is not transitive, this is because the relation is 'asexual parent of' and not 'asexual ancestor of', thus the relation might be immediately terminal and relate only three of our amoebae.

The second part of the definition excludes, in our protozoan example and elsewhere, the spontaneous generation of amoebae. All of the amoebae of our clone, those alive now, their ancestors and their future descendants form a class or set described as the Field of the Relation.

Consider the members of the Field, or the individual amoebae of our clone, we can clearly group together all of those that have divided, or are about to do so. This set forms a part of the Field called the Domain of the relation. The beginner of our clone, our protozoan Zeus, will be a member of this set, and so will every other member of the clone except those alive at the time the clone is destroyed. We can form another group consisting only of those members of our clone that have been derived by binary fission. This set is called the Converse Domain of the relation. It includes those members of the clone that are alive at extinction time, together with all of the members of the domain but for Zeus.

It is as though the Field of the Relation were a ladder and the Converse Domain were formed by slipping one rung down the ladder from the Domain.

Consider again the Hierarchical Relation, the relation in which Zeus stands to his sexless daughters is called the First Power of the Relation, he stands in the Second Power to his four grand-daughters, in the Third Power to his eight great-grand daughters, and so on.[1] The same can be said of any member of the Domain of an infinite hierarchy, so that any such member can be regarded as the beginner of a sub-hierarchy, any of us can become a grand-parent.

Woodger coined the apt word 'Assemblage' for the members of the converse domain that stand in the first power of the converse of the relation to some other one member of the Field. Thus the daughters of Zeus form an Assemblage, his grand-daughters two different Assemblages and so on.

Rules of hierarchy formation can be added which are only indirectly dependent on a Hierarchical Relation. For example we might require that the two members of an Assemblage would divide at the same time, after assemblage formation. This would result in a hierarchy in which all members standing in the same power of the Converse Relation, that is those formed by division, divide contemporaneously. One can picture such a hierarchy as ripples or concentric waves spreading equidistant from each other, each ripple would comprise all of the members of the field to which the beginner of the relation, Zeus, stood in a given power of the relation. Such a ripple is called a Level.[2] The two daughters of Zeus, to whom he stands in the First Power of the Relation, comprises the members of level one, the eight grand-daughters to which he stands in the Third Power of the Relation comprise the members of Level three, and so on. If, on the other hand, we arrange our hierarchy so that no two members of an Assemblage divide at the same time, we may have a hierarchy in which the members of the levels get progressively further out of step.

Assymetrical relations can be introduced as ancillary ordering relations in the field of a hierarchy in several ways. Firstly they can be used to assist in naming the individual members of the hierarchy in the way that will now be described.

Woodger started exploring the manner in which the natural number series could form the Field of a $1 \rightarrow$ many hierarchy. He calls such hierarchies 'Number Hierarchies'. Number Hierarchies can be used to name, or uniquely identify the members of the Field of Division Hierarchies. This can be done by ordering the assemblage partners of the Division Hierarchy by an asymmetrical relation, for example the spatial relation 'left of'. And by identifying members[3] of the Converse Domain of such a spatial relation by say odd numbers, and members of the Domain by means of even numbers. If the spatial relations are multiplied then so are the identifying groups of numbers.

In unpublished work, Woodger introduced a three-dimensional co-ordinate system whose three axes were related to the idealised planes of morphological symmetry of a developing embryo. These spatial relations were described as 'Directly Anterior', 'Directly Dorsal', and 'Directly Left'. These relations are symbolised here by A, Do and L. Within this system Woodger worked out techniques specifying the positions and numbers allotted to cells after a given number of divisions, the planes of the divisions alternately orientated by the relations A, L and Do.

It was these preliminary experiments on formulating a hierarchy that would describe the spatial relations between cells forming simple shapes such as cubes, that led me to attempt the description of such a hierarchy for more complex systems [2].

The above three very specific relations restricted the application of the system, and made it extremely difficult to describe division hierarchies applicable to embryological processes.

Another way in which spatial relations can be introduced as ancillary relations between members of the Field of a hierarchy avoids asymmetry between assemblage partners, for example we may only require that assemblage members touch. If the stipulation is made that every member but two belonging to levels two and above of a binary fission hierarchy, touched two other members of the level: the members of each level would describe a pattern of cells that was unbranched, might be straight or sinuous but cannot be circular and which doubled its length at every division.[4] With this insight you can follow the manner in which a model of animal or plant development, or some other, can be set up in an abstract hierarchy. Such a hierarchy has none of the limitations of diagrams, which mirror real or idealised[5] division stages of organisms. However, in many stages and perhaps always, it is desirable to identify the particular progeny of any cell in the tradition of the study of cell-lineage. This, as has been seen, can be done for the members of the field of a hierarchy by utilising the natural number series to construct a simply numberable hierarchy.

The system at present in use by students of cell lineages is primarily concerned to make diagrams intelligible by naming concisely the cells so represented. This is a very primitive and inadequate method, nearly pre-hieroglyphic in its development. It is not possible to represent intelligibly both the position and history of every cell in a developmental stage of even a few thousand cells, by means of diagrams supplemented by notations such as those introduced by Whitman [3] and Boveri [4]. The designations of the individual cells of an embryo so described, rapidly become excessively cumbersome.[6]

The description of lineages in very complicated stages of division, which would be beyond the scope of two-dimensional pictorial representation, is made possible by the use of number hierarchies. This is because the spatial and temporal relations of the cells or blastomeres are written explicit into the axioms and postulates that control the manner in which the numerical designations are allotted.

Of course this doesn't do away with the complexity inherent in the developmental process, nor with the complexity inherent in the description of later stages of development. This method of description does have the extreme advantage that, as the theory of sets is applicable to it, it is accessible to programmed computation.

In the remainder of this essay I propose to discuss some features of number hierarchies and then proceed to a discussion of the rules that permit the application of a simply numberable hierarchy to the development of a sea-gooseberry or ctenophore from the zygote to a stage beyond gastrulation.

The following description of a cleavage system involves the three entities that have been mentioned previously. In the first place, an idealised cleavage system, in the second, a number hierarchy, and third a set of rules by means of which the number hierarchy is generated in a manner that permits it to act as an adequate model of the cleavage system.

A growing idealised filamentar alga will serve as a simple example to which a number hierarchy can be applied[7]. This biological fiction originates from one cell and grows by a series of regular and synchronous divisions of each of its constituent cells in a manner more military than biological. The number zero is allotted to the beginner of the number hierarchy[8] and corresponds to the zygote that begins the time extent of the alga. Then we can have the following correspondences between numbers and cells of the synchronously growing alga:

Algal Filament	Corresponding members of the field of the number hierarchy	level
	0	0
	1 2	1
	3 4 5 6	2
	7 8 9 10 11 12 13 14	3
	15 16 ⋯ 19 20 ⋯ 29 30	4
... and so on	... and so on	... and so on

Obviously each level of the number hierarchy contains 2^n numbers.

For instance zero level contains only 2^0 numbers, that is only one number, the beginner. The third level contains 2^3 or eight numbers, namely 7, 8, 9, ..., 14. The numeral members of each level of the field are easily specified. For example the twenty second level contains 2^{22} or 4194304 members. These members are the numbers between and including 4194303, $(2^n - 1)$ to 8388606, $(2^{n+1} - 2)$. 'n' represents the number of the level.

Further, given the designation of any member of the hierarchy we can easily determine the designations of its daughter cells. So far then this system of designating the cells of such an idealised, totally synchronised, alga; has worked simply and consistently. One problem has not yet been explained. This is how we correlate the numbers of the numberable hierarchy with the individual cells of the alga. To do this we can either assign the numbers purely arbitrarily[9], or we can introduce another feature into the set of characteristics of our imaginary alga. Let our alga cleave asymmetrically so that its filaments are orientated in the earth's magnetic field in such a way that one member of each assemblage is always directed to the North whilst the other is always directed to the South. If you consider the four daughter cells derived from such a N–S assemblage pair, you notice that the pair of daughters derived from the northfacing assemblage partner are both north of the pair derived from the south-facing assemblage partner. In this way the numbers of any appropriate level of the numberable hierarchy can be applied to a corresponding filament if the assemblage pairs are similarly asymmetrically and uniquely related. Then the lowest number of any level, n, the number $(2^n) - 1$, names the north-most filamentar cell, the number 2^n names the next north most cell, the number $(2^n) + 1$ names the third cell down the filament from the north and so on down to the most southern cell which will be named by the number $(2^{n+1}) - 2$. If the assemblage relation chosen to allot the numbers of the numberable hierarchy is one such as 'directly north of' then the numbers of the numberable hierarchy will at each level fit a filament whose arrangement is that of a particular straight line. But we yet require other rules. So far we have no information on whether or not the cells named by the numberable hierarchy are close together or far apart. We therefore introduce a rule that any two cells to which consecutive numbers are assigned must touch each other except when the two members belong to different levels of the corresponding arithmetical hierarchy. These each name a cell touching only one cell. We also require an additional rule relating to the generation of the members of subsequent levels of the field. In order to model our alga we need a numberable hierarchy in which all the members of any level are generated instantly and at a time subsequent to the generation of all the members of the previous level. This can be done by stipulating this as a requirement for generating the hierarchical relation.

The type of hierarchy we have been discussing is conveniently identified by means of a name, I call it a P· hierarchy [2]. In this name the letter P indicates that the assemblage partners are ordered by means of a spatial[10] relation and the dot indicates that this relation is held between the assemblage pairs of every level of the field of the hierarchy. A slightly more complicated hierarchy of this kind is illustrated by a (P, Q)· hierarchy. Here the assemblage-orientating relations are repeated alternately at every other level. For example, the assemblage members of the first level might be orientated by a relation, 'P'. those of the second level by a relation, 'Q', of a different kind, those of the third level by 'P' again those of the fourth level by 'Q' again, and so on. The dot again means 'and so on'. If P and Q both represent asymmetical linear spatial relations and each is always in the same plane but the two are at right angles to one another then the members of each level of the field of the hierarchy will form a pattern of regularly growing rectangles. If P is taken as the spatial relation ordering assemblage partners in rows across the page and Q the relation similarly ordering columns down the page, and if the hierarchy is numberable having the lowest numbered member of any level naming a point or spot or cell in the uppermost left hand corner of the pattern and the highest numbered cell forming the lowest right hand corner of the pattern then the first four levels will be arranged and numbered as indicated in the diagram on page 242.[11]

In the diagram the letters indicate the spatial relations holding between assemblage partners. The numbers adjacent to the letters represent the assemblage partners. It is not necessary that the letters, or for that matter the numbers, be represented diagrammatically once the rules for the hierarchy have been laid down.

A third type of simple hierarchy is given by introducing a third linear and asymmetrical spatial relation to form a repeating triplet of relations. Such a hierarchy is Woodger's A, L, Do hierarchy, or in the terminology we are using a (P, Q, S)· hierarchy. Such a numberable hierarchy results in a pattern of regularly growing solid rectangles whose numbered members can be represented as indicated on p. 243.

		Members of Hierarchy			level
		0			0
1		(P)		2	1
3				5	2
(Q)				(Q)	
4				6	
7	(P)	8	11 (P)	12	3
9	(P)	10	13 (P)	14	
15		17	23	25	4
(Q)		(Q)	(Q)	(Q)	
16		18	24	26	
19		21	27	29	
(Q)		(Q)	(Q)	(Q)	
20		22	28	30	

Within any of the preceding three diagrams identical hierarchical lines, or cell-lineages if you will, can be picked out. For example we have in each case $O R$ (1, 2), followed by 1 R (3, 4), followed by 3 R (7, 8), 7 R (15, 16) and so on: R indicates that the hierarchical relation holds between the members of the field that it separates. [12] In the different cases illustrated the actual spatial distribution of the members of the field named is very different indeed.

Having looked at some of the characteristics of number hierarchies let us now examine the spatial relations that we are about to use to describe development in a member of the Phylum Ctenophora. Earlier, it was pointed out that the simple linear spatial relations will not do to relate assemblages in the complex situations presented to us by morphology because they lack flexibility.

Members of Hierarchy	Level
0	0
- - - - - - -	
1 (P) 2	1
- - - - - - -	
3 5	2
(Q) (Q)	
4 6	
- - - - - - -	

7 ———————— 11
(S) →8 ———————— 12
9 - - - - - - - 13
(S) →10 ———————— 14

Level 3

15 ←(P)→16 ——————— 23 ←(P)→24
17 ←(P)→18 ——————— 25 ←(P)→26·
19 ←(P)→20 - - - - - - 27 ←(P)→28
21 ←(P)→22 ——————— 29 ←(P)→30

Level 4

and so on

Consider the two members of the first level of any 1→2 number hierarchy. They will of course be the members of an assemblage. Whatever the members are, and whatever the relation assembling them it is possible to represent them as points related by a line. If the field-members are cells then this first cell pair may in fact be related by some such morphological relation as antero-posterior, left-right or dorso-ventral. Whatever it is this gives us our bsae-relation.[13] Now we define a relation 'A' such that any member of the field is 'A' to any other member of the field as long as it is not posterior to it. Thus such a relation will describe equally a pair of cells one of which is directly anterior, or anterior and left, or directly left, or directly dorsal, and so on to the other. In our numberable hierarchy we distinguish between such members of the field by saying that the member of the domain of the relation has the smaller number. Similarly we can define a general notion of the other fundamental morphological relations, dorsal, 'Do' and bilateral. The last of these is most conviently defined like the others as an asymmetrical relation, left of 'L'. Using these general relations we can now define other special relations such as directly dorsal. A cell will be directly dorsal to another if it is dorsal to, and neither left or its converse nor anterior or its converse, to another. And so on.

With these notions clarified we can now go on to describe the early development of a sea-gooseberry. The cleavage of ctenophores [5] results in complete cleavage of the egg, it is strongly unequal however, resulting in micromeres and macromeres that are remarkably different in size. It is determinate. That is to say that if a blastomere is destroyed then those tissues to which the blastomere in question would have contributed fail to develop. The early cleavage is perfectly regular. Bearing in mind the terminology previously used to describe hierarchies such a hierarchy can be called a $(P, Q, P, S\cdot)$ hierarchy at least up to level six, a stage in which gastrulation has commenced.

Reverberi [6] described the cleavage briefly as follows: The fertilised egg divides first meridionally into two equal blastomeres; the second segmentation is also meridional and equal, and four cells of the same size are formed. The third segmentation is again meridional but unequal, the four outer cells being smaller than the median ones; these eight blastomeres are disposed in a circle, though not exactly in the same plane. The fourth cleavage takes place latitudinally, and eight micromeres are budded off by the macromeres. The fifth segmentation ... (gives rise to) ... the 32 cell stage, ... every octant having three micromeres and a macromere. The sixth segmentation is again asynchronous; the micromeres divide first, and six of them are now present in every octant, then the macromeres divide meridionally, and a 64 cell stage is reached with 48 micromeres and 16 macromeres. The successive segmentations are not synchronous for all cells; the macromeres remain quiescent for a certain time, and then give rise again to some micromeres, which are supposed to represent the presumptive mesoderm. An epibolic gastrula is formed by the expanding of the micromere fields to form eight rows of comb plates, and also the epidermis'.

The following account achieves two tasks. It traces the individual cell lineages, not referred to in Reverberi's account, uniquely and in so doing provides a non-geometrical apparatus for describing morphology – and in fact any space. The use of numberable hierarchies enables far more information to be given than in Reverberi's brief account. In order to make the approach clear the description will have to be given in plain English thus reducing considerably the economy of the description. I shall first give eight rules which define the simply number hierarchy, R_y, that describes this type of development and having done simply that, discuss them.

I. The R_y is a member of the class of division hierarchies.

II. R_y is regular. Each cell is related to only two others. That is, it is a 1–2 hierarchy.

III. The beginner of R_y is named zero and for any three cells, x, y, z, if x is the parental cell of y and z and y is anterior or dorsal or left of z then the number allotted to y is twice that allotted to x, plus one, and the number allotted to z is the same as that of y, plus one.

IV. Any pair of assemblage partners belonging to level one or level three of the hierarchy is related by the relation anterior. Similarly the relation left of holds for assemblage members of level two. For any level above level three the relation ordering the assemblage partners is that of dorsal to.

V. Any tier is a cyclically touching set and so are the members of level three.

VI. For any level, say level n, above the third there are 8 sets [14] of $2^{(n-3)}$ cells whose members are related dorso-ventrally. If x and y are any two members of one of these sets and x Cnv R^p/R^p y then $p < n - 3$. [15]

VII. If, in the sets of cells defined by VI there are two cells, one of which is a ventral-most cell, then they touch one another.

This last rule is perhaps the most difficult to follow. Its intention is that the cells that are descendants of the eight members of level three do not form a layer within the tier formed by the most ventral cell.

VIII. Let X and Y be two of the sets of cells defined by VI and x and z

represent cell members of X and y a cell member of Y. Let x and y be ventral most cells. Then if x is anterior and left of y then z is anterior or left or dorsal to x. Conversely, if x is posterior or right of y then z is posterior or right or dorsal to x. Thus the following two situations are possible,

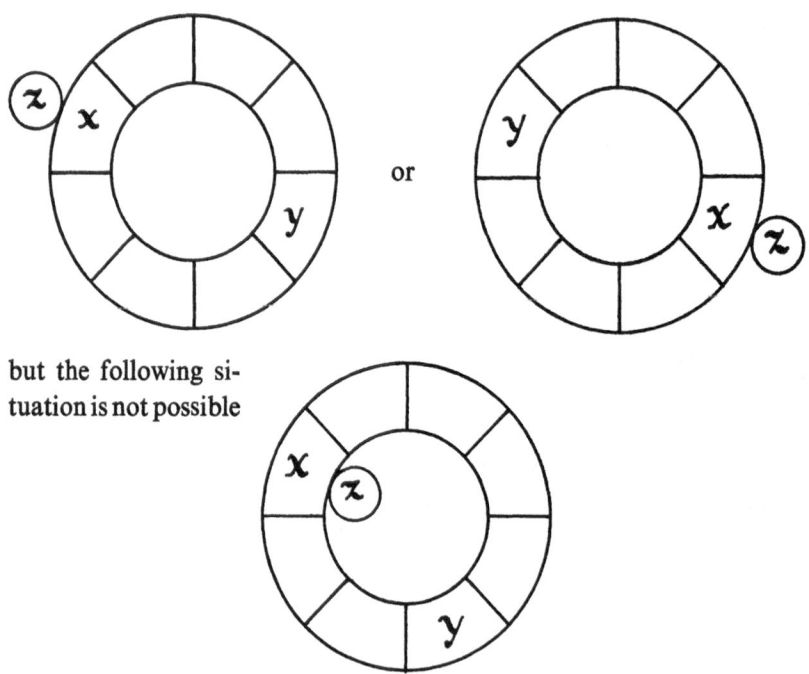

but the following situation is not possible

The first three of these postulates are necessary, in some similar form, for any 1–2 number hierarchy. The third and fourth postulates set up an ordinally similar correspondence between the members of the field of the division hierarchy exemplified by Ctenophora and the members of a 1–2 number $(P, Q, P, S\cdot)$ hierarchy. The rules for the formation of such a hierarchy can of course be formulated separately. They are very similar to the above named postulates.

The fifth postulate is only slightly less general than the first three in that it is necessary in defining any hierarchy that models a circle or sphere. This fifth postulate thus ensures that the hierarchy does not describe a bilaminar sheet as would happen with the similar $(A', L', Do'), (A', Do')\cdot$

hierarchy. It is important to make some such restriction because such special relations are included in the general relations *A. L, Do.* The fifth postulate does not of course limit the description to geometrically perfect examples of cyclically touching things. Anything that fits it will do, whether the circle is corrugated or indented. Far more weakly it merely fails to exclude perfection, what more can most of us hope for? Perhaps you will excuse me if I use the importance of this as licence to labour the point further. Embryologists, I work cheek by jowl with them, and biologists at large are famed – some might say selected – for their suspicion of 'mathematical' tools when applied to biology. The statistical treatment of variability is now a familiar tool just because it describes this variability without fitting it into a mould. Postulate five here gives an example of the manner in which a tool can be used to describe certain basic features of living morphology without having to take the variability into account. It does this by being able to ignore variability due to the manner in which it takes the essence of variety into account, or if you will, the common theme which makes variability recognisable. The fifth postulate then can be seen to invoke a general characteristic of the morphology of many living things. It is on the other hand also very specific. It appears to be peculiar to Ctenophora and Porifera in incorporating the requirement that the cell-members of level 3 form a cyclically-touching set.

The fourth, sixth, seventh and eighth rules are probably adequate to distinguish the type of cleavage shown by Ctenophores and Poriferans from all other types of cleavage. This is again a very important point. The rules for the hierarchy not only permit the spatial relations between any two cell-members of a level to be calculated, the shape formed by sets of the members and the cell lineages to be described but it also provides a means of classifying cleavage types.

The seventh rule, or postulate, in conjunction with the fourth, ensures that the cell members of any level, or cleavage stage if you will, stay in contact with one another as an entity and don't stand apart.

The eighth rule ensures that cell descendants of the cell-members of the third level do not form a cylinder or a sphere of cells but a mantle of cells about eight ventral cells. Let me emphasise again that there are no geometrical concepts within this system. Nevertheless if one assumes that the cells of the system have spatial dimensions, as cells do, then necessarily the ventral-most cells of each of the eight groups will have to be larger than the other members of the groups. This is because all of the other members touch the ventral-most cell as required by rule seven.

No provision has been made in the rules to specify the touching relations between tiers. It is required that all of the members of each of the eight groups of cells touch the most ventral. In the earlier stages of cleavage there are gaps between the tiers forming the ectodermal and comb-plate cloak, later on however the gaps become obliterated. If it is necessary to allow for this explicitly it can be done by saying so in an addition to rule six [2].

No provision, *to take into account asynchrony in cleavage*, has been made in this elementary exposition of the rules for ctenophores. Actually the divisions of the cells are not synchronous in the later stages of development. Such differences are, in common with differences in size, of great physiological and biochemical interest in themselves. Within obvious limits they do not affect the outcome of the development described here and have for simplicity been disregarded.

Similarly there has been no attempt to indicate any variation in size between the members of any one tier. Certainly such differences in size do exist but they have no important effect on morphogenesis before the tenth level, which completes gastrulation. Certain important size differences are however an outcome of, that is would be deducible as theorems from, the rules as they stand. For example, rule VII requires that all members of the eight sets of cells[16] touch their respective ventral-most cell. Thus necessarily beyond the fifth level, in which each of the eight groups of cells would consist of four cells, the ventral most cell must be larger than the others. The three dorsal cells are restricted to one half of the ventral-most cell by rule eight and not more than four equal sized cells can be so arranged.

I hope that enough has been said to show that this method of hierarchical numbering in association with non-hierarchical relations is of interest as a useful language for embryology and that in common with this example of its subject-matter shows promise of development.

Department of Biology Applied to Medicine, Middlesex Hospital Medical School, London

NOTES

1. The concept of power is similar to that of exponent in algebra.
2. The metaphor is misleading in that such a ripple originates from the beginner and not from the preceding ripple.
3. More precisely the identification involves a simply enumerable hierarchy such as that defined by J. H. Woodger in his paper in *Colloques de logique mathématique*.
4. Such a hierarchy is of course perfectly general and applies to any subject matter
5. All diagrams of embryonic development are idealised in that, if they are to be applicable to other instances of the same kind of development, they must be diagrams and not drawings.
6. For example the cells of an embryo resulting from the fifteenth division are designated, in Whitman's system, by names composed of a selection of the following: the first four letters of the alphabet in upper and in lower case, the first ten members of the natural number series as prefixes and the numbers 1 and 2 in 12 combinations of from 2 to 8,192 permutations as suffixes. On the other hand utilising a numbered hierarchy to describe the developmental states of animals is only as cumbersome as arithmetic, the designations corresponding to the above being merely the 65,536 numbers from 65,535 to 131,070.
7. The relation that correlates with the allocation of the numberable hierarchy, as a model, to the alga is that of ordinal similarity.
8. Zero is allotted to the beginner because it results in a simplification of the rules for allocation of numbers when the assemblage number of the hierarchy is greater than two.
9. This does not result in logical fun alone. Such a model allows the numberable hierarchy to be used to explore what happens when only stated cells of the filament divide, or divide more rapidly.
10. Of course 'P' does not have to represent a spatial relation, it can represent, for instance, the relation 'containing more information than'.
11. This does provide a method of generating an infinite number of very unlikely series. The series themselves, for some of which Woodger and I have found general terms [2], do give a method for working out the spatial position of a designated member of the field of a numberable hierarchy.
12. Strictly this should be written 0R1, 0R2; 1R3, 1R4, etc.
13. There are two different sets of difficulties here. One concerned with the initial assumptions involved, the other with the technicalities of embryological cell-lineage studies. Neither is insuperable but space does not permit their discussion.
14. 8 sets of cells are chosen in place of 16 into which they in fact divide at level 6. 8 sets serve adequately for this illustration.
15. Converse is abbreviated by Cnv, and power by ^{po}.
16. See note 13.

REFERENCES

[1] J. H. Woodger, 'The Axiomatic Method in Biology', *C.U.P.* 1937, 39.
[2] F. T. C. Harris, London University M. Sc. Thesis, 1958.
[3] C. O. Whitman, 'The Embryology of Clepsine', *Q.J.M.S.* **18** (1878) 276.

[4] Th. Boveri, 'Entwicklung von Ascaris megalocephala', *Fests. zum Carl von Kupffer*, 1899.
[5] C. Dawydoff, *Traité d'embryologie*, Paris, 1928.
[6] G. Reverberi, *Acta Embryologiae et Morphologiae Experimentalis*, **1** (1957) 134.

ANALOGIES IN BIOLOGY

When I was no more than a boy and beginning to show some interest in living creatures I can remember being sternly warned by my elders to beware of the dangers of analogies. It was said in the same tone one might tell someone not to eat a certain kind of mushroom. I do not believe it ever seriously curbed my instinct for analogy; I went on comparing as any person will whether it be consciously or unconsciously, but the feeling of slight guilt has always remained and is no doubt now responsible for my writing this short essay.

The difficulty with analogies, I was told, is that they are always rough and inexact and therefore could easily be misleading. Something may be like something else in many ways, but not in all ways and the difference makes any talk about the similarities dangerous, aside from being inexact and sloppy. Loose analogies are the stuff crackpots and quacks are made of; it is a game of no more substance than reading tea leaves. And indeed this can be true for analogies can easily be senseless.

It was therefore a considerable shock to me when I first came upon an essay of William Morton Wheeler, on "The ant colony as an organism", Here was an eminent and respected scientist openly making a large and comprehensive analogy and even advertising his sins in his title. But the success of the essay did not escape me; by making the analogy which is so striking, many facets of ant behavior and of the organization within an organism are neatly emphasized. Of course Wheeler is not the only biologist who has succumbed to the temptation of comparing insect societies to other things, and A. E. Emerson, and C. R. Haskins among others have effectively continued the tradition. In fact it would be a bit too much to expect of any man of imagination not to be reminded of other phenomena as one observes social insects go through their fascinating and fantastic behavior patterns.

However it must be admitted that although these discussions make good stimulating reading they are analogies, still bristling with the old dangers. The answer seems to be that if the author is sufficiently wise and

251

sophisticated he can express his analogy in unobjectionable terms, but with the greatest of ease some less cautious individual could draw conclusions that would be totally unpalatable. There is nothing wrong with saying that division of labor among social insects is similar to division of labor among the organs of our body, but we would shudder at the thought of trying to compare totalitarian and democratic political systems with the social behavior of insects. Playing with analogies would seem to be a dangerous sport only to be attempted by the careful. But even then we may ask have the analogies done more than prick the imagination and provide pleasant reading; since the correspondence between the two things one compares is incomplete does this contribute to a significant advance into the insight of either? This is a hard question to answer, but it is sufficiently pointed so that one might wonder if the cautions about the dangers of analogies are indeed not justified.

There is a simple and old way out of this difficulty. It is the way of the comparative anatomists who flourished during the latter part of the last century, although they did not always use the tool at their disposal. If one compares two structures and makes the proposition that the functions of the structures are identical, then one has achieved perfect correspondence, at least as far as the functions are concerned and furthermore one may learn something of the different ways a particular function is carried out. To give an ancient and obvious example one need only think of a wing as a structure designed for the function of flying. The anatomists placed the emphasis on homology versus analogy because they were primarily concerned with phylogenetic relations. The fore arms of a swimming seal, a walking deer, a brachiating monkey and a flying bird were homologous, while the insect wing was merely analogous to that of the bird and their ancestry quite unconnected. Nowadays we attach greater importance to the flying rather than the blood relation. This shift in emphasis is largely because we understand, through natural selection and modern genetics, how the phylogeny could have come about, although a hundred years ago that was the compelling mystery.

The bat, the bird, and the insect flies yet each has solved the problem in his own way. The stiffening structure of each wing is different, the covering material, the shape, the musculature, and many other properties. By comparing the three it is possible to see, through selection, the variety of different ways a challenge can be met, a function fulfilled. We need no longer say the wings of different types of animals are analogous and cringe at the thought that someone will emphasize their many differences; we simply say that their functions are identical and the object of our interest is how this one function can be performed by mechanically different structures.

Students of evolution have, in a restricted way, been interested in this "functional" approach for some time in their study of convergence. The fact that two unrelated animals such as a marsupial and a placental sabre-tooth tiger, produce similar structures, implies a similarity of function, in this case of eating habits. There are of frequent occurrence. The interest of the anatomist and the paleontologist in these cases is, of course, primarily descriptive. The bones of two convergent forms will be a matter of close scrutiny, for such a comparison is considered a particularly advantageous method of understanding the influence of function on anatomical form.

The evolutionary ecologist also has an interest in convergence for often the examples of convergence leave the clear implication that any particular environment has a certain number of environmental niches and that certain specific types of animals may eventually fill these niches. For instance, to take another illustration among marsupials, the fauna of Australia has the same general ecological types as the placental fauna of North America; there are marsupial equivalents of many of our major animal types. Or another illustration, which is always close to the heart of biologists, may be found in the finches of the Galapagos Islands described by Charles Darwin which have shown "adaptive radiation". One stock of finches came to these relatively birdless islands, and a whole series of new finches, with new diets and new habits blossomed as the variations of the offspring of successive generations were selected in different ways.

The point is that any particular environment can support a certain group of animals and plants which possess certain specific functional characters. The whole environment is a complex maze of these opportunities one interlocking with and dependent upon the other. Wood borers are dependent on trees, and woodpeckers, on the larvae of the wood borers. If we consider the living activity of each of these species, that is, its feeding, its movement, its dwelling habits and so forth, as their functional aspects, then we can see that each environment has room for a particular set of such functional niches. In fact we could, with a little thought and care, describe the whole environment in terms of these needs and actions – these functions. In other words, there may be some merit in thinking of the outside world in terms of specific groupings of functions. It is conceivable in this way that it might be possible to see some order to the activity of living organisms, and if this is so it would show that the method of making analogies by seeking common functions has not been entirely useless.

Turning now from the environment to within an organism we see that similar problems exist and that any individual performs a group of functions, such as feeding or reproduction, and it is quite impossible for the organism to continue to exist unless it keeps all of its living activities. This is so obvious it is hardly worth the words to say it, and for many years now "life" has most conveniently been defined by a series of functions which are bound together. The motor car can eat (transform chemical into mechanical energy) but it cannot grow or reproduce. The idea that the functions come in groups, in packages, is important for by being contained there is automatically a self sufficiency and a stability. And it might be added that this packaging is no doubt the result of natural selection, for stability is a property of high adaptive value.

The interesting point, however, is that the units, or groups of functions are not solely those that make up individual organisms, but there might be larger or smaller units that also contribute to stability. For instance the grouping of functions within cells, among cells, among organs are examples of units within an organism, and family groups or animal societies provide good examples of larger units. It is important to emphasize that the same functions may be performed at each of these hierarchial levels. For example there are methods of communication within a cell, among cells, within organs, within an individual among the organs, and between individuals in a society. These methods differ at different levels; sometimes diffusion will be involved, sometimes polar conduction of impulses, sometimes visual and auditory stimuli, but in each case signals are passed and received. So if we wish to compare insect societies with the internal workings of a single organism, we can do so without difficulties or even cautions, if we examine the different methods by which a particular function, such as communication, is carried out. In this way comparisons lose their dangers and may become fruitful without hindrance.

254

To end with a larger notion, let me point out that I have, during the course of these few words, outlined a manner of looking at the world of life that may be useful of pursue. The environment both without and within the organism along with the inter-relation of parts of an organism and organisms one with another should all ultimately be described in terms of a complex scheme of interlocking activities or functions. Through reproduction, inheritance mechanisms, and selection (with the help of time) these functions are fulfilled during the course of evolution. By the careful examination of the variety of methods of this fulfilment we will have a comprehensive catalogue of how living substances can be marshalled to perform certain jobs. In essence it would be a catalogue of protoplasmic capabilities. It is difficult to predict the usefulness of such a list, but as with any system of descriptive classification, the chances are excellent that it might provide some further clue as to the order or the system of Nature.

Princeton University, Princeton, New Jersey, U.S.A.

PROBABILITY MODELS AND THOUGHT AND LEARNING PROCESSES

I

One of the most interesting phenomena of recent years has been the growth of Cybernetics [1] and the application of logical and probability techniques to the study of thought and learning processes. Models, both of a theoretical and practical kind have been constructed in order to simulate human and animal behaviour.

Craik [2], one of the first writers in this field, took as his criterion of a model that it must bear a proper relation-structure to the steps in the process simulated. By this he meant that it must work in a similar sort of way. One of the aims of this paper will be to see how far such models satisfy this criterion.

Take, for example, Craik's account of thinking in which psychological processes are compared to physical ones. Starting from the position that thinking is an operation performed upon symbols, he notes that one of its most important features is the ability to predict the future development of events in the external world. For Craik three essential processes are involved here: (1) the "translation" of some external process into words, numbers or other symbols; (2) the arrival at other symbols by a process of reasoning, deduction, inference, etc.; (3) the "re-translation" of these symbols into external processes, as in building a bridge according to some previously thought out design.[3] Human thought thus provides a small scale symbolic model of processes going on in the external world. We can, in this way, by trying out in our heads alternative plans of action, predict their consequences in practice. Thus if we are engineers we can calculate how much stress and strain the bridge we are planning can stand. It was Craik's contention that many mechanical devices, such as calculating machines and tide predictors also show the same predictive ability.

Since this notion of imitation or simulation attains great importance in Cybernetic theory, let us look more closely at Craik's conception of a model. He conceives a model in a very general way as any physical or chemical system which has a similar relation-structure to that of the process it imitates. And, by having a similar relation-structure he means, as we have already noted, "that it is a physical working model which works in the same way as the process it parallels".[4] This, of course, is the fundamental point at issue: does the model work in the same way as the process copied?

Craik's hypothesis, that thought parallels reality, that its essential feature is symbolism, and that this symbolism is largely of the same kind as that which is familiar to us in mechanical devices which aid thought and calculation, is, he would then argue, based on the notion of identity of relation-structure. However, to say that two things have the same relation-structure may not always be very illuminating, if one interprets this notion in a purely formal sense. The relationship owned in common may be a trivial one. Take the notion of group-structure (i.e. the structure possessed by a mathematical group). All sorts of things in the world to which the laws of logic, geometry and arithmetic apply manifest such a structure, and between which otherwise there may be very little in common.

In psychology, at least, it is not very helpful to say that two systems have the same structure, unless a particular concrete interpretation can be specified. Because of this, concrete models, however erroneous they may be, often have a higher suggestive value in psychology and biology than abstract mathematical ones. When Craik talks of two processes having a similar relation-structure, he does not adequately bring out this difference between an abstract system and its interpretation, possibly because he himself seems to have thought rather in terms of concrete models. Craik does not therefore seem to use the notion of relation-structure in its precise logical sense.[5]

Granting that a model can be constructed which can simulate, say, the intellectual behaviour of a human being, the crucial question is then not, for example, whether they produce the same solution to a logical problem, but whether they produce it in the same sort of way. Both human beings and digital computers can supply a correct solution to a logical problem, but they arrive at it in very different ways. If we look into the way such a computer operates, though it produces a correct solution we would say that it goes about its job very unintelligently. What it does is to try out every possible solution until it hits on the correct one. An intelligent human being, on the other hand, usually acts selectively, rejecting certain solutions and accepting others. Psychologists seem to be aware of this fact nowadays when they talk about strategies in thinking.

II

Because of this difference in operation, the digital computer model has fallen into some disrepute and probability and feedback models have been introduced. It is argued that however closely a digital computer may in principle parallel brain and thought processes, it executes most of these in a quite different manner. What the probability and feedback models then claim to do is to reproduce not only the end-results but also the mode of functioning. Goal-aiming mechanisms using negative feedback have therefore been introduced, which, since they are error-correcting, resemble inductive rather than deductive mechanisms.[6]

One of the assumptions of the feedback model is that random trial-and-error behaviour is a fundamental feature of human and animal learning. When, for example, an animal has to learn a specific task, its activities at first exhibit a random character, but as it begins to learn each success increases the possibility of the successful action recurring, and each failure decreases this possibility. The errors which occur on every trial are progressively eliminated until the specific task is completely learned. Learning on this view is then a process which can be imitated mechanically (or electronically), and mechanical (or electronic) rats and rabbits have been constructed to work on the feedback principle.

It is further claimed that not only is the feedback model applicable to learning behaviour, but also to thinking and problem solving. In these cases too, there is a continuous feedback of information by means of which we correct our tentative intellectual endeavours. Hence, it is argued, there is no good reason for believing that solving a problem by insight involves a fundamentally different process from its solution by trial-and-error behaviour. In solving an intellectual problem we usually try out one solution after another, correcting them until the problem is finally solved. In a similar way a feedback operated anti-aircraft gun will automatically correct its firing until it achieves its goal.

The above view has been criticised on the grounds that we need something more than a feedback device to explain adaptive behaviour at its higher levels in human beings. What is left out of such goal-seeking activities (no doubt as irrelevant to an objective analysis) is the power to anticipate the results on a conceptual level, and to evaluate our actions with reference to ideal aims and values.

Even in the field of animal behaviour, most psychologists would agree that learning is not simply the gradual reinforcement of responses originally made at random. We know that learning is often discontinuous. The number of errors occurring during learning may suddenly fall off without warning, and the kind of error made one day may be quite different from that made the next day.

The more sophisticated learning theorists would agree that the assumption of randomness is not necessary for trial and error behaviour. Motivation has therefore been introduced as a secondary principle, so that we find Hull and others speaking of need-reduction, etc. As we have seen, something similar to this occurs when we deal with human intellectual activities; reference has to be made to some value system in terms of which possible courses of action may be evaluated.

Because of this feature of human behaviour, namely our ability to contemplate ideal aims and values, and to make decisions according to them, our responses are not predictable with the same accuracy as the responses of a simple goal-seeking mechanism. Nevertheless, it has been argued by some writers that the comparative unpredictability of our individual actions can be simulated by an artifice, if we include within it some sort of randomising device. We are told, for example, that the definite introduction of randomness in a machine with suitable corrective feedback can give a flexibility of action and a power of response not seen before.

It has therefore been claimed that an artifice shows randomness in that domain which in a human being is that of "free-will", and that behaviourally there is no reason why the two should be distinguishable. However, the word "randomness" as used in this context is ambiguous. We need to distinguish between (1) its popular everyday usage, and (2) its mathematical meaning. In (1) we concern ourselves with the cause and not merely the result of a performance, as when we refer to an event which occurs by chance or accident, and contrast it with events brought about by design or purpose. In (2) as used in statistics, we are primarily concerned with the result obtained, with certain irregular series, but not with their origin.[7]

When it is suggested that an artifice can simulate "free-will" by having a randomising device put into it, "randomness" in sense (1) is undoubtedly referred to. It is assumed that the physical causes which determine the artifice's behaviour are unknown, but if they were known its behaviour would be seen to be causally determined. And when it is further asserted that human behaviour, insofar as it manifests "free-will", exhibits a random character, the implication is that our decisions may ultimately be explainable in physical or physiological terms.

Nevertheless, the parallelism between "randomness" in an artifice and "free-will" in a human being does not take us very far. Two series of events may both exhibit "randomness". But in one case it may be due to design, as when we deliberately shoot at a target so as to produce a random distribution of hits. In the other, it may be the outcome of a complex interplay of physical causes, for example, the series of digits produced by some randomising device. Further, though to an external observer a man's actions may appear to exhibit a "measure of randomness", they immediately acquire a significance for the observer when he becomes aware of the motives underlying these actions.

Even those writers who contend that there is an objective similarity between voluntary behaviour in a human being and the behaviour of an artifice exhibiting a "measure of randomness", would probably agree that human activities do not normally exhibit the essential characteristics of a random process. Even in throwing a die we would find it difficult to produce a series at all resembling that produceable by an ideal thrower, namely, a six turning up once in about six times. Similarly, when an individual is asked to choose between several alternatives, for example, a number between zero and nine, his answer is usually quite different from what might be expected from a randomising device. Though he might feel that he selected the number at random, his choice would not be random in the formal sense. The numbers thus selected are usually numbers such as three or seven. This would seem to indicate that the selection has proceeded according to some unconscious choice pattern.

III

It is clear that there is a certain orderliness and predictability about human and animal behaviour which lies somewhere between the logical certainty of a deductive process and the unpredictability of a random one. Certain refinements in probability theory have therefore been brought in by cyberneticians and psychologists to deal with the systematic temporal uniformities exhibited in behaviour. These are, (1) the concept of a stochastic process, where we are concerned with the probability of a series of events occurring in time, rather than single events, and (2) the allied notion of dependent or conditional probabilities, where we consider the probability of an event B as conditional upon the probability of an antecedent event A.

Looked at more generally, any system which produces a sequence of symbols according to certain probabilities is called a stochastic process. The special case in which the probabilities depend on the previous event is called a Markov process (after the Russian mathematician who introduced this technique for the study of word distributions). As an example, take the case of the English language to which the concept of a stochastic process has been applied.[8] Suppose when writing we put the definite article "the" down on paper, then the probability that the next word selected will be an article or a verb is small. Starting from the known word frequencies of the English language it is assumed that similar considerations apply to any English word sequence. From this it is concluded that there are probabilities which exert a certain degree of control over the language.

The above approach has been objected to on a number of grounds. These merit consideration as somewhat similar objections occur when stochastic models are applied to psychological processes. It has been argued that it is difficult to say what is a representative sample of English; different speakers of the language may make different selections from it. Overnight, as it were, radical changes may occur in the linguistic probabilities of certain words, and these may have little relation to their previous frequencies. Further, the English language is not a substantial thing capable of generating sequences of symbols in the way in which a tossed coin generates a sequence of heads and tails. And finally, in such studies psychological considerations relating to thought and meaning are usually left out of account.

However, this has not prevented the study of stochastic word sequences from being more or less bodily transferred to an analysis of human and animal learning behaviour. The concept of dependent probabilities has been applied to explain the fact that when a response occurs in a learning situation the organism does not return to its original state, since it has learned by experience.

But even if learning behaviour can be described in terms of time-dependent probabilities, there are many other things in the world around us which can also be thus described. To give some examples: (a) velocity of a given point in a turbulent field; (b) temperature of a room at a time t; (c) the number of bacteria in a culture at a time t; (d) yield of wheat in a given year. All the above deal with situations which are not amenable to a simple causal analysis, as they are due to a large number of variables. In quantum theory and statistical mechanics all the variables associated with physical systems are also defined by stochastic processes. But what is of some importance is to know exactly how these processes differ from each other. In other words, what characteristics mark off, for example, biological and psychological systems from physical ones. Further, it has not yet been demonstrated that causal laws have outlived their usefulness, at least as far as psychological studies are concerned.

With regard to the debate as to whether the fundamental laws of physics are ultimately causal or statistical, physicists are not entirely in agreement amongst themselves. The followers of Einstein, though in a minority, continue to accept a causal position.[9] They believe, as did Einstein, that the essentially statistical character of contemporary quantum theory is to be ascribed solely to the fact that this theory operates with an incomplete description of physical systems. And even if it should turn out that this view is incorrect and that causal laws in physics are nothing more than extreme cases of statistical dependency, this still does not mean that the situation is similar in psychology.

IV

In their book *Stochastic Models for Learning*[10], Bush and Mosteller (hereafter referred to as BM) have applied probability methods (i.e. stochastic models) to learning theory. We intend to discuss this book in some detail, as it is the first extensive systematic attempt to present a probabilistic analysis of the data obtained in learning experiments.[11] We are given a mathematical model which it is claimed is applicable to learning situations. It is interesting to note that though their account is worded in probabilistic terms, it is based on the learning studies of Hull and Guthrie [12], and employs the classificatory categories of these avowedly causal theories. This is reflected in the selection of the variables to be sampled and correlated, i.e., in the choice of the system's units (stimuli, responses, environmental events, etc.). In a sense, then, though we are given a probabilistic analysis of behaviour, it is behaviour as seen from the rather restricted viewpoint of the learning theorist.

We are told by BM that many experiments in conditioning and learning involve choice situations where the animal or person has to choose between alternative courses of action. For example, a rat can turn right or left in a maze, and a person can choose one of several answers in a questionnaire as the correct one. Learning experiments are then characterised by a set of possible courses of action between which the subject has to choose, and certain events (punishment, reward, etc.) which alter the subject's tendency (or probability) to choose between these courses of action. For example, the experimenter may reward one response and punish another, and this will alter the probabilities of the various responses.

What BM then try to do is to describe the change in probability occurring after each performance. As a measure of behaviour they choose the probability p that the instrumental response will occur at a specified time. After each occurrence of the response, this probability will be increased or decreased by a small amount, and will thus change during conditioning and extinction. The amount of change is determined by the environmental events (punishment and reward) and the work or effort the organism expends in making the response. BM express this change in terms of the probability immediately prior to the occurrence of the response, since it is assumed to be independent of the earlier values. As far as the mathematical system itself is concerned, we are told that the physical basis of the probability is irrelevant. For all that it matters, the organism might very well employ a random number table or a roulette wheel to make its decisions.[13] BM merely assume that the organisms behave *as if* they possessed such probability mechanisms.

It is perfectly true that when we refer to probabilities in a mathematical system we are only concerned with an abstract set of elements, which can be manipulated according to the laws of combinatorial analysis. In

263

such a system, when we talk of the sequence of *H*'s and *T*'s involved, for example, in coin-tossing, the *H*'s and *T*'s merely become names given to our symbols without reference to the mechanics of any actual process of coin-tossing. One might say the same for the sets of alternatives, outcomes and environmental events in the BM model. They are merely names for the mathematical variables and operators of the system. Nevertheless, BM do assume that to the variables of their system correspond "classes of responses of real organisms" in the experimental situation. If this is the case, the mode of production of these responses is not entirely without interest.

A fundamental assumption of this mathematical model, as we have seen, is that in order to assign a probability to some future event, we need only to know the probability of the event just preceding it. We can remain in complete ignorance of the past history of the system. Or as they put it, the probability on trial $n + l$ depends upon the probability on trial n and not upon how it got there, i.e., upon the earlier values of the probabilities.[14] This they term the *independence-of-path-assumption*. It has the advantage that it enables them to apply Markov processes to specific learning procedures, since in such a process "the transition probabilities from a particular state do not depend upon how that particular state was reached".[15]

Now the fact that the model is non-historical is a weighty argument against it. One of the most important characteristics of organisms (as well as some physical systems) is their historicity. Their present behaviour is often affected, as Freud has shown in the case of human beings, by very early happenings. One of the difficulties inherent in all psychological experiments is that the human subject "remembers" stimuli and responses over relatively long periods. This "memory" effect may come in to bias significantly the responses of the subject, and it may not always show itself in the $n + l$ response of trial n. Like genetic inheritance, its manifestations may be discontinuous and only appear at selected points in the succession of events.

That there is a certain amount of arbitrariness in setting up such classes of responses is well brought out by Miller and Frick when they state: "Most psychological experiments involve the classification of the organism's responses and the tabulation of the number of responses in each class. There are many difficulties concealed in the summary sentence

– the decisions as to which responses to record, which environmental changes to use in their classification, how large a sample is necessary, etc. In what follows it will be assumed that the psychologist has been able to limit his interest to certain portions of the behaviour stream that he feels are particularly diagnostic, and that these portions can be recognised whenever they occur. They have the status of 'responses' or 'response units'. The psychologist's problem is to describe the relative frequencies of occurrence in responses in each of the classes he has set up." [16]

It is clear that those portions of the behaviour stream which the psychologist singles out as being particularly diagnostic and to which he gives the status of response units, will vary with his own particular approach and interests.

BM outline possible objections to their path independence model, namely, that it does not seem to provide for "memory", for a "practice" effect or for long-range effects of "trauma". They believe that the appropriateness of such objections depends upon the specific event identifications which are made. "In principle", they tell us, "we can always make identifications that impose as long a memory as we want. Suppose, for example, we have a sequence of successes and failures *SFFSSSFSS*, we could call *S* and *F* the two events, or the sequence could be considered to contain four kinds of events, *SS*, *SF*, *FS*, *FF*, or triplets of events etc." [17] They do not therefore think that their general model is seriously restricted by the path independence assumption.

It is not denied that on a purely formal level this approach might work. But the question is, can such an empirical identification be made? It is obvious that once the sequence has occurred, we can arrange the symbols recording the results into longer or shorter sequences without changing their value. Further, the above view seems to treat the successes and failures of the experimental trials *as if* they were the result of coin-tossing. It assumes that the actual time-order of the occurrence of events is irrelevant; that the *S*'s and *F*'s are independent of each other, and will not change as a result of their association into triplets, etc. But can one really call such a classification of symbols into longer or shorter sequences the "imposition of a memory"? Has "memory" any significance at all in this context, unless we deal with an organism having a definite history?

There is a difference between a set of abstract elements related according to the laws of combinatorial analysis, where the elements remain the same under any transformation, and a temporal sequence of events produced by an organism. If we study the organism in the process of generating such successes and failures, then the order in which they occur will certainly make some difference to the final outcome. In actual practice different types of sequences of S's and F's in the earlier part of such a series produced by an organism, might lead to radically different outcomes. Further, in the case of a human being, one cannot neglect his attitude towards the series which is being produced. If he suspects it to be random, or looks for a problem, or believes that he is being deceived, the final outcome may in each of these cases be very different.

There is also the empirical question as to the optimum length of an event as experienced by us. How long can the memory span of an event be? There seems to be a psychological time limit to the duration of a perceived event. The present moment has, for example, been metaphorically described as tinged with the memory of the past and the anticipation of the future. In historical discussions, too, we assign certain arbitrary limits to the duration of events. Though we might describe the Battle of Waterloo as an historical event, we would probably think twice before describing Napoleon as one.[18] On the other hand, in theoretical physics an event could have any length and might indeed include the whole universe. But all that we would be dealing with here would be simply a set of abstract space-time co-ordinates.

V

Another feature of the BM model is that it is linear, i.e., the system is reversible. Or as they put it, a linear operator R satisfies the two relations $R(u + v) = (Ru + Rv)$; $Rcu = cRu$.[19] To this one might object by pointing out that in highly organised biological systems having a Gestalt character, where the whole is more than the sum of its parts, linearity cannot be the rule. The system is irreversible. For example, the time-order of the presented stimuli may play an important role in determining the kind of response produced. The linearity assumption is suspect when a human subject is required to make a number of responses, since it assumes that the order in which the stimuli are given is indifferent. In the case, for example, of two stimuli A, B, the result may be different when the order is changed, or if they are given at different times t^1, t^2, t^3, t^n. Further, in an organic system the nature of the responses depends not only on the present stimulus, but also on past stimuli and their responces.

This view also assumes that there is an intrinsic measure of the applied stimulus existing independently of the system to which it is applied. In the above context this would mean that "reward" and "punishment" are objective factors independent of their evaluation (or assessment) by the subject. Such an approach, however, overlooks the fact that the stimulus may have different values for different subjects and lead therefore to different responses. To take account of such differential outcomes we may, as we have seen, when dealing with intellectual activities require to bring in a valuational system.

BM's reason for introducing their linear assumption is that only linear operators enable them to cope with their "combining of classes restriction", which is a fundamental feature of their model. As an illustration of what they mean by this, they consider a bar-pressing experiment (in the rat) where three classes of responses can be defined, (1) pressing the bar from the left, (2) pressing the bar from the right, and (3) not pressing the bar.[20] Now we should be able, they tell us, "to combine the first two classes into a single class of bar-pressing and thereby obtain predictions equivalent to those which would have been obtained by defining only two classes in the first place." [21]

Such an exercise seems quite innocuous on a purely formal level. Putting it a little differently, in r classes of responses, it should be possible to combine any two classes into a larger class and thereby obtain predictions equivalent to those obtained by the two classes taken separately. Though we can always define classes of behaviour which can be included in a larger, more comprehensive class, what we have to ask is whether the predictive capacity of the larger class remains the same. One can certainly see that the combining of classes restriction will apply to some features of behaviour, for example, the formal characteristics which left and right bar-pressing may possess in common. It neglects, however, the qualitative differences in behaviour which are usually of vital importance for predictive purposes. Even left bar-pressing may have a certain asymmetry, and somewhat different characteristics and concomitants from right bar-pressing. The rat might, as it were, be "left-handed".

What we need to remember is that with an increase in the comprehensiveness of a class (i.e., the greater its extension) there is usually an increase in the generality and often triviality of its defining characteristics. Instead of the general class being richer in predictive power, it may be poorer than either of its component sub-classes. If this is the case, the combining of classes restriction will not always be of much help as far as the more complex kinds of behaviour are concerned. It will leave out of account the more important qualitative characteristics belonging to the included sub-classes, and which may have a high predictive value.

One can undoubtedly apply probability laws to human activities, but it ought never to be forgotten that they possess many other properties than those needed for the making of probability judgments. Venn has pointed out (*Logic of Chance*) that by thus applying probability theory we no more put human actions on the same footing as games of chance, than the historian neglects the distinction between vice and virtue when he applies the same rules of arithmetic to reckon up, say, the numbers who in any country have been burnt at the stake as martyrs, or hanged on the gallows as thieves. And if we are interested in the distinction between vice and virtue, the laws of arithmetic will not help us very much.

In any case, it is doubtful whether most kinds of behaviour can naturally be expressed in terms of clear-cut alternative courses of action which are mutually exclusive and exhaustive. One of the difficulties in dealing with human activities is that they naturally shade off into each other. Like words, they are also ambiguous and may have different significance in different contexts.

BM take up the position that mutually exclusive and exhaustive classes of behaviour can always be defined in any experimental problem.[22] This may be the case when dealing with rats running mazes (where the nature of the experiment may already predetermine the alternatives), and other simple forms of experiment where the alternatives are controlled by the experimenter. However, not all behaviour can be thus controlled, and even a rat running a maze and a dog being conditioned will show differential outcomes in accordance with the "temperament" of the animal.

As far as their use of linear operators in their model is concerned, BM try to reassure us, "if it is any comfort to the reader, we hasten to point out that the whole of quantum physics is based upon an assumption of linear operators".[23] This is cold comfort indeed when physics itself is showing nowadays more interest in irreversible processes, in non-linear systems which have in this respect certain similarities to biological systems.[24]

Further, as we have seen, BM are not concerned with the insides (or causal mechanisms) of their organisms; they merely conceive that every organism has a true probability at the start of each experimental trial. Indeed, as far as their mathematical model is concerned, it does not matter to them whether behaviour is intrinsically statistical or whether it only appears to be so. As we have already noted, they assume that the organism might use a random number table or a roulette wheel to generate its decisions.[25] However, it is doubtful whether "randomness" has any significance without a reference to the process which generates it. There are no such things as random sequences in themselves, since as soon as such a sequence is written down it becomes predictable. Randomness seems to be ephemeral, having meaning only during the activity of generating the numbers. As soon as we are able to specify a random process it ceases to be random.

Further, though the mathematician may not be interested as to how the series is produced, whether by design or accident, the psychologist, at least, ought to be concerned with its manner of production. To say that the organism can be thought of as if it had a random number table inside it [26], by which it makes decisions, is simply to evade the issue. It also overlooks that the production of such tables by statisticians involves a considerable amount of design and ingenuity to prevent certain sequences (regularities) from occurring. Indeed, it is more difficult to construct a really random sequence than a sequence with well-defined regularities.

As we have seen, BM in their mathematical model regard the organism after the manner of a probability device such as a roulette wheel. They therefore compare some forms of learning behaviour in the rat to sequences fashioned on what is called the Monte Carlo method – where as a result of successive plays the series converges to a given frequency. To such a sequence they give the name of a *stat-rat* [27], which is, we are told, a sort of theoretical organism – a mathematical robot. By this method employing random number tables, they claim they can generate artificial sequences of responses, similar to the sequences produced by a real-life rat whilst learning.

It may very well be that rats when running mazes do generate such simple statistical sequences. But as we have already suggested, what differentiates the rat from that major instrument of probability theory, the tossed coin, is precisely the mechanism which generates such a sequence. This might, if we were dealing with an extremely intelligent rat – a super-rat, be due to design; he might perhaps be deceiving the experimenter.

If we reduce the behaviour of a rat or a human being to that of a tossed coin, where the only behaviour exhibited is that of turning left or right, it is not surprising that their behaviour takes on such a statistical character. If the two series, a tossed coin and the behaviour of a rat in a maze, show similar statistical properties, is this not perhaps due to our having in both cases imposed restrictive conditions which limit the type of behaviour observed? In the case of rat-learning, it has been pointed out that the specific features of learning behaviour exhibited in the "learning curve" [28] may arise from the experimenter rigidly maintaining uniform conditions throughout the training period. The experimental situation may perhaps force the learning behaviour of the rat to exhibit such features.

As we have observed in connection with feedback mechanisms, though in a particular experimental situation a human being may act like a feedback device, it must be remembered that his behaviour is of extreme flexibility. Put into an appropriate experimental situation, a man can be made to behave in a great many different ways, and he can quite easily be made to simulate the behaviour of a rat when running a maze.

It is sometimes also forgotten that the frequency theory of probability, upon which stochastic learning models are based, does not tell us anything about future single events. It only applies to individuals as members of classes of events. Economists such as Shackle [29] have therefore pointed out that decisions have often to be made in life concerning novel and unpredictable ventures. There are many kinds of decisions of extreme importance which are virtually unique – choosing a career or a wife – and to these any frequency theory seems inapplicable.

270

What moral are we to draw from all this? It would appear that we can apply probability and statistical concepts to physical and in some cases biological data, without unduly bothering about the nature of the concepts used. In psychological studies, on the other hand, the assumptions implicit in probability theory may come in to bias our results.

VI

For practical purposes, we may divide empirical probability series into two classes. (1) Those containing the results of games of chance in which the conditions of production and the laws of statistical occurrence are largely fixed, and to which the abstract probability model approximately applies. When pennies are tossed they give heads and tails equally as often as they did when they were first tossed, and we believe they will continue to do so. Further, this series tends to a fixed numerical proportion. In (2), which covers the bulk of ordinary statistical enquiries, the conditions of production may more or less vary, and so may the results. In most cases of human behaviour in which we find statistical uniformity, as seen, for example, in vital statistics, the observed phenomena are generally the product of very numerous and complicated antecedents. Unlike games of chance, there is not always a steady approach to a limiting value as is required by Bernouilli's theorem. In such statistical series (for example, the number of marriages, births, accidents, etc., occurring in a year) it will be found that after perhaps a period of uniformity they will tend to fluctuate and change.[30]

Not only do such changes occur in the field of social behaviour, but the individual too does not as a rule show a static type of ordered behaviour. A specific behaviour pattern may be elicited repeatedly by certain stimuli over a long period of time, and then quite suddenly these stimuli may cease to elicit the familiar pattern. It seems to be a fundamental feature of most organisms that they manifest a dynamic state of order which permits sudden, irreversible, unpredictable changes.

On the other hand, in the case of statements relating to games of chance, we can, as we have seen, usually assume with some certainty that the future behaviour of such systems will have a well-defined character, namely, tend towards certain limiting values. We are confident that the physical structure of these games will not change significantly. Indeed, we take great care in the manufacture of such devices as dice, roulette wheels, etc., to see that the outcomes are equally probable. The roulette wheels at Monte Carlo are continually being tested and adjusted, so that they come close to an ideal randomisation system.

University of Manchester, Manchester, England

REFERENCES

1. The science of control and communication in animal and machine.
2. K. J. W. Craik, *The Nature of Explanation,* Cambridge University Press, London, 1952.
3. *Ibid.* cf. p. 50.
4. *Ibid.* p. 51.
5. Russell tells us (*Introduction to Mathematical Philosophy,* pp. 53–4) "We may define two relations P and Q as 'similar', or as having 'likeness', when there is a one-one relation S whose domain is the field of P and whose converse domain is the field of Q, and which is such that, if one term has the relation P to another, the correlate of the one has the relation Q to the correlate of the other, and *vice-versa*." Compare the above to Craik's assertion (*Ibid.* p. 51) "By a 'relation-structure' I do not mean some obscure non-physical entity which attends the model."
6. An example of the use of negative feedback as a controlling device may be seen in the early radar-controlled anti-aircraft fire. After successive approximations resulting from "fed back" error signals, the gun is finally enabled to hit its target. Such mechanisms have been termed goal-aiming, since they tend to simulate the behaviour of animals when actuated by goal-seeking tendencies – such as hunger or sex.
7. Though as we shall see later, the agency by which the series is generated cannot be entirely separated from the notion of randomness.
8. It has also found an application in those branches of physics where causal laws have been ousted by statistical ones.
9. For an account of this approach, cf. David Bohm, *Causality and Chance in Modern Physics,* Routledge and Kegan Paul, London, 1957.
10. Robert R. Bush and Frederick Mosteller, *Stochastic Models for Learning,* John Wiley, New York, 1955.
11. It also covers in a more systematic and abstract way some of the questions regarding the application of probability models to learning behaviour discussed in the earlier part of this paper. As it treats of the way learning behaviour may be described by mathematical models, its descriptions are much more general than those obtained from a study of the functioning of concrete mechanical and electronic models. One might say, however, that the general principles explicitly formulated in the mathematical model are implicit in the functioning of the more concrete type of model.
12. *Ibid.* cf. p. 333. "We recognize that we have been strongly influenced by the Hullian and Guthrian schools of thought."
13. *Ibid.* cf. pp. 15–16.

14. *Ibid.* cf. p. 20.
15. *Ibid.* p. 78.
16. G. A. Miller and F. C. Frick, 'Statistical behaviouristics and sequences of responses', *Psychological Review*, 1949, 311.
17. *Ibid.* cf. p. 331.
18. Though some philosophers such as Whitehead would have done this.
19. *Ibid.* footnote p. 28. Where u and v are arbitrary operants and c is a constant.
20. *Ibid.* cf. p. 39.
21. *Ibid.* p. 39.
22. *Ibid.* cf. p. 14.
23. *Ibid.* p. 20. Nevertheless, they add cautiously, "Although linear operators are basic to the mathematical machinery which follows. linearity is not essential to the general approach". This is amplified on p. 331, "if the linearity axiom proved to be unsatisfactory, nearly everything we have said in this book would have to be modified". They go on to suggest a possible remedy in which an "intervening variable could change according to linear operators". And they point out that Estes and Burke use such a strategy in their general model of stimulus variability.
24. Cf. the work of I. Prigogine in thermodynamics, *Étude thermodynamique des phénomènes irréversibles*, Liège, 1947, and I. Prigogine and J. M. Wiame, 'Biologie et thermodynamique des phénomènes irréversibles', *Experientia* 2 (1946).
    Bertalannfy and others have also made the distinction between closed and open systems. In the former, to which most physical systems belong, the processess are self-contained and continue without intervention from outside; in such systems there is usually an increase in entropy. In the latter, the system is open to external influences and there may be a decrease in entropy. Most biological systems are of this type as they reorganise themselves towards states of greater heterogeneity and complexity. An interesting feature of such systems is that they may attain a steady state and continue to do work. (Cf. L. von Bertalanffy, *Problems of Life*, Watts, New York, 1952).
25. Cf. pp. 15–16.
26. *Ibid.* cf. p. 16.
27. *Ibid.* cf. pp. 129–31. The asymptotic increase in probability thus becomes identified with the frequency curve of learning.
28. In a learning curve the time taken (or errors made) in successive performances or trials is plotted against the number of trials. The curve rises at first with comparative rapidity until it becomes practically horizontal.
29. G. L. S. Shackle, *Expectation in Economics*, Cambridge University Press, London, 1949.
30. Cf. John Venn, *The Logic of Chance*, 3rd ed., Macmillan, London, 1888, pp. 91–95.

R. C. LEWONTIN

MODELS, MATHEMATICS AND METAPHORS*

In exposing the process by which scientific knowledge is accumulated, the epistemologist deals primarily with two problems. First, what is the *logical structure* of the *statements* of science, and second, what is the *experimental content* of the *entities* in these statements? Essentially, these are the problems of the syntax of scientific language on the one hand, and the vocabulary, on the other.

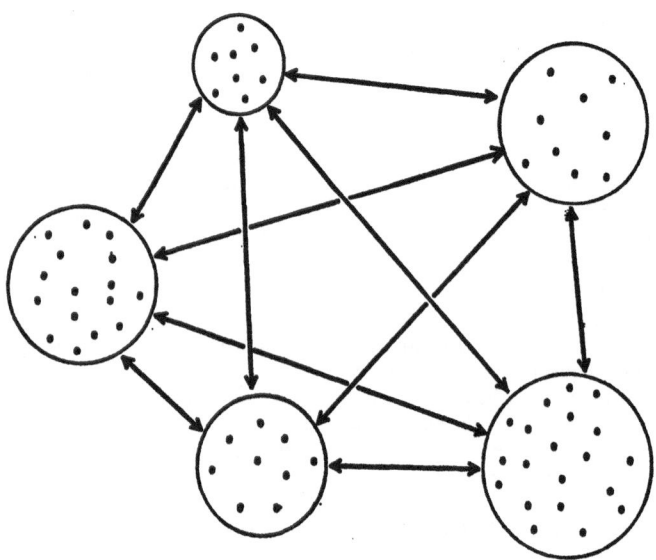

Fig. 1.

* This essay is based in part on a paper given at the American Society of Zoologists Refresher Course in Evolution in August 1959 and in part on a paper read to the University of Rochester Philosophy Discussion Group at its December 1959 meeting. I am indebted to that latter group and especially to Prof. Lewis Beck for their unstinted criticism of my somewhat unorthodox view of models.

It is in dealing with this latter question that the philosopher of science and the practicing scientist must concern themselves with the notion of models, their relation to theories, and the ways in which they serve as useful scientific devices.

Now it is obvious, nay, trivial, that scientific knowledge is more than a collection of experientia. Rather, science is, to appropriate a term from psychology, a *structured* collection of experiences, and any inquiry into the nature of science is really an examination of this structure and its making. How are the raw data of the senses juxtaposed, connected, related to produce what we so innocently call scientific knowledge?

In Figure 1 is depicted the crudest sort of structure that might be imposed on raw experience. Each dot in the figure represents an individual spatio-temporally defined observation, while these data are enclosed in circles representing sets of observations delimited by some common property. In the language of modern probability theory the observations become "occurrences" of the "events" represented by the sets.[1]

Second there are lines drawn between the events, lines representing rules of correlation between occurrences of one event and occurrence of another. As it is shown in the figure this is the simplest sort of structured experience. It does not include sets of predicted but as yet unseen phenomena, nor is there any provision in it for inferred unobservable entities like genes or protons. Moreover the lines connecting events are considered to be only correlative statements, rather than causative or explanatory ones. The grammar of such statements would be paratactic rather than hypotactic and for this reason the arrows indicating the connections point in both directions. Despite the simplicity of this structure it is sufficiently general as a basis for discussion for it already contains entities outside of direct experientia. The classes themselves represent a degree of abstraction although the symbols for these classes may be the same as those used for individual occurrences, as "man" is an abstraction from individual men. The correlative rules even more obviously cannot be framed entirely in the language of the experiences, even at the most elementary level. In the very process of erecting the structure, then, it has become necessary to introduce entities not directly experienced. As Benjamin[2] puts it, science "talks about things which do not obviously exist in order to convey information about things that obviously do exist". Finally we must observe that irrespective of whether a "real"

275

structure exists for the universe of observables, Figure 1 is unlikely to represent the real structure. The question of mind as a unique force is not at issue, but only the obvious fact that a limitless number of different structures can be imposed upon a limited raw experience. It is for this reason that I have said that the observations become occurrences of events in the process of construction of the figure. Some conscious selection of possibilities is involved in making Tom, Dick and Harry members of the set "man" rather than of the set "150 pound objects".

From such a simple beginning the structure may be developed in two directions. One leads to "scientific theory" or "scientific hypothesis" while the other leads to "model". Both directions require some elaboration of the structure, but not to equal degrees or in the same way. The development of a theory requires the introduction of *as yet unobserved* but *observable* entities connected to previous observation by lines of causality and of explanation. That is, we demand both prediction and explanation from a theory and especially prediction of new phenomena, not simply "more of the same". Looked at in another way, a theory is distinguished by its *content* and *intent* from other kinds of scientific structures. The content of a theory, especially its predictive aspect, must render the theory falsifiable or verifiable as a result of new accumulations of experience. Thus, a theory is contingent as a whole and must contain at least one contingent statement. The intent of a theory, on the other hand includes explanation. Theories are designed to provide answers to "why" questions. It is not possible here to explore the criteria of satisfactory answers to the question, "Why"?, but there is no doubt that the satisfaction of such inquiries is one basis on which the scientist judges the usefulness of a theory.

Models differ from theories both in their content and intent. First, models are not contingent in the way that theories are. The rules of correlation between variables in a model may be (indeed should be) derived from a theory and to the extent that theory is incorrect, the model will relate observables in a way which does not correspond to nature. But the model is not thereby incorrect, it is simply wrong to say that it is a model of a given natural situation. Because models are not contingent themselves but entirely *analytic*, the model can never be said to be true or false in an empirical sense. This important distinction between models and theories can be illustrated by a similar situation in art criticism. In his book "Art and Illusion", Gombrich raises the question of whether a painting can be said to lie. As he quite correctly concludes, the concept of truth cannot be applied to the painting alone, but only to the combination of the painting and statements about it. As an illustration Gombrich points to two identical views of cities in the Neurenberg Chronicle (made in fact from the same woodcut) one titled Damascus and the other Mantua. The titles are in fact lies but they might have simply stated, "The artists concept of what a city looks like" and this would be true. Pictures are one species of model and like all models they are not by themselves statements about the real world, but only deliberate juxtapositions of arbitrarily chosen signs. That is, they are not contingent.

Models differ from theories too in the purpose which they are intended to serve. A model is essentially a calculating engine designed to produce some output for a given input. It is precisely for this reason that electronic computers are so useful to the model maker. The concept of a model as a calculating device is completely in accord with what I have said about the analytic nature of models. Because they are not contingent, models cannot produce really "new" knowledge, but can only demonstrate what is entailed by the theory from which the model is built. For this reason, successful use of models demands a preexistent theory and in biology, at least, it has been only those disciplines with a well developed theoretical structure, which have had any notable success with models. This point will be looked into more closely in the later discussion of the metaphorical element in models.

THE STRUCTURE OF A MODEL

A model will be defined as a system made up of the following components: 1) *A set of entities*. These entities may be "events" as previously defined or they may be inferred unobservables like protons or finally (and here I diverge most strongly from usage) they may be entities with *no imaginable* correspondence to observables. It is those models that contain some of the last, which may give pain to the epistemologist, but which scientists insist upon using. The measure theoretical probability of Von Mises and the quantum theory of Heisenberg and Schrödinger contain entities of a totally non-experiential character. The photon in quantum theory is not an inferred entity like the old fashioned atom or proton[3], because

277

unlike them it is invested with qualities that are experientially contradictory. A photon can for example be in two places at the same time.

2) *Quantity*. To each entity there corresponds one or more variables which may be number, time, spatial location, color, etc. Whether these quantities are subject to mensuration or not depends upon the entities to which they refer. Thus, in Figure 1, only one step removed from experience, such quantities as weight and color might be appropriate. But in probability models we may speak of the Borel measure of a set without for a moment imagining that any Borel tape measures exist.

3) *Rules*. The final essential elements of a model is the set of rules. Rules connect entities in such a manner that the quantity variables associated with entities are made dependent upon each other. A set of rules will, for example, relate the velocity of a baseball at any instant, with the velocity of the bat that struck it, the elastic modulus of the bat, the density of the air and so on. The rules may be in the form of a logical or mathematical calculus or they may be rules of physical behaviour. Again, the form of these rules will depend upon the experimental content of the entities. It is important to distinguish between the rules of the model and, for example, mathematical formulae which may be logically deduced from them. The distance d, traversed by a freely falling object during a time t is given by the formula

$$d = \tfrac{1}{2} g t^2$$

where g is the acceleration of gravity. This is not a rule of the Newtonian model but rather a deduction from the assumption that the body began at rest and that it is subject to a constant acceleration, g. The formula may of course, be a rule in another model, but here the question arises of how to choose a model, a question to which I have already alluded and have postponed.

The way in which the rules of the model interconnect the entities and their associated quantity variables, determines the input and output of the model. Values are arbitrarily imposed upon certain quantity variables and these values, by means of the rules, result in values for yet other quantity variables. A model is generally not isotropic in this respect. Because of the escapement mechanism in a watch it is not possible to make it go by applying pressure to just any gear. So, in a model, the assignment of values to some variables may have no effect on the others.

The rules of a model are a unique aspect of it and, so to speak, make the model. The entities and their associated quantities are "dummy" variables. In this sense a model is like a mathematical transformation, an operator whose characteristics are given by the forms of the equations and not by the names of the variables being transformed. That is, we are free by an appropriate thesaurus to replace one set of entities by another and usually this will require a change in the quantity variables as well. This operation, *per se*, does not change the model because the rules have not changed, but unfortunate side effects may occur. A thesaurus is a set of one-many relations and an incautious use of it can result in the transformation of a perfectly sensible phrase into complete gibberish. This arises because words carry with them, the Mad Hatter not withstanding, a set of rules of their own and these rules intrude themselves upon the consciousness. There are very few true synonyms. So too, physical entities (but not formal ones) carry a set of physical rules known to us from previous experience and so long as these rules, which are exogenous to the model, intrude themselves upon the model, they change it.

The particular change of variable in a model which involves the substitution of a *physical entity with an already associated set of observed or inferred attributes different from the replaced variable*, I shall call a *metaphorical* change. It may be objected that the whole point of a metaphor (Man is a wolf, for example) is to replace one entity with another of the same characteristics, but more distilled, leaping more quickly to the inner ear and eye. The answer is that although it may be the purpose of metaphor to carry over certain attributes (as the etymology of the word suggests), it cannot be doubted that extraneous attributes are also carried over, else we would have not metaphor, but identity. "The best material model of a cat", say Rosenblueth and Wiener [4] "is another, or preferably the same, cat".

In most discussions of models, the metaphorical element is taken to be an integral part of their structure.[5] In fact a model is often thought of as being physical, as a model airplane, or a billiard ball model of a gas. In this view a "mathematical model" is impossible for "metaphor" and "model" are virtually synonymous. The argument that I have developed in the previous paragraphs, however, is in sharp contradiction to the

279

idea of model as metaphor. The essential structure of a model is given by the rules, while the metaphor, if any, is chosen so that the physical entities involved in it behave as much as possible in accordance with these given rubrics.

What, then, is the purpose of the metaphorical elements in a model? The first may be said to be *didactic*. If we can but relate the elements of a model to our familiar experience, we "understand" it better. It is for this reason that metaphorical substitutions must be steps toward *experientia*, or at least not away from them. Metaphors in this sense are concessions to the sense – bound human mind and even such a genius as Lord Kelvin insisted that he could not "understand" a theory unless he could make a physical model of it.

Second, there is the *experimental* value of metaphor. It is easier to breed fruit flies than elephants, easier to put model airplanes in wind tunnels than jet bombers. This advantage of metaphors, unlike their didactic function, is not unique to them and as I shall show shortly, a move in the direction of formalism is sometimes superior in this respect.

The problems introduced by the injection of metaphors have already been alluded to and the *reductio ad absurdum* of Rosenblueth and Wiener points them out better than any lengthy discussion. It is necessary, however, to say something about the appropriate choice of a metaphor. While a cat is the best material model of a cat in terms of safety because it avoids the exogenous qualities inevitably introduced into a model by metaphor, such a model is obviously the worst possible from the didactic standpoint and of on advantage operationally. Moreover a model is, above all, an abstraction from *experientia* which means precisely that certain aspects of the real objects of enquiry are deliberately ignored as being irrelevant. A model airplane is a valid model of a jet bomber not *because* it is smaller and made of different materials but *in spite* of these differences. The result of wind tunnel tests are useful only insofar as the size and material of the model are irrelevant to the answers sought. We are concerned only that for a given input into the model the rules (including rules of translation from model to real object) will provide us with an output which can be compared with experience. So long as the input-output relation is consonant with our experience we do not reject the model. Continually, the range of input is increased and the model adjusted, if necessary, so that it continues to put out acceptable results.

Whether the limit of this adjustment is finally "reality" rather than a model or even a "true model" is open to speculation. The notion that such a convergence does occur must assume that on a finite universe of observation one and only one structure can be imposed that is both necessary and sufficient to relate the observations in a consistent way. Such a faith seems, at the present stage of development of scientific knowledge, unwarranted. Whether or not models can "reveal" nature, their primary function is to order it. The best metaphorical model is then one which has a high didactic value and at the same time in which the exogenous physical attributes of the metaphorical entities can be absorbed into or made consonant with the rules of the model. In this way we both concretize and more important *unify* our picture of the universe.

This latter notion reveals the third purpose of metaphor in models. It is the hope of all science, I suppose, to find a sort of unified field theory, but at the same time to make this theory in its working detail a part of experience. In an effort to do this the model builder introduces metaphorical elements not as metaphors *per se*, but as articles of faith in the unity of experience. But this faith seems no longer tenable, if it ever was, and scientists, especially physicists, seem now content to accept purely formal, non-iconic models of physical phenomena, because common sense has been so often confounded. No physicist can explain quantum theory in its entirety in terms of believable experience, but this does not make the wave equation any less of a model, only less metaphoric.

In sum, what is to be said about the choice of metaphor is that the model should be chosen *before* the metaphor and not by means of it. The metaphor is to be chosen by virtue of its elements of similarity to the pre-existent structure of rules.

KNOWLEDGE FROM MODELS

How do models, aside from the advantages which accrue from their metaphorical content, work in the accumulation of knowledge? Can they predict new phenomena, or only more of the same? One claim for the metaphorical element in models is the very one that I here rejected. It is supposed that it is the metaphor alone which is capable of generating new propositions because it is the metaphorical elements which bring to the model "exogenous qualities". Thus, if we say that particles in a gas are like billiard balls we might look for certain qualities of a gas which would derive from elastic collisions, from conservation of momentum and so on. But billiard balls have color, they make a noise, they break when dropped and they are made of clay. Ought we to ascribe these properties to gas molecules too? Obviously not and if we infer these qualities by analogy and fail to find them should we discard the billiard ball as an appropriate metaphor? Again, no. In fact, the billiard ball analogy cannot really provide a *constructive* tool for future inquiry because we are too sophisticated for that. The attribution of properties to an entity by means of metaphor may *accidentally* reveal a new phenomenon and thus advance knowledge, but it is always accidental and as a systematic approach to knowledge it is an extraordinarily clumsy technique. Such accidents do happen because there are some threads connecting apparently unrelated phenomena so that by chance metaphor and object may be knots in the same thread. Worse yet, metaphor may lead to propositions which interfere with the accumulation of new relevant experience. "The price of metaphor is, indeed, eternal vigilance".

The construction of models *can* lead to the discovery of new phenomena in a systematic way if the metaphorical element in them is ignored (but not only if). This is true despite the fact that models are entirely analytic and in no way contingent upon experience. A calculating engine, an analytic model, provides new knowledge because knowledge is after all, a state of the human mind. In Leibniz's terms, an increase in knowledge may come from the conversion of *perception* into *apperception*. This is what models do. Although models as here defined contain no relationships that have not been built into them, they contain much *knowledge* that has not been put into them. As a kind of meta-model we might consider the series of numbers 0, 25, 36, 9, 1, 4, 16. These make little sense as they stand but if they are ordered in a increasing series, the series becomes

$$0 \ 1 \ 4 \ 9 \ 16 \ 25 \ 36$$

which after a certain reflection can be seen to be the squares of the successive integers. But this is no increase in knowledge. True, but on further reflection it is clear that the differences between successive numbers are simply

$$1 \ 3 \ 5 \ 7 \ 9 \ 11$$

and so on. We thus learn that the squares of the integers differ by the successive odd numbers, a discovery which of course could have been made algebraically *if one knows algebra*. The relationship between successive squares was contained in the series all the while, but the knowledge of this relationship proceeded from the imposition of an ordering.

Now it may be objected that this empirical procedure demonstrates only a relationship among the first seven integers and does not prove a general case. This objection is, of course, true and leads to an attempt to prove that the relation holds for any pair of successive integers by other means. That is, a problem is suggested by the ordering, a tentative answer given, and an impetus for further research is provided.

The triviality of the example ought not conceal the importance of the general result. Models can contribute knowledge in two ways. First, they can suggest, or in the case of some models even prove, general results that were contained in but not apparent from collections of observations. An important aspect of this use of models is that observations that seemed inexplicable may appear as the inevitable concomitants of a theory, although before the application of the model the connection was unknown.

Second, models may give an impetus and direction to future inquiry and thus to the discovery of truly new phenomena not hitherto sought for. This function of a model arises when input and output of the model fail to correspond to observations from nature. If the model contains the entire theoretical structure relating the observations and if the output of the model does not conform to expectation the scientist is driven to look for new phenomena by which a reconciliation can be achieved. Again, it is the theory, not the model, that is incomplete but the analytic power of the model is a tool by which the completeness of the theory may be tested.

In the succeeding sections of this essay, I shall attempt to illustrate these properties of models by reference to some problems in evolutionary and population genetical theory which have been solved by the use of mathematical models.

THE PROBLEMS OF EVOLUTIONARY THEORY

The whole notion of evolution rests upon three basic observations of neontology and paleontology, observations which induced naturalists of the eighteenth and nineteenth centuries to formulate evolutionary theories. First, at any instant of time and within any population of a species there is phenotypic variation from individual to individual. Second, the extent of this variation differs from population to population within a species. Finally, the phenotypic composition of populations and species changes in time. The contribution of Darwin and Wallace was to perceive the relationship between intrapopulational variation and variation in time and space. Their theories amount, in short, to the realization that *intrapopulational variability is converted into spatial and temporal differentiation. The process of this conversion is the process of evolution.*

The importance of their theories is in no way diminished by the twentieth century discovery (or rediscovery) that the mechanism of inheritance is quite different from what Darwin supposed it to be, or that natural selection is not the only mechanism by which variation is controlled. On the contrary, the two modern problems of evolutionary study are the same as they were a hundred years ago: What is responsible for the origin and maintenance of intrapopulational variation? What are the processes by which this variation is converted to the observed differences between populations in time and space? Observational, experimental, and theoretical studies all are concerned with these problems.

There is some doubt in the minds of biologists that mathematical theory can be at all important in an understanding of nature. This doubt has been fed by some overly elegant and essentially vacuous mathematical redescriptions of biological phenomena. Genetics and evolution, however, are singularly susceptible to useful, even essential mathematical model building.

Observational and experimental studies are primarily concerned with the discovery and measuring of all possible natural mechanisms affecting phenotypic and genotypic variation. How high are mutation rates? What factors affect recombination of genes? How do genes interact among themselves and with the environment to determine fitness? What are the rates of evolution of characters, of speciation? These are questions of observation and experiment. In addition, there are the correlative questions. Do species lower in the food chain differ in their breeding systems from those higher up? Do annuals tend to self-comptability? Is the degree of polymorphism of a population related to environmental stringency?

It is after the answers to those questions have been given that theoretical and especially mathematical constructions become important. Given the mutation rates, the breeding system, the interactions of genes in determining fitness, what are the directions and rates of change of phenotypic and genotypic characteristics of populations and species? Only mathematical treatment can give quantitative predictions which in turn are the only tests of the adequacy of the structure of evolutionary theory. While natural selection can be demonstrated to occur and while it is an attractive hypothesis that it is responsible for evolutionary change, mathematical treatment can demonstrate that a given force of natural selection is sufficient to produce a given amount of alteration in population structure. It is mathematics that has been responsible for the destruction of the naive notion that natural selection can and will produce any and all "desirable" changes. For example, it is known that mutation rates are under the control of genes at loci other than the mutating ones. Further, it may be argued that under some circumstances it is of advantage to a population or species to change its mutation rates. Mathematical analysis shows, however, that selection for genes controlling mutation rate is a second order process whose efficiency is so low as to make it doubtful that it would occur at all.

In addition to testing generalizations about evolutionary forces, mathematical models serve a creative role in evolutionary studies. General relationships that are not obvious from observational data emerge from models, and these relationships can then be tested by critical experiment. One illustration of this is the discovery that heterosis can lead to balanced polymorphism. This is by no means obvious on the face of it, nor would it be a simple matter to arrive at this result by inference from masses of experimental data. That heterosis is in fact an important factor in maintaining polymorphism has been demonstrated by experiment and can only be so demonstrated, but that heterosis should be looked for in polymorphic systems we owe to mathematical treatment.

The possibility of entirely new forces in evolution also emerges from mathematical models. The principle that subdivision of a population and restriction of the breeding size of a unit can in themselves lead to spatial differentiation by genetic drift we owe to the theoretical insights of Wright and Fisher. Whether random drift plays an important part in evolution is a matter for experimental verification, but that it *can* is a mathematical truth.

THE STRUCTURE OF EVOLUTIONARY MODELS

The relationship among the elements of an evolutionary model: entities,

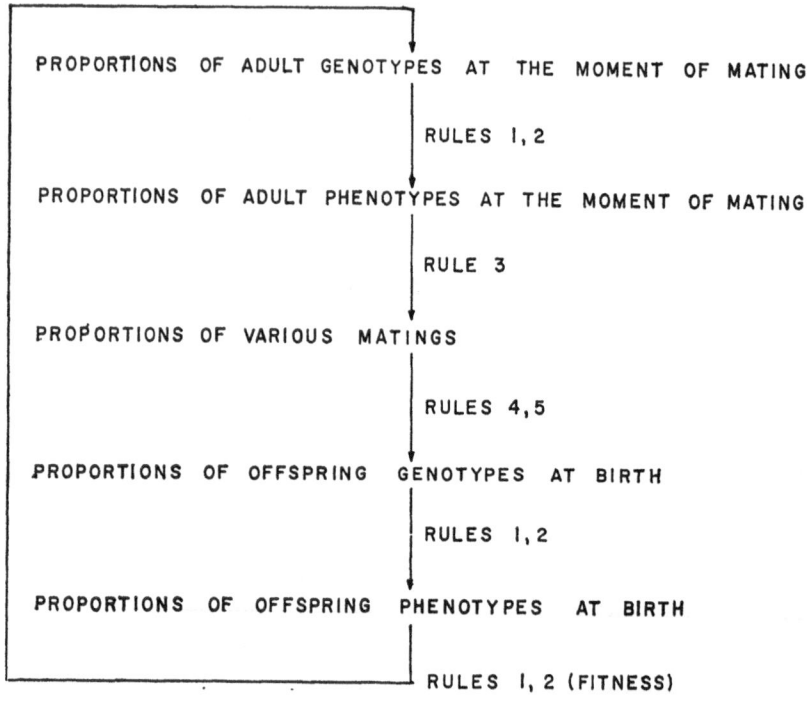

Fig. 2.

quantity variables and rules, is a critical one. Thus, if the concept of Darwinian fitness refers to differential reproduction of diploid genotypes, the entities in the model must be genotypes and not genes, although it may be possible to derive expressions for changes in gene frequency from the genotypic model. It is for this reason that models using populations or species as the basic physical entities are so difficult to construct, for there are no rigorous rules comparable to Mendelian laws and mating rules, which apply to populations or species as wholes. The failure of population ecology and demography to establish really useful mathematical models is a consequence of the lack of known rules of behavior for population growth and competition with other populations. Attempts have been made to relate the average fitness of individuals within populations to the fitness of the populations relative to each other. The lack of any mechanical model of competition between populations, however, makes the parameter "average population fitnes" useless *in this context*, whatever may be its importance for understanding intrapopulational change. This does not mean that nothing can be done with the notion of population fitness, but rather that at the present time the critical ingredients of the model are lacking.

For intra-populational models, however, the general laws of inheritance, of mutation and of natural selection are of such a form that they can be used in the construction of models. Figure 2 is a general diagram of a model for changes in the composition of a population due to known evolutionary and genetic phenomena. The figure shows a single generation, but its form is repeated again and again in successive generations as indicated by the path from the last step in the sequence back to the first. The rules referred to in the diagram are: (1) Relation between genotype and phenotype (dominance, pleiotropy, norm of reaction especially of fitness, epistasis); (2) Environmental fluctuations; (3) Mating rules (panmixia, assortative mating, inbreeding, apomixis, etc.); (4) Mechanisms of gene transmission (segregation ratio, number of alleles, number of loci, linkage, polyploidy, etc.); (5) Mutation.

The entities (genotypes, phenotypes, matings) and their associated quantity variables, in this case proportions, are connected one to the other by the rules much in the manner of Figure 1. By an appropriate choice of initial values of the quantities and of the particular forms of the rules pertinent to the investigation, the general model of Figure 2 can be turned into a specific calculating device for the prediction of the quantity variables in each generation.

A DETERMINISTIC MODEL

One example of the way in which the general structure of Figure 2 can be turned into a specific model is an attempt to understand the polymorphism in an Australian grasshopper, *Moraba scurra*.[6] Populations of *Moraba* are polymorphic for two independent genetic factors on two different chromosomes, with the result that there are nine different genotypes: *AABB*, *AABb*, *AAbb*, etc.

The proportions of the various genotypes are different from population to population but are apparently constant from year to year. That is, the polymorphism is stable. The problem is to explain the stable polymorphism. Specifically, can the observed composition of the population be explained by the known forces of natural selection, forces whose magnitude can be estimated in nature?

Since the populations of *Moraba* are known to be large, it can be assumed that the rules in Figure 2 are of a deterministic rather than probabilistic nature. That is, if a process like mutation of A to a is said to occur at a rate of 10^{-6} per generation, then precisely one millionth of the A genes will be converted to a genes each generation. The result is that the successive steps in the model can be related by exact algebraic functions and these in turn can be reduced to a single equation for the change in gene frequency per generation. If x is the frequency of A genes among A and a, and y is the frequency of B genes among B and b, one form of this equation is the differential equation

$$\frac{dy}{dx} = \frac{y\,(1-y)}{x\,(1-x)}\frac{\dfrac{\delta W}{\delta y}}{\dfrac{\delta W}{\delta x}}$$

where W is the average fitness of the population. This average fitness is a function of the frequencies y and x of the genes at the two loci. This equation is extremely suggestive in that it is of the same form as the equation of motion of a particle in a two dimensional potential field where x and y represent the positions of the particle along the x and y axes and W is the potential of the field at any given point x_0, y_0. The analogy is not a strict one because the factors $y(1-y)$ and $x(1-x)$ do not usually appear in the analogous physical equations but the essential feature, the condition for no further change in x and y, is the same in both the genetic and physical model. This condition simply stated is that the gene frequencies will reach a state of stable equilibrium at a maximum point in the potential field. The analogous statement for a physical particle would be that stable equilibrium is the point of minimum potential

288

energy. The metaphor of the particle moving in a potential field suggests a simple way of determining whether the known state of a population corresponds to the expected stable equilibrium condition. A map of the

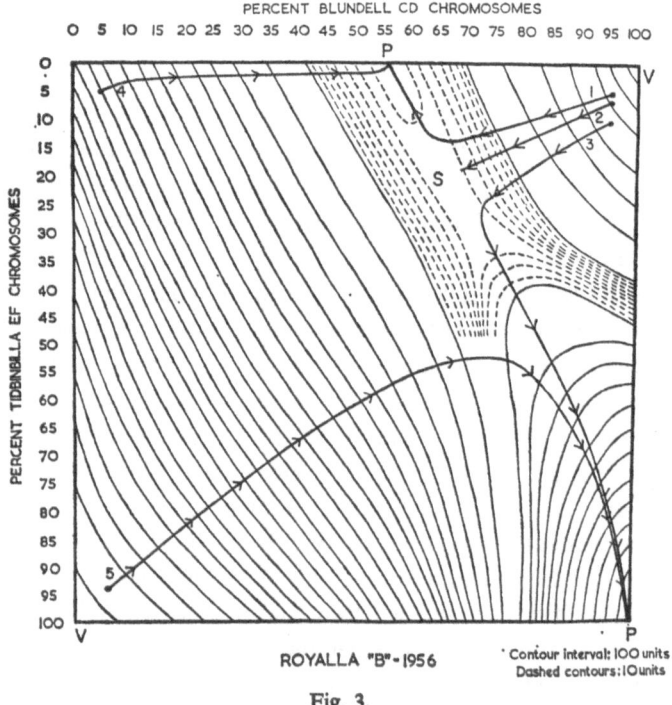

Fig. 3.

potential field can be drawn, the point or points of maximum $\bar{W}$ can be seen by inspection and compared with the known composition of the population. Figure 3 is an example of such a map. The two axes represent the frequencies x and y of the genes A and B. The lines in the field connect points of equal $\bar{W}$ and it can be seen that there are two maxima marked by the letters P and two minima marked by V. In addition there is a minimax point or saddle point indicated by S. The numbered trajectories represent the paths of gene frequency change which the

289

population would follow if it were assumed that the initial composition of the population were at points 1, 2, 3, 4 or 5. The expectation is that the gene frequency composition of the population should correspond to one of the two points *P*. In fact the population is at point *S* and every population of *Moraba* investigated shows a similar map with a similar discrepancy between expectation and observation.

Another way to look at Figure 3 is to imagine it as a topographic map with two peaks at *P*, two valleys at *V* and a col or saddle at *S*. The population can then be analogized with a mountain climber ineluctably drawn upwards by that mysterious impetus common to a Mallory and a Hillary. The rules of the genetic model specify that the climber must go always upward, but they also specify that he be myopic. Thus, he will climb the peak nearest him without being able to see the more distant, albeit higher, peak.

Several points about this model are pertinent to our general discussion of model and metaphor. The two metaphors, the physical particle in the potential field and the mountain climber in the adaptive landscape are introduced *after* the equations of gene frequency change have been derived, and are *suggested by* the *equation*. In fact the physical particle has certain attributes not shared by the gene frequency composition. An important one is that physical particles have momentum and reach a stable equilibrium by an oscillating path, overshooting the equilibrium point on either side in a damped oscillation around the equilibrium point. The genetic rules do not predict such an oscillation and it is not observed in experiments. The only advantage of the metaphor, or of the mountain climber analogy, is a didactic one. It enables us to picture, indeed to draw on paper, the evolutionary situation and to "understand" the evolution of the population in a way not possible from the equation alone. In addition to this didactic function what has the model contributed? It has shown that the real composition of the populations does *not* correspond to expectation. Several explanations are possible, but their exact content is not relevant. What is relevant is that each explanation involves the search for a phenomenon not considered in the original explanation. A direction has been given to future experimental research on these populations and an impetus provided for further study. Without the model it would have been impossible to say whether the populations were, in fact, behaving according to expectation.

290

A STOCHASTIC MODEL

In the previous model, the rules were formed in terms of algebraic equations relating successive states of the population. It often happens, however, that because of important random events, the rules cannot be so formed but must be regarded as *probabilities* of transition from one state to another.

Certainly real populations of plants and animals are not infinitely large and the number of effective parents of each generation may be much smaller than the total population, as in social bees. Moreover, environment is not usually constant but undergoes a certain amount of random fluctuation. A more realistic group of models is then one in which the role of chance in determining which individuals will leave what sort of offspring is taken into account. These are the *stochastic* models.

The structure of stochastic models is essentially the same as shown in the diagram for deterministic models but the form of the rules is different. Each rule in a stochastic model gives rise to a set of probabilities that particular events will occur. As an illustration we may take segregation in a heterozygote. Mendel's law leads to the expectation that half the offspring of the mating $Aa \times aa$ will be Aa and half will be aa. In a deterministic model this proportion is exactly realized. If only 5 offspring are produced from such a mating, however, it may be that the ratio of Aa to aa is 5:0, 4:1, 3:2, etc. To each of these events a probability can be assigned and from any particular mating any one of these ratios might be realized. Since each step in a genetic process depends upon the outcome of the previous step, each of the possible offspring ratios will have a different effect on the next generation. In a deterministic model the history of the population can be represented by a single line, but in the stochastic case, it will appear as a branching scheme with a multiplicity of paths to each of which can be assigned a probability. Thus, stochastic models do not allow of the exact prediction of the outcome of evolutionary processes. Even if all the parameters *and the initial state* of the population are given, the result is not sure. What can be done, however is to form a probability distribution of results. The kind of statement that can be made is of the form, "The probability that the population will have a gene frequency of A equal to 0.725 in the n th generation is one quarter". Another way to state this is that if 1 000 000 populations all began with

291

the same gene frequency, 250 000 of them would have the gene frequency 0.725 in the nth generation. One way in which the probability, one quarter, can be arrived at is purely mathematical, using the theory of

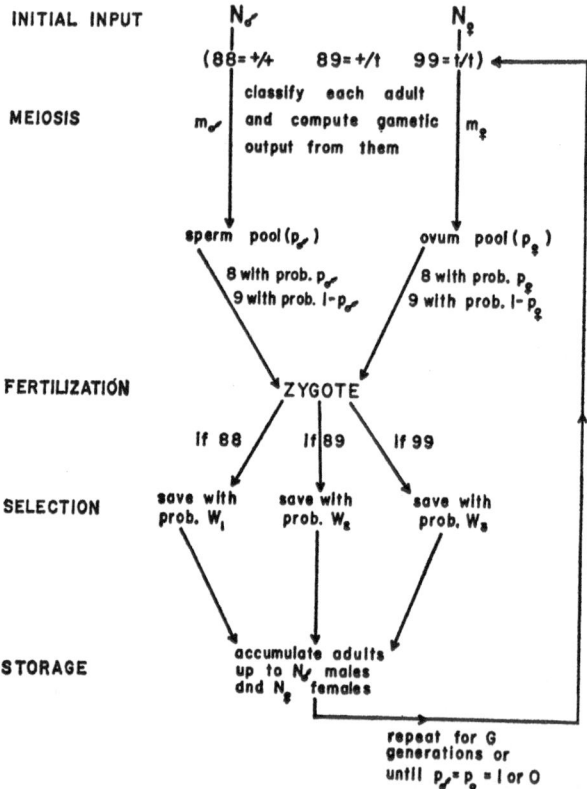

Fig. 4.

probability distributions. Another is to make a mechanical analogue of a population with the rules and random processes built into it and then to allow this pseudo-population to run for n generations over and over again.

An example of such an analogue is shown in Figure 4, a model constructed to explain the polymorphism in certain populations of *Mus musculus*.[7] The figure looks very much like Figure 2 and is, as a matter of fact, the general diagram of a program for a digital computer based on Figure 2. The numbers 8 and 9 are symbols for genes and are operated on *logically, not numerically* by the computer. The various quantity variables, m, N, p, W etc. are treated as numbers to be interpreted arithmetically and the rules represented by the arrows are either logical or arithmetic operations analogous to the rules of Figure 2. What has been constructed is a truly numerical and symbolic model of a genetic process, devoid of any metaphorical content. It is a model which has moved toward formalism rather than away from it. In no way can the numbers or the vacuum tubes of the machine be thought of as metaphors for genes and Mendelian laws. They have no didactic content, the model having been reduced to the barest logical bones.

The biological problem posed is the following. In natural populations of *Mus* there is a high frequency of a mutant gent t, the frequency being different in different populations. Homozygous t individuals invariably die before birth, but heterozygous males produce an excess of t bearing sperm. There is then a balance between the loss of t through death of homozygotes and the gain of t through sperm formation in heterozygous males. Although the t allele differs in different populations in the degree of meiotic abnormality, all are favored in meiosis to a high degree (between 90 % and 99 % of sperm from heterozygotes bear the t allele). This is in contrast to the known behavior of newly arisen t mutants which may have a much smaller meiotic advantage. The problem is to explain the observed frequency of t alleles in natural populations. A deterministic model of this situation predicts a *higher* frequency of t alleles than is actually observed and it does not explain in any way the absence in natural populations of alleles with a more moderate meiotic advantage.

As it turns out, both the observed degree of polymorphism and the concentration of t alleles with high meiotic advantage are predicted by the stochastic model of Figure 4. By assuming that the numbers of male and female parents are small each generation, we can show by means of the computer model that there will be eventual *loss* of the t alleles rather than a stable equilibrium and second that the rate of loss of t alleles is quite sensitive to the degree of meiotic advantage of a given allele. We than expect to find in nature only those alleles still remaining in populations which have an extreme advantage and these should be in frequencies less than that predicted in a deterministic model. Thus, the second function of models is exemplified, the possibility of making apparently inexplicable observations flow from a theoretical structure. That is, a phenomenon has been "predicted" and then found in nature (although the historical process in this case was reversed) quite unexpectedly and without a deliberate attempt to build it into the model. New knowledge has certainly been gained, and this without recourse to any metaphor.

INDETERMINATE MODELS

While events cannot be predicted exactly in a stochastic model, probabilities of different outcomes can be specified. For some phenomena even this is not possible. Probably much of environmental variation is of a non-recurring sort, or of such periodicity that probabilities cannot be assigned. What is the probability of a glacial epoch? How often will a series of dry days cause an ephemeral pond to dry up? What is the chance than an epidemic will occur? One may say that organisms play a game against a capricious nature, a game in which they are certain to become bankrupt eventually, but in which some strategies will enable the population to hold out for a longer time. At present there are no models which are capable of dealing with the problem of population and species survival in a completely indeterminate universe. One way of attacking the problem may be through the theory of games and decisions. A game theory model of evolution would have the following form

	N_1	N_2	N_3	....	N_k
S_1	P_{11}	P_{12}	P_{13}	....	P_{1k}
S_2	P_{21}				
S_3	P_{31}				
$\vdots$	$\vdots$				
S_r	P_{3r}				

The N's represent the possible *states of nature* (hot, cold, wet, dry, etc.) The S's stand for *strategies* manifested by the population or species. These strategies are in the form of general rules of genetic and phenotypic behavior. Degree of heterozygosity, amount of inbreeding, recombination index, mutation rate all are strategic problems. To each combination of strategy and state of nature corresponds a value of P, the *pay-off* or *outcome*. These may be measured as rate of increase of the population,

294

mean adaptive value, replacement ratio, among others. Some strategies will result in uniformly poor outcomes in all states of nature. Others, very few, may be uniformly good, but the vast majority will be advantageous in some states and disadvantageous in others. Simply, by a process of historic elimination, only those populations will survive and spread whose strategies are in some sense optimal over the set of states of nature.

Models of this sort are only beginning to be constructed and whether they will prove to be useful depends primarily on how the values of P are determined and how optimal strategies are defined. This in turn is contingent on the existence of a rigorous set of rules connecting strategies with outcomes, and another set connecting outcomes with observable and measurable evolutionary change. These two requirements are not easy to fulfil. For example one pay-off measure might be the probability that a population will survive a given number of generations. This fulfils the second requirement, since the survival of a population is directly observable, but it is difficult to make rigorous rules connecting, say, mutation rate with probability of survival. Alternatively, the mean adaptive value of the population might be the pay-off. This is simple to compute given the strategy and the state of nature, but what information of evolutionary interest does it convey. It must be emphasized that at the moment the mathematical problems of this sort of model are subordinate to a proper identification of its components with evolutionary phenomena.

It is in game theoretical models of evolutionary processes that the dangers inherent in metaphor become important considerations. Game theory with its notions of "pay-off", "utility", "strategy" smacks strongly of purposive thinking. It is not only the words and their connotations which are a danger in this direction. Game theory has grown out of research in economics and psychology and a number of the concepts are *defined* purposively. If a game theoretical attack on evolutionary problems is to be fruitful it will be necessary to strip the metaphor of its aspects of will and purpose, to discard what I have called the "exogenous qualities" of the metaphorical elements.

Above all, such a model does not differ from deterministic or stochastic models in what it attempts, but only in the entities with which it is dealing (populations and species), and in the rules of operation. It cannot disclose what strategies a population *ought* to adopt (a meaningless notion in a mechanistic universe) nor what strategies a population *will* adopt, nor even what strategies *have been* adopted. It is, like other models, a mechanism for determining the evolutionary outcome of a given set of natural phenomena.

University of Rochester, Rochester, New York, U.S.A.

REFERENCES

1. In *The Logic of Scientific Discovery* K. R. Popper uses the concepts of "events" and "occurrences" in this same way. They are obviously related to Russell's "sensibilia" and "constructions" but I wish to avoid, as far as possible, the issue of phenomenalism versus realism.
2. A. C. Benjamin, *The Logical Structure of Science*, 1936, Chapter XI.
3. Dudley Shapere, in an unpublished paper, has suggested that even the "old fashioned" atom implied experientially contradictory qualities and that this is an ineluctable aspect of scientific explanation.
4. A. Rosenblueth and N. Wiener, *Philosophy of Science* 12 (1945).
5. A. C. Benjamin, *The Logical Structure of Science*, 1936, Chapter XI; P. Franck, *Philosophy of Science* 4 (1937); H. Meyer, *Philosophy of Science* 18 (1951).
7. R. C. Lewontin and M. J. D. White in *Evolution* 14 (1960).
8. R. Lewontin and L. C. Dunn, *Genetics* 45 (1960).

OLAF HELMER

THE GAME-THEORETICAL APPROACH TO ORGANIZATION THEORY

Organization theory and game theory both are concerned with the interactions of the decisions of a group of people acting under given constraints. It herefore is reasonable to expect the existence of an approach to organization theory – or at least to some sector of organization theory – via the theory of games. As it happens, it appears that the application of the game-theoretical apparatus to the study of organization promises to be of substantial benefit to both parties concerned, – to game theory as well as to organization theory. One can go even further, it seems to me, and in fact assert (a) that both game theory and organization theory are in real trouble today, (b) that organization theory can be viewed as a very natural extension of game theory as far as applications are concerned, and (c) that by giving proper recognition to this intimate relationship between the two fields they are both likely to overcome their present difficulties.

These difficulties are of a very different nature in the two cases. The trouble with organization theory, briefly and bluntly, is its non-existence. There has been a lot of talk about organization theory in the last decade as the up-and-coming thing, and there have been numerous sporadic studies in this general area. But there has, as far as I can see, been no serious and successful attempt to build up an adequate conceptual framework within which to construct a unified theory of organizations. Even the basic concept of an organization, and hence of the subject matter of this supposed theory, is extremely vague. The notions which people working in this general area have with regard to this concept seem to be quite uncertain and frequently at variance with one another. For this reason alone it would be most fortuitous if game theory had to offer a ready-made conceptual framework which might be exploited and expanded appropriately to establish a satisfactory conceptual basis for a theory of organizations.

The trouble with game theory is of a very different kind, since its concepts are exact enough to permit axiomatization. Despite considerable advances in the mathematical theory, especially regarding two-person zero-sum games, the theory of games in my opinion has reached a state of near stagnation with regard to its applicability to the real world. This is true not only of the non-zero-sum and n-person parts of the theory but applies equally to the zero-sum two-person case.

I am not referring to the failure, to date, to establish a satisfactory theory of differential games, which would open up a whole new realm of applications, especially in the military field, but rather to the inadequacy of the basic concepts, precise though they are, for grasping the realities of conflict situations among people.

In order to establish a framework within which to discuss these matters, let me recall the epistemological apparatus for applying a mathematical theory to the real world. A mathematical theory of necessity takes as its starting point some basic assumptions, which possibly may be a set of

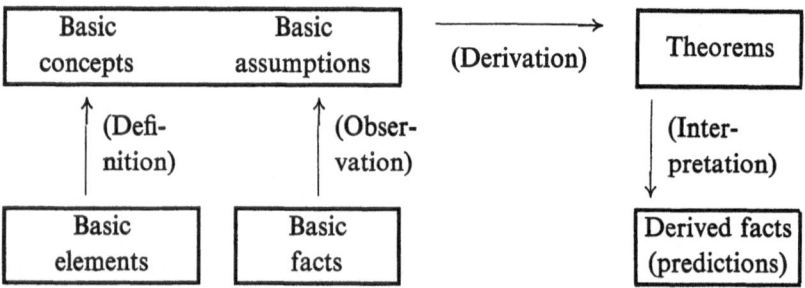

formal axioms and which are formulated in terms of certain basic concepts. From this basis the theory will derive certain theorems. The application to reality takes place by establishing what are sometimes called correlative definitions, that is by interpreting the basic concepts as standing for certain basic elements in the real world. This correlation associates certain basic facts with the basic assumptions, and if the basic facts have been established by observation, we may predict – as derived facts – the statements about the real world corresponding to the theorems of the theory.

This of course is trivial. Applied to two-person zero-sum games, the theory usually assumes merely that the players have a utility preference scale which is linear in the payoff amounts and in the probabilities involved in probability combinations of outcomes.

Let us note in passing in what sense we may speak of predictions in game theory: Usually the normative formulation is preferred, namely that, if a zero-sum game is played by two persons each having a linear utility scale with respect to the outcomes (let us call them briefly "linear players") then each ought to play an optimal strategy as prescribed by game theory. But this is equivalent to predicting that each will on the average be most satisfied with the outcome if he plays an optimal strategy. A slightly different twist can be given to the predictive element by the following formulation: A linear player, if he is rational and if he assumes his opponent to be rational, will select an optimal strategy. By thus formally adding the player's rationality and his appraisal of the opponent's rationality to the basic assumptions, we can now shift from a prediction of the player's relative satisfaction with the outcome to a prediction of the player's behavior during the play of the game. This formulation will greatly facilitate the transition to predictions in organization theory.

Now, to defend my statement about the conceptual inadequacy of game theory in point of applicability, let us remember that the principal fields of application to date are economic, where the payoff is monetary, and military, where the payoff is in terms of some combat advantage. In the economic case, the assumption of linearity, as is well-known, is usually absurd, especially if the amounts involved are sizable compared to the player's assets; and by the time we go over from the ostensible monetary payoffs to linear utilities, the game almost invariably turns out to be non-zero-sum. As for the military applications, the situation is even worse: the payoff there is usually in terms of some vaguely conceived military advantage, and the mathematical model has to introduce some completely fictitious measurable utility in order to make the situation amenable to game-theoretical treatment. Take for instance the well-known Blotto-type game, where two opposing commanders commit their battalions to fight over the possession of several positions of military advantage. The customary payoff is the difference in the number of surviving battalions plus so many credit points for each position held. Even if such a scoring system were to reflect reasonably accurately the military utility of specific outcomes to one side, there is little reason to assume either that this utility is linear with respect to probability combinations of such outcomes or that the military utility to the other side should be assessed by the same formula.

Thus in order to do justice to the areas of application for which game theory was supposedly developed, it is necessary, firstly, to create an applicable theory of non-zero-sum games, and secondly, to go rather thoroughly into an analysis of the sensitivity with which optimal strategies depend on the choice of a particular payoff function when no such specific function is unambiguously implied by the real situation.

The next thing to look at then is the status, with regard to applicability, of the non-zero-sum theory. Here, aside from rather inconclusive attempts to deal with special negotiatory situations, only the two extreme basic assumptions have been considered in great detail, namely of complete cooperation and of complete non-cooperation, in each case coupled again with the assumption of linear utilities. The former, of course, is the original von Neumann-Morgenstern theory, the latter the equilibrium point theory.

Among the two principal areas of application, the economic and the military, the von Neumann-Morgenstern theory of course may at best be applicable to the economic one, since the assumption of complete cooperation among military opponents is clearly absurd. In the economic case, we no longer are confronted with the difficulty of non-linear utilities as in the zero-sum case, because – since the game is non-zero-sum anyway – we may as well assume that the ostensible monetary payoffs have for each player been replaced by linear utilities. But here a new difficulty arises: The linear utilities associated by the players with the monetary outcomes are no longer commensurable and transferable, and therefore the prescription, which the theory offers, of maximizing the expected return to the coalition of the players becomes meaningless.

The equilibrium point theory, on the other hand, is not quite so badly off. It requires no commensurability of utilities and is at least in special cases able to predict the behavior of rational players. It does presuppose complete non-cooperation, however, and for that reason is rarely applicable directly to a real-life situation. Let us consider two trivial examples. The game in which the players have the respective payoff matrices

$$\begin{pmatrix} 1 & 11 \\ 0 & 10 \end{pmatrix} \text{ and } \begin{pmatrix} 1 & 0 \\ 5 & 10 \end{pmatrix}$$

(the first player choosing between rows, the second between columns) has a unique equilibrium point* with payoffs of one to each player (first row, first column). Yet the mere agreement among the players to let the first player go first will cause both to switch to their second strategies and give each a tenfold return. Notice that in this case it is merely a question of whether the constraints of the situation permit transition to this slightly different game; if they do, a pair of rational players will automatically accept it. Next look at the game obtained from the previous example by changing '5' to '11':

$$\begin{pmatrix} 1 & 11 \\ 0 & 10 \end{pmatrix} \begin{pmatrix} 1 & 0 \\ 11 & 10 \end{pmatrix}$$

The equilibrium point is the same as before, but the corresponding strategies are now strongly dominant** for both players; therefore under strict non-cooperation the theory rightly recommends that the equilibrium point be played. Again, a small amount of cooperation would cause each player to switch to his second strategy, giving each a payoff of 10. But in this case more than just a transition to a slightly different game, acceptable to any rational player, is involved. The cooperation here presupposes a relationship between the players extraneous to the immediate game situation. This relationship may be one of several: They may be playing within a social environment having the institution of enforceable contracts and avail themselves of this opportunity; or there may be an attitude of mutual trust which would warrant the entering of a purely informal agreement; or, more specifically, there may be an experience pattern derived from a series of plays of the same or similar games aginst the same opponent, which will lead each to expect the other to play his second strategy.

The point which I am hoping to make with these trivial examples is that the equilibrium point theory often suggests what appears to be an unreasonable solution, in the sense that there may be a different outcome which is preferred by all players, and that this preferred solution can often be achieved by a rather minute and obvious amount of cooperation. What happens when such cooperation takes place may of course formally be

* An equilibrium point is a pair of strategies such that neither player would wish to change his stragey unless the other did.
** That is, each is preferred by its player, no matter what he knows of the opponent's choice of strategy.

interpreted as the embedding of the given game in a larger game, which can then be regarded as a strictly non-cooperative game and for which the equilibrium point solution is reasonable, in the sense that no other outcome is preferred by all players.

But this shows clearly the limits to which game theory as such can be expected to take us. Given rational players with known utility preferences, it can – at least in many cases – predict the players' behavior in a given non-cooperative game. Moreover, given a game with certain cooperation options, it may tell us if that game can be embedded in a larger, non-cooperative, game having a reasonable equilibrium point solution (in the earlier sense). But what it cannot generally do is to predict the behavior, even of rational players with known utility preferences, with regard to their cooperative options, insofar as these depend on the players' attitudes toward their fellow players and toward any behavior patterns which they may have observed in past plays. At this precise point, therefore, in order to be in a position to predict decision-making behavior in cooperative situations of this kind, it becomes mandatory to go beyond the limits of game theory proper. An it is in this context, I think, that organization theory appears as the natural extension of game theory.

Like game theory, organization theory is somehow concerned with an interacting group of people in a decision-making situation. But in two decisive respects it has to allow for greater sophistication than game theory: firstly, it must assume that the members of the group, in addition to certain utility preferences, have other recognizable characteristics of a psychological nature (such as attitudes toward fellow members and toward risk, and possibly learning ability); and secondly, instead of dealing with just one fixed competitive situation with well-determined outcome possibilities, it must be capable of handling a continual recurrence, a flux, of situations calling upon the participants to make decisions in the light of their preferences, capabilities, and attitudes.

Applied to the simplest possible case of a two-person game, game theory should be able to predict, on the basis of the rationality and the known utility preferences of the players, how the game will be played once non-cooperatively. But if we wish to predict what would happen if this game is played repeatedly under certain cooperation options, we have to turn to organization theory, – which however will require additional assumptions on the players' attitudes.

302

What I am saying does not amount to a definition of the subject matter of organization theory. I am merely trying to state, from the game-theoretical point of view, some minimum conditions as to what real-life situations a theory of organizations should be capable of handling.

If this proposal, of establishing organization theory as a natural extension of game theory, should find acceptance then this entails a whole program of research in this joint area of interest, which is the study of human interactions in decision-making situations under given constraints.

Let me then take a brief look at the prospects of such a joint enterprise, – a theory of games directed somewhat more deliberately toward applicability, and a theory of organizations built on a game-theoretical foundation.

Conceptually, game theory would profit, because by relegating motivations other than utility preferences to organization theory, it would confine game theory to that area in which it can reasonably be expected to show results, and remove the uneasiness as to why game theory seems to be unable to cope with cooperative games (in the sense of suggesting, or predicting, rational behavior). As for organization theory, the advantage of being able to latch on to the existing vocabulary of game theory, as I pointed out before, would be considerable. But it also is clear that an immediate effort would have to be made to introduce formally the additionally needed, typically organizational, concepts having to do with the players' attitudes of various kinds.

As for specific areas of research suggested by this approach, let me list a few, without the slightest attempt at completeness:

One. An analysis of the sensitivity with which optimal strategies depend on slight changes in the payoff function, both for zero-sum and non-cooperative non-zero-sum games.

Two. An examination, both analytical and experimental, of ostensibly zero-sum games, that is, of games with ostensible zero-sum monetary payoffs, which in fact are non-zero-sum when realistic utility preferences are introduced. For example, the game

$$\begin{pmatrix} \$ & 9 & \$\,0 \\ \$ & -90 & \$\,1 \end{pmatrix}$$

has a theoretical "value" of 9 cents to the row player. If I had to play this game as the row player, I would gladly settle for 0 by always playing the

303

first strategy. In general it would be interesting to find out under what circumstances the associated non-zero-sum game has a reasonable equilibrium point solution, and how in fact people play such games.

Three. A complete theory of negotiation games under the assumption of the availability of enforceable contracts.

Four. A systematic examination of the following type of embeddability problem. Given a pair of payoff matrices for a non-zero-sum game, assume that the equilibrium point theory has no reasonable solution to offer for non-cooperative play. In that case, by what cooperative devices (such as passage of information, contractual agreements) could the given game be embedded in a larger game which both players would like to play at least as well as the given game and which does have a reasonable equilibrium point solution?

Five. The definition, and subsequent analysis, of what for lack of a better name I will call games with time-series payoffs. What I have in mind, in the simplest case, is a two-person interaction involving a time sequence of moves, where a payoff occurs after each move. This clearly is a borderline case between game and organization theories. A situation as described involves planning over time, and decisions may well depend on attitudes other than just the direct utility preferences with regard to the ostensible payoff.

Six. A consolidation of the various efforts which have been made at reducing non-linear utility complexes to a single linear utility, with special emphasis perhaps on non-monetary rewards.

Seven. The isolation of suitable attitudinal predicates, in terms of which people can be characterized and their actions in competitive decision-making be predicted. A first step in this direction might possibly be taken by sujecting subjects to a series of standardized non-zero-sum games and classifying them according to their choices of strategy.

Eight. An analysis of the mutual-influence structure within a group as a function of the members' attitudes.

Nine. An empirical study of the performance of small organized groups in playing games. Given n people (where n at first may well be 2 or 3) with certain utility preferences and attitudes, with a given communications structure and a preestablished mutual-influence structure, how well will they each perform when the group is subjected to a suitable series of standardized test games?

These few examples of what seems to be in need of being done will suffice, and I would like to finish with a plea regarding methodology.

Several of the examples which I just gave call for what is primarily psychological research. Yet it would be disastrous if the mathematicians used this to rationalize themselves into doing nothing about it. Here if ever seems to me to be a clearcut and perhaps even crucial case calling for the much-advertized application of mathematical methods to the social sciences. Paradoxically, psychology can possibly receive the greatest help by being steered away from its traditional method of carefully controlled series of laboratory experiments and toward what amounts to more of an engineering approach to its problems. That means the construction of possibly a multitude of tentative mathematical models, the evaluation of which need not be through detailed experimentation, since we are at this stage interested not so much in minute exactness but in gross workable ideas. There are in fact two devices which we can use to trade unneeded over-exactness for badly needed speed. One is a cautious use of expert judgment wherever it is available, the other an exploitation of high-speed computing facilities, which can be used in Monte-Carlo fashion to examine the implications of mathematical models for a multitude of values of their input parameters. There is some real hope, I think, that a concerted effort along these lines may help to produce within the foreseeable future the foundations of a highly applicable theory of organizations.

The Rand Corporation, Santa Monica, California, U.S.A.

B. DUNHAM/D. FRIDSHAL/R. FRIDSHAL/J. H. NORTH

DESIGN BY NATURAL SELECTION

There is an old philosophical argument based upon the assumption that a good design presupposes an intelligent agency. If we encounter efficient machines, we must assume an intelligent planner. In the nineteenth century, biologists discounted this assumption. Suppose, they argued, structural changes do occur in living organisms by chance only. Nevertheless, the changes which register will be mostly in a given direction, since those less efficiently designed will not survive. In this way, highly efficient and extraordinarily complicated designs will "evolve" without any intelligent planning simply because (1) the capacity to survive is a "scoring" mechanism and (2) in a long interval of time there will be many generations.

In building equipment with extreme requirements as to size, weight, or efficiency, we must often resort to automatic methods of optimization. Unfortunately, many design problems are difficult to formulate in workable mathematical terms. Even though, as is sometimes the case, a satisfactory answer is easily recognized, we cannot always by "intelligent agency" compute such an answer. Consequently, there is a definite need for alternative methods of design; and something like the biological process of "natural selection" is worth considering.

The modern computer has both speed and precision. If one design followed from a slight change in its predecessor, a computer could scrutinize many generations. Further, a way of "scoring" competing candidates is available in most design situations. The very requirement of "efficiency" implies a basis for telling one design from another. With no knowledge of internal combustion engines, we can still measure which car gives more miles per gallon, and so forth. Hence, it would appear both possible and profitable to design by "natural selection" using a computer. Certainly, the everyday world of complicated living objects around us is proof that such a method does work.

As a matter of fact, we have already used the technique successfully a number of times, and it may be pertinent to cite one or two case histories.

The first such occasion was in connection with a somewhat abstract investigation to discover logically efficient building blocks for computers.[1] Suppose we regard a computer as composed of many input-output devices hooked together in various sequences. It is well known these devices can often be correlated with truth-functional particles. For example, a two-input, one-output element which registers ON when both inputs are ON, but otherwise OFF, functions much the way the connective AND operates. Hence, it is often possible to reduce an engineering problem to a logical one.

Suppose we found, for two otherwise comparable devices, that fewer units of A than of B were sufficient for assembling desired hookups. We would judge A "logically" more efficient, but would still lack a straight-forward test of that supremacy. It turns out, however, that the efficiency of an element can be equated with its "versatility", and the latter is easily measured. Let us see how this is done.

By certain external adjustments, a logical element can be made to operate in a variety of ways. For example, inputs can be "biased" (so fixed as to be always ON or OFF) or "duplicated" (so tied together as to receive identical signals). Consider the element defined by the following table:

ROW	X_1 X_2 X_3	OUTPUT
1	1 1 1	1
2	0 1 1	1
3	1 0 1	0
4	0 0 1	1
5	1 1 0	1
6	0 1 0	0
7	1 0 0	0
8	0 0 0	0

The numbers ONE and ZERO symbolize the ON and OFF states, respectively. X_1, X_2, and X_3 represent the inputs; and the eight rows of ONEs and ZEROs under them, the possible input conditions. Suppose X_3 were biased OFF. Then, as the bottom half of the table shows, the device would simply AND together X_1 and X_2. On the other hand, suppose X_1 and X_2 were tied together to receive the same signal. Rows

1, 4, 5, and 8 reveal the device would OR that signal to the input at X_3, and so on.

Because different adjustments sometimes produce equivalent outputs, one device can be made to operate in a greater variety of logical ways than another. Because, in turn, this added versatility will result in greater efficiency, we can "score" a logical element simply by counting the number of effectively different operations it can perform.

It is easy to identify the more efficient four-input, one-output elements. We need only score all possible cases and pick out those with higher scores. The five-variable problem is not so simple, however. The total domain of individuals is in the billions. At the time we studied the matter, we were not yet wise enough to construct examples with high scores, even though we could evaluate easily given ones. Hence, the problem in question was well suited to the "natural selection" approach.

The actual case history can be stated briefly. First, the class of five-input elements was correlated with the binary numbers having 32 digits or "bits". For example, the number

0 0 0 1 1 0 0 1 1 0 1 0 0 0 1 0 1 1 1 0 1 0 1 1 0 1 1 0 0 1 0 1

represents the output of an element where the 32 possible input conditions are ordered in a standard way. The computer then operated as follows. A "random" 32-bit number was generated and scored. A second number was obtained from the first by changing one bit. If the second had a higher score, it was retained and the first discarded. Otherwise, the original number was reinstated and a different bit changed. This process continued, giving numbers with higher and higher scores until the computer became "stuck", that is, no one of the 32 possible changes produced a number with higher score. At this point, we might either (1) record the result and make another "random" start, or (2) try for a brief period to become "unstuck" by accepting small changes not producing a higher score. Eventually, we could start over. As it turned out, the latter approach was preferable; and the over-all method, most successful. The given run on the 704 computer produced in a few minutes elements with far higher scores than had been obtained by other means.

Another occasion when "natural selection" proved helpful was in the sizewise reduction of a 1400-terminal "black box". In outline, we were presented with the following problem, which had already frustrated much effort at solution by "intelligent agency". A black box was to be built sandwich style in four layers. Each layer, or board, would contain ten columns of seven "slugs" each. A slug was a four-input, one-output device. Information was to be carried by lines printed vertically on one side of the board and horizontally on the other. The board could be pierced at any point, but all lines must be separated by at least 0.03 inches. The inputs and outputs for all 70 slugs were given, different slugs having various inputs in common. Now, between two adjacent columns of slugs, there would always be room for ten printed lines. Should more than ten be required, however, the board must be made larger; and therein lay the difficulty. The over-all size of the box was critical, and no extra lines could be permitted. Hence, a way of positioning and wiring the slugs on the board must be found with all the connections made and no more than ten lines between columns.

The approach we suggested was the following. First, a systematic method should be worked out for wiring any given arrangement of slugs with reasonable economy. Since the size of the board is a function of the wiring, this would enable every arrangement of slugs to receive an explicit "score". The method in question need not be especially subtle nor sophisticated, but only fast. We did not regard it as important that the automatic wirings be near optimal, since the difficult problem was to find a workable arrangement of slugs, not to wire it once found. It seemed better to run through many "generations" with only approximate scores indicating progress than to manage a very few "evolutions" with rather exact statements of position. The computer would then operate as follows. A "random" arrangement of the 70 slugs on the board would be wired and scored. A second arrangement would be obtained from the first by interchanging two slugs and scored. The "selection" process would continue, as in the problem above, until the machine became "stuck". At this point, we thought it best that a strong effort be made to "unstick" the machine before starting over. As before, a number of changes not producing higher scores could be accepted.

The results of the actual computer run were again most successful. We were notified subsequently that, in 80 minutes on the 704, usable designs were produced for all four boards. The black box was manufactured from these.

Now, of course, trial-and-error problem solving is at least as old as man himself, and we are not making an especially original observation in saying it can be used for design purposes. The matter is one of emphasis. We feel the "natural selection" approach is more powerful than commonly recognized, and we hope to see it more widely exploited. It should be noted, however, that the succes of its application is in many cases a function of the skill with which it is used. Let us consider some of the points upon which success of failure may hang.

First of all, the scoring mechanism should be as fast as possible. The essence of the method is that many "evolutions" will take place from a variety of starts. Suppose a complicated problem is encountered. It may be best to use two scorers, one fast and the other slow. The former might well be sufficient for general progress in the right direction; the latter could be called upon in the closing stages when refinement matters.

Secondly, an effective "traffic-jam" mechanism is required. Generally speaking, if we are transported from one point to another by some conveyance based upon a scoring of individual positions, we will often become "stuck" upon a relative peak.[2] In the two case histories cited, our approach was to suspend the scoring mechanism temporarily, while permitting slight changes. In this way, without losing the total progress to date, we edged (so to speak) into another line of traffic. The best way of doing this varied slightly in the two cases. Suppose, however, a second way of scoring were at hand with as much over-all validity as the first, but with somewhat different relative peaks. When the first mode of transportation proved ineffective, we could switch to the second, and vice versa, thus very much reducing the incidence of traffic jams. In point of fact, our experience has been that such second scorers are frequently available.

Thirdly, restraints must often be introduced. It sometimes happens that small changes in certain critical conditions will have substantial effects. Such conditions may well have a best setting, which should be "guaranteed" throughout the entire selection process. Because of their pronounced influence, such best settings can many times be predicted either by judicious statistical survey or truant reflection. In general, trial computer runs are most helpful.

IBM Research Laboratory, Poughkeepsie, New York;
New York University, New York

REFERENCES

1. For additional details of this investigation, see B. Dunham and J. H. North, 'The Use of Multipurpose Logical Devices', *Proceedings of an International Symposium on the Theory of Switching,* 192–200 Harvard University Press, 1959, Part II, and 'The Multipurpose Bias Device, Part II', *IBM Journal of Research and Development* **3**, (1959) 46–53.
2. For a related discussion of this problem, see R. M. Friedberg, B. Dunham, and J. H. North, "A Learning Machine, Part II', *IBM Journal* **3** (1959) 282–287.

PART IV

ANALYTIC BIOLOGY

MARIAN PRZEŁĘCKI

ON THE CONCEPT OF GENOTYPE

Woodger's book: "Biology and Language" constitutes a great achievement in his attempts at constructing a precise and adequate language for genetics. In the conceptual structure outlined there the concept of genotype occupies an important place and most of the remaining concepts are introduced with its help. The definition of genotype as proposed by Woodger is interesting not only for geneticists. It deserves special attention of all concerned with problems of general methodology. Genotype is usually considered a "theoretical concept" appearing at the highest level of genetical theory. It is said to denote the genetic constitution of an organism, specified by reference to genes – entities which are not accessible to direct observation. Woodger treats genotype as an "elementary concept" belonging to the lowest level of genetical system. His definition of genotype is couched in terms of directly observable things, properties and relations. It does not make use of the much disputed concept of gene. On the contrary, the latter is defined with the help of the former. Woodger's concept of genotype – a theoretical concept defined explicitly on the observational level – calls for a careful examination, which may throw light on some difficult problems connected with the relation between theory and observation. In what follows I am not trying to give a full analysis of Woodger's definition. I should like rather to point out one peculiar feature which it seems to share with all definitions of the same kind, i.e., definitions of theoretical concepts in terms of observable entities.

The lowest level of genetical theory, called in Woodger's terminology the zero level, consists of generalized observation records about parents and offspring and the environments in which they develop. Woodger does not separate these concepts. He uses a single primitive term which is explained in the following semantical rule:

'$F_{X,Y,Z}(W_1, W_2)$' denotes the set of all offspring x such that for some u and some v, $u \in W_1$ and is a parent of x, $v \in W_2$ and is a parent of x, and u has developed in an environment belonging to the set X, v in an environment belonging to the set Y and x in an environment belonging to Z.

Let us put this explanation into a symbolic form:

$$F_{x,y,z}(W_1, W_2) = (\hat{x}) \left[(\exists u)(\exists v)(u \neq v \,.\, u \in W_1 \,.\, u \, Ps \, x \,. \right.$$
$$\left. .\, v \in W_2 \,.\, v \, Ps \, x \,.\, en(u) \in X \,.\, en(v) \in Y \,.\, en(x) \in Z) \right],$$

where 'Ps' denotes the relation of sexual parenthood[1] and '$en(x)$' – the environment of x [2], both interpreted as expressions belonging to zero level. If we chose these terms as our primitive signs, the explanation of 'F' might serve as its definition. But, defined or undefined, it is also an expression at zero level referring to observable things and relations. In most of its uses the three environmental sets: X, Y, Z are not distinct. Therefore the following simpler form of this expression is introduced:

$$F_x(W_1, W_2) = F_{x,x,x}(W_1, W_2).$$

'F' may be said to denote the first filial generation. A notation for the second generation is easily provided:

$$F_x^2(W_1, W_2) = F_x(F_x(W_1, W_2), F_x(W_1, W_2)).$$

What is needed now, is the classification of environments, parents and offspring. Environmental sets are specified by reference to the composition of the soil, water or air, the amount and composition of food, etc. Two kinds of classification of parents and offspring are involved in genetics. First we may specify sets of parents or offspring by means of observation records which state to which species they belong and how they are distinguished from other members of this species. For example, the matrices:

x belongs to Pisum sativum and x has yellow cotyledons,
x belongs to Pisum sativum and x has green cotyledons

specify such sets. Sets specified in this way are called phenotypes. But a classification of parents and offspring into phenotypes is not sufficient for genetical purposes. We require a further classification of members of phenotypes into sub-sets called genotypes. These sets are usually specified by reference to certain unobservable entities postulated by the highest level genetical hypotheses. Woodger defines them without resorting to those somewhat mysterious concepts. His definitions are based solely on the restricted observational vocabulary explained as above. The meaning of the word 'genotype' which these definitions are intended to grasp is characterized by the following quotation from Haldane: "a class (of organisms) which can be distinguished from another by breeding tests is called a genotype."

Let us describe a simple breeding test, e.g. Mendel's experiments with garden peas. Two kinds of statements will be needed to that purpose:

(1) $F_X(Y, Z) \subset W,$

(2) $F_X(Y, Z) \in p_1 W_1 + p_2 W_2 + \ldots + p_n W_n.$

(1) states that every member of the set $F_X(Y, Z)$ is a member of the set W, (2) that the set $F_X(Y, Z)$ contains members of W_i in the proportion p_i for each i from 1 to n. We will denote by 'Y' the set of all garden peas with yellow cotyledons, by 'G' the set of all garden peas with green cotyledons, by 'A' and 'B' the actual sub-sets of Y and G which Mendel used in his experiments, and by 'E' the set of environments existing in Mendel's garden. His observations can be recorded as follows:

I. (1) $A \subset Y$ II. (1) $B \subset G$
    (2) $F_E(A, A) \subset Y$ (2) $F_E(B, B) \subset G$
    (3) $F_E^2(A, A) \subset Y$ (3) $F_E^2(B, B) \subset G$
III. (1) $F_E(A, B) \subset Y$
    (2) $F_E^2(A, B) \in \frac{3}{4} Y + \frac{1}{4} G$
    (3) $F_E(A, F_E(A, B)) \subset Y$
    (4) $F_E(B, F_E(A, B)) \in \frac{1}{2} Y + \frac{1}{2} G.$

The three sets of organisms: A, B and $F_E(A, B)$ behave genetically in different ways and, consequently, belong to three different genotypes: the homozygous genotype of $Y : H(Y)$, the homozygous genotype of $G : H(G)$, and the singly heterozygous genotype of Y and $G : Ht(Y, G)$. $H(Y)$ is defined as the set of all peas which behave genetically like members of A, $H(G)$ – as the set of all peas which behave genetically like members of B, and $Ht(Y, G)$ – as the set of all peas which behave genetically like members of $F_E(A, B)$. The precise definitions of these concepts are formally rather complicated. I am going to present only one of them – the definition of $H(Y)$ – and I will do it in a somewhat simplified way.

$H(Y)$ contains all sets Z which satisfy the following conditions:

(1) $Z \subset Y$
(2) $F_{E(Y)}(Z, Z) \subset Y$
(3) $F^2_{E(Y)}(Z, Z) \subset Y$
(4) $F^2_{E(Y)}(Z, Z) \neq \wedge.$

Conditions (1)–(3) correspond to observation records about members of A. Condition (4), which implies:

$$F_{E(Y)}(Z, Z) \neq \wedge \text{ and } Z \neq \wedge,$$

guarantees that conditions (1)–(3) are not satisfied vacuously. $E(Y)$ is a certain maximized environmental set which contains, loosely speaking, all environmental sets similar to E, i.e., sets in which some organisms do behave genetically like members of A. If we write '$K(Z)$' as an abbreviation for conditions (1)–(4), the definition of $H(Y)$ may be formulated as follows: $H(Y)$ is the sum of all sets Z such that $K(Z)$.
This is equivalent to the following formula:

$$H(Y) = (\hat{x})\,(\exists Z)\,(K(Z)\,.\,x \,\epsilon\, Z).$$

In an analogous manner we can define $H(G)$ and $Ht(Y, G)$.

The definition of $H(Y)$ seems to be intuitively correct. An organism belongs to the homozygous genotype of Y if and only if it belongs to a class of organisms which are Y and whose offspring in the first and second generation are Y too. A closer examination shows, however, that the definition of $H(Y)$ leads to some undesirable consequences.

Suppose that x_1 is a garden pea with yellow cotyledons which has not produced any offspring:

$$x_1 \,\epsilon\, Y\,.\, \sim (\exists y)\,(x_1 \, Ps \, y).$$

It can easily be shown that x, will belong to $H(Y)$ provided there is a class Z such that $K(Z)$. Let us assume that Z_0 is such a class, i.e., that $K(Z_0)$. Then the class $Z_0 \cup \{x_1\}$ must also satisfy the condition K, i.e., $K(Z_0 \cup \{x_1\})$. The assumption: $K(Z_0)$ amounts to the conjunction of the following conditions:

(1) $Z_0 \subset Y$
(2) $F_{E(Y)}(Z_0, Z_0) \subset Y$

(3) $F_{E(Y)}(Z_0, Z_0) \subset Y$

(4) $F^2_{E(Y)}(Z_0, Z_0) \neq \wedge.$

We will show that the corresponding conditions for $Z_0 \cup \{x_1\}$ hold as well.

The condition:

(1) $Z_0 \cup \{x_1\} \subset Y$

is true because, under our assumption, $x_1 \in Y$.

In order to show that the condition:

(2) $F_{E(Y)}(Z_0 \cup \{x_1\}, Z_0 \cup \{x_1\}) \subset Y$

is satisfied, it suffices to demonstrate that:

$$F_{E(Y)}(Z_0 \cup \{x_1\}, Z_0 \cup \{x_1\}) = F_{E(Y)}(Z_0, Z_0).$$

This is quite obvious since x_1 has no offspring. If we add it to the class Z_0, we do not change any class of offspring which members of Z_0 have produced. This may be shown in a more formal way. The class: $F_{E(Y)}(Z_0 \cup \{x_1\}, Z_0 \cup \{x_1\})$ is, according to the definition of F, identical with the class:

$$(\hat{y})[(\exists u)(\exists v)(u \neq v \,.\, u \in Z_0 \cup \{x_1\} \,.\, u Psy \,.\, v \in Z_0 \cup \{x_1\} \,. \\ .\, v Psy \,.\, en(u) \in E(Y) \,.\, en(v) \in E(Y) \,.\, en(y) \in E(Y))].$$

Since: $u \in Z_0 \cup \{x_1\} \equiv u \in Z_0 \vee u = x_1$ the above formulation may be expanded as follows:

$$(\hat{y})[(\exists u)(\exists v)(u \neq v \,.\, u \in Z_0 \,.\, u Psy \,.\, v \in Z_0 \,.\, v Psy \,. \\ .\, en(u) \in E(Y) \,.\, en(v) \in E(Y) \,.\, en(y) \in E(Y) \vee \\ \vee\, u \neq v \,.\, u \in Z_0 \,.\, u Psy \,.\, v = x_1 \,.\, v Psy \,. \\ .\, en(u) \in E(Y) \,.\, en(v) \in E(Y) \,.\, en(y) \in E(Y) \vee \\ \vee\, u \neq v \,.\, u = x_1 \,.\, u Psy \,.\, v \in Z_0 \,.\, v Psy \,. \\ .\, en(u) \in E(Y) \,.\, en(v) \in E(Y) \,.\, en(y) \in E(Y) \vee \\ \vee\, u \neq v \,.\, u = x_1 \,.\, u Psy \,.\, v = x_1 \,.\, v Psy \,. \\ .\, en(u) \in E(Y) \,.\, en(v) \in E(Y) \,.\, en(y) \in E(Y))].$$

Now the last component of the above disjunction is false for every y for purely logical reason: $\sim (\exists u)(\exists v)(u \neq v \,.\, u = x_1 \,.\, v = x_1)$. The second and third components are false for every y since, under our assumption

319

concerning x_1 : $\sim (\exists y)(\exists u)(u = x_1 . u \, Ps \, y)$. Thus the whole disjunction is equivalent to its first component and the above class is identical with the following one:

$$(\acute{y})[(\exists u)(\exists v)(u \neq v . u \, \epsilon \, Z_0 . u \, Ps \, y . v \, \epsilon \, Z_0 . v \, Ps \, y .$$
$$. \, en(u) \, \epsilon \, E(Y) . en(v) \, \epsilon \, E(Y) . en(y) \, \epsilon \, E(Y))],$$

which is nothing else but the class: $F_{E(Y)}(Z_0, Z_0)$.

The identity of the first filial generations:

$$F_{E(Y)}(Z_0 \cup \{x_1\}, Z_0 \cup \{x_1\}) = F_{E(Y)}(Z_0, Z_0)$$

implies identity of the second generations:

$$F^2_{E(Y)}(Z_0 \cup \{x_1\}, Z_0 \cup \{x_1\}) = F^2_{E(Y)}(Z_0, Z_0),$$

since

$$F^2_{E(Y)}(Z_0 \cup \{x_1\}, Z_0 \cup \{x_1\}) =$$
$$= F_{E(Y)}(F_{E(Y)}(Z_0 \cup \{x_1\}, Z_0 \cup \{x_1\}), F_{E(Y)}(Z_0 \cup \{x_1\}, Z_0 \cup \{x_1\})).$$

Thus the conditions:

(3) $\quad F^2_{E(Y)}(Z_0 \cup \{x_1\}, Z_0 \cup \{x_1\}) \subset Y$
(4) $\quad F^2_{E(Y)}(Z_0 \cup \{x_1\}, Z_0 \cup \{x_1\}) \neq \wedge$

are satisfied too. It follows then that: $K(Z_0 \cup \{x_1\})$. But if $K(Z_0 \cup \{x_1\})$ then $(\exists Z)(K(Z) . x_1 \, \epsilon \, Z)$, and hence, according to our definition of $H(Y)$, $x_1 \, \epsilon \, H(Y)$.

That paradoxical consequence holds not only for those members of Y which have not produced any offspring at all. If an organism belonging to Y has produced some offspring but not with members of the class Z_0, it will belong to $H(Y)$ for similar reasons. The same is true of those members of Y which have not developed in an environment belonging to $E(Y)$ or whose offspring have not developed in such environments. Generally speaking, if $x_1 \, \epsilon \, Y$ and $\sim (\exists y)(\exists v)(x_1 \neq v . x_1 \, Psy . v \, \epsilon \, Z_0 . v \, Psy .$ $en(x_1) \, \epsilon \, E(Y) . en(v) \, \epsilon \, E(Y) . en(y) \, \epsilon \, E(Y))$ then $x_1 \, \epsilon \, H(Y)$. In other words, if $x_1 \, \epsilon \, Y$ and $F_{E(Y)}(\{x_1\}, Z_0) = \wedge$ then $x_1 \, \epsilon \, H(Y)$. The point is that, in so far as x_1 is concerned, conditions (2)–(4) of our definition of $H(Y)$ are satisfied vacuously. The same goes for conditions (3)–(4) if $F^2_{E(Y)}(\{x_1\}, Z_0) = \wedge$.

This is, undoubtedly, an undesirable consequence. The class $H(Y)$, as defined above, is quite different from what it was meant to be. It is unduly large and contains, in addition to "genuine" members of the homozygous genotype of Y, all members of Y which have not produced offspring with members of some class satisfying the condition K. Similar consequences hold for the definition of the heterozygous genotype of Y and G constructed along the same lines. Consequently, the sets: $H(Y)$ and $Ht(Y, G)$ are not exclusive. They have some members in common. For example, all "childless" members of Y belong both to $H(Y)$ and $Ht(Y, G)$.

If we want to avoid these paradoxical consequences of our definition, we shall have to strengthen the condition K so as to exclude from the class $H(Y)$ all members of Y which belong to it only due to their lack of offspring. We can do it by adding to conditions (1)–(4) the following clause:

(5) $\quad (x)(x \, \epsilon \, Z \supset F^2_{E(Y)}(\{x\}, Z) \neq \wedge)$.

It will guarantee that the class Z contains only such organisms which have produced some offspring (in the first and second generation) with other members of Z. Let us write $K'(Z)$ as an abbreviation for conditions (1)–(5). Now the statement: $(\exists Z)(K'(Z) \, . \, x_1 \, \epsilon \, Z)$ will imply that x_1 is Y, that it has produced some offspring (in the first and second generation), and that these offspring are Y too. If one of these facts does not hold, the statement cannot be true. And so, if x_1 has not produced any offspring, it will not belong to any class satisfying the condition K'. The definition:

$$H(Y) = (\hat{x}) \, (\exists Z) \, (K'(Z) \, . \, x \, \epsilon \, Z)$$

will thus exclude from the class $H(Y)$ all those "illegitimate" members which belonged to it according to its former definition. But the present definition is not satisfactory either. All "childless" members of Y will now belong to $\overline{H(Y)}$. Strictly speaking, all members of Y which have produced no offspring with members of any class satisfying the condition K' will belong to $\overline{H(Y)}$. This is certainly as unacceptable consequence as the former one.

The above-mentioned defects seem to be incurable. Every explicit definition of $H(Y)$ based on our restricted vocabulary will result in classifying all "childless" members of Y as belonging either to $H(Y)$ or to $\overline{H(Y)}$. We have no ground for making any distinction between them since all our decisions are based on observation records about phenotypes of the given organism and its offspring. As our "troublesome" organisms belong all to the same phenotype and have produced no offspring, they will all be put into the same class: either $H(Y)$ or $\overline{H(Y)}$. Both decisions are equally arbitrary and unjustified. Moreover, they can be found, later on, to conflict with some experimental evidence. Genetics in its further development may – and, in fact, does – widen its observational basis including, e.g., some cytological evidence. We can formulate certain new criteria for $H(Y)$ in terms of these cytological observations. They will enable us to classify some of the "childless" organisms as $H(Y)$, some others as $\overline{H(Y)}$. But such decisions will clash with our definition of $H(Y)$, which has arbitrarily classified all these organisms as $H(Y)$ or $\overline{H(Y)}$.

Genetics at its "elementary", Mendelian stage can tell us nothing about the genotype of "childless" organisms. Therefore the definition of genotype formulated at this stage must also leave that question open. It cannot then be an explicit definition. The concept of genotype may be defined adequately only by some more "liberal" procedure. I mean what has often been referred to as partial or conditional definition (or reduction). This kind of definition was introduced by Carnap[3] and then generalized by some other authors.[4]

The simplest form of partial definition introducing the term Q may be rendered as follows:

(1) $Px \supset Qx$
(2) $Rx \supset \sim Qx.$

The main difference between partial and explicit definition lies in the fact that the defining conditions: P and R are neither logically exhaustive nor exclusive, whereas in the case of explicit definition they amount to: P and $\sim P$. Since they do not exhaust all possibilities, there may be individuals which are neither P nor R. With regard to such individuals the meaning of the term Q is not determined. We have no criteria of application for Q in such cases. Q remains an "open" concept. It may be determined closer by some additional criteria.

From the foregoing analysis it can be seen that genotype is an "open" concept which in some cases must remain undetermined. Therefore, as a zero level concept, it can be defined only partially. The following sentences may serve as its partial definition based on the adopted vocabulary:

(1) $(\exists Z)\,[K'(Z)\,.\,x \in Z] \supset x \in H(Y)$

(2) $(\exists Z)\,[K'(Z)\,.\,(\sim x \in Y \vee \sim F_{E(Y)}(\{x\}, Z) \subset Y \vee$
    $\vee \sim F^2_{E(Y)}(\{x\}, Z) \subset Y)] \supset \sim x \in H(Y).$

Sentence (1) formulates a sufficient condition for the membership in $H(Y)$, sentence (2) – a necessary one. I have already tried to show that the condition expressed by (1) is an adequate one. The condition formulated in (2) excludes from the class $H(Y)$ all organisms which are not members of Y or whose offspring which they produced with members of some class satisfying the condition K' do not belong to Y. This seems adequate too. The conditions formulated in (1) and (2) are not logically exhaustive. There may be organisms which do not satisfy either of them. All organisms considered previously belong just to that kind. If an organism has produced no offspring with members of any class satisfying the condition K' it does not satisfy the antecedent of (1). If, in addition, it is a member of Y it does not satisfy the antecedent of (2) either. (It will be noticed that the clause: $\sim F_{E(Y)}(\{x\}, Z) \subset Y$ implies: $F_{E(Y)}(\{x\}, Z) \neq \wedge$.) With regard to such an organism we have no criteria of application for the term $H(Y)$. We cannot apply either $H(Y)$ or its negation. But that is just what was wanted.

The meaning of $H(Y)$ would be completely undetermined, if there were no class satisfying the condition K'. We may declare (as did Woodger) that in such case $H(Y) = \wedge$ by adding to condition (2) an additional clause:

(2) $\sim (\exists Z)\,[K'(Z)] \vee (\exists Z)\,[K'(Z)\,.\,(\sim x \in Y \vee$
    $\vee \sim F_{E(Y)}(\{x\}, Z) \subset Y \vee$
    $\vee \sim F^2_{E(Y)}(\{x\}, Z) \subset Y)] \supset \sim x \in H(Y).$

The conditions expressed by (1) and (2) are not logically exclusive. Since our definition implies their exclusiveness, the statement asserting it must have a "factual content". It is a matter of experience whether or not an object satisfying the antecedent of (1) may also satisfy the antecedent of (2). The observations seem to exclude that possibility and, thus, to confirm the empirical consequence of our definition. But empirical evidence can never be conclusive. Logical exclusiveness of conditions (1) and (2) may be guaranteed by a suitable modification of the second condition:

(2) $\sim (\exists Z) [K'(Z) . x \in Z] . (\exists Z) [K'(Z) . (\sim x \in Y \lor$
$\lor \sim F_{E(Y)}(\{x\}, Z) \subset Y \lor \sim F_{E(Y)}(\{x\}, Z) \subset Y)] \supset$
$\supset \sim x \in H(Y).$

The partial definition of $H(Y)$ consisting of statements (1) and (2) seems to supply that term with the intended meaning. It avoids the shortcomings of the explicit definition mentioned previously. At the same time, it is based on the same "elementary" vocabulary as our original definition. The price of its adequacy is a rather complicated and awkward form. It seems possible to introduce the concept of genotype by means of a simpler partial definition formulated in the same language. As this concept differs in some respect from $H(Y)$, let us denote it by $H^*(Y)$. The definition of $H^*(Y)$ reads as follows:

(1) $\{x, y\} \subset Y . F_{E(Y)} \{x, y\} \subset Y . F^2_{E(Y)} \{x, y\} \subset Y .$
    $. F^2_{E(Y)} \{x, y\} \neq \land \supset \{x, y\} \in H^*(Y)$

(2) $\sim \{x, y\} \subset Y \lor \sim F_{E(Y)} \{x, y\} \subset Y \lor$
    $\lor \sim F^2_{E(Y)} \{x, y\} \subset Y \supset \sim \{x, y\} \in H^*(Y),$

where $F_{E(Y)} \{x, y\}$ is simply a shorter formulation of $F_{E(Y)}(\{x\}, \{y\})$. The difference between $H(Y)$ and $H^*(Y)$ lies in the fact that $H^*(Y)$ is not a class of individuals but a class of (unordered) pairs of individuals. The definition of $H^*(Y)$ enables us to say whether or not a pair of organisms belongs to a given genotype, but it does not enable us to classify individual organisms with regard to their genotypes. Apart from this important difference, the meaning of $H^*(Y)$ seems to be very much like that of $H(Y)$. The conditions specified by the definition of $H^*(Y)$ are analogous to those formulated in the definition of $H(Y)$. They imply therefore similar consequences. $H^*(Y)$ is also an "open" concept. A pair of organisms which belong to Y and have not produced any offspring does not satisfy either the antecedent of (1) or the antecedent of (2). Consequently, neither $H^*(Y)$ nor $\overline{H^*(Y)}$ can be applied to it. This seems to be in accordance with what has been said of the situation in classical genetics. Genetics, in so far as it is founded purely on genetical experiments, i.e., on breeding tests, can tell us nothing definite about the genotype of such organisms. That is why the concept of genotype can be defined only partially.

One reservation should be made in connection with that problem.

There are situations in which a determination of the genotype of a "child-less" organism can be made at the elementary level. This is possible if such a statement can be deduced from some known statements about genotypes of other organisms. Let us assume that x_1, which belongs to Y, has not produced any offspring but its parents: y_1 and y_2 have produced some offspring in the first and second generation. Then we are able to determine the genotypes of both parents and – in some cases – also the genotype of x_1. If y_1 and y_2 belong to $H(Y)$, x_1 will belong to $H(Y)$ as well. If one parent belongs to $H(Y)$ and the other to $H(G)$, x_1 must be $Ht(Y, G)$. The same will hold if the genotypes of parents are: $Ht(Y, G)$ and $H(G)$. But if both parents belong to $Ht(Y, G)$, or one of them to $Ht(Y, G)$ and the other to $H(Y)$, x_1 may be $H(Y)$ as well as $Ht(Y, G)$. Only an examination of x_1's offspring could decide the question. Situations as described above are quite conceivable. Our definitions of $H(Y)$ or $H^*(Y)$ do not imply anything that would exclude them.

We have seen that the shifting from explicit definition of $H(Y)$ to partial definitions of $H(Y)$ or $H^*(Y)$ makes it possible to avoid some undesirable consequences of the original definition. But against that definition there may be raised objections which cannot be avoided in that way. There are defects which the original definition of $H(Y)$ shares with all partial definitions of this concept. Suppose that an organism x_1 belonging to $Ht(Y, G)$ has, by chance, produced the first and second generation offspring all belonging to the phenotype Y. This is quite probable, especially if the number of x_1's offspring is not very large. In that case, all definitions of $H(Y)$ constructed along the above lines – the explicit as well as partial ones – will lead to the conclusion that x_1 belongs to $H(Y)$! This is certainly inconsistent with the meaning intended for $H(Y)$. The point is that the connection between the genotype of an organism and the phenotypes of its offspring is of a probabilistic nature. Consequently, an adequate observational definition of $H(Y)$ must also be a "probabilistic" one. Such probabilistic partial definitions have been proposed by some authors[5] but they are still in need of further elaboration. I shall not deal with that problem in the present paper which is concerned only with one aspect of the analyzed concept: the "openness" of its meaning.

This "openness" of meaning characterizes not only the concept of genotype but also several other concepts belonging to the conceptual apparatus of genetics as constructed in *Biology and Language*. First of all, terms defined with the help of $H(Y)$ inherit the "open" character of the latter. But, in addition to it, some of them acquire that character of their own. Their definitions may be analyzed in the same way as the definition of $H(Y)$. And the same remedies in the form of partial definitions can be found for their defects. One of those concepts deserves, however, special attention. The definition of $Gm(P)$ leads to certain paradoxical consequences for reasons closely similar to those analyzed before. Woodger suggests a way out not by abandoning explicit definition but only by modifying it slightly. His interesting suggestion calls for a closer examination.

'$Gm(P)$' is to denote the set of all gametes such that the zygotes which result from their union in pairs develop into members of $H(P)$. In order to formulate the definition of $Gm(P)$ we require two additional signs:

(1) $dlz\,(x, y, z)$, which may be read 'the zygote x develops in the environment y into the life z' and

(2)   '$U\,(\alpha, \beta)$', which denotes the set of all zygotes formed by the union of a gamete belonging to the set α with one belonging to the set β.

We write an abbreviation '$L(\alpha)$' for the following conditions:

(1) $\alpha \neq \wedge$

(2) $(x)(y)(z)\left[dlz(x, y, z)\,.\,x \in U\,(\alpha, \alpha)\,.\,y \in E(P) \supset z \in H(P)\right].$

$Gm(P)$ is defined as follows:

$$Gm(P) = (\hat{u})(\exists\alpha)(L(\alpha)\,.\,u \in \alpha).$$

Against this definition Woodger himself raises certain objections. "It is always possible to add to the set α in this definition terms which would make the antecedent of the conditional which is implicit in this definition false and would thus still satisfy the definition. The set α would thus become unduly large and would contain all manner of objects (*e.g.* some which were not gametes at all) which are not wanted in the set $Gm(P)$". This is quite true and can easily be shown by a procedure analogous to that used in the case of $H(Y)$.

Let us assume that a class α_0 satisfies the condition L and that u_1 is an object which has not formed any zygote with a member of the class α_0:

$\sim (\exists x)\,(x \in U\,(\{u_1\}, \alpha_0))$. The class: $U\,(\alpha_0 \cup \{u_1\}, \alpha_0 \cup \{u_1\})$ will then be identical with the class: $U(\alpha_0, \alpha_0)$, since we do not change the class of all zygotes formed by the union of members of α_0 by adding to α_0 an object which has not fused with any member of α_0 to form a zygote. This identity implies that: $L(\alpha_0 \cup \{u_1\})$. But if $L(\alpha_0 \cup \{u_1\})$ then $(\exists \alpha)(L(\alpha)\,.\,u_1 \in \alpha)$ and hence, according to our definition of $Gm(P)$, $u_1 \in Gm(P)$. Thus $Gm(P)$ will contain all objects which have not formed zygotes with members of some class satisfying the condition L. Moreover, it will also contain all gametes such that the zygotes which result from their union do not develop in an environment belonging to $E(P)$. Strictly speaking, if a class α_0 satisfies the condition L then all objects u such that:

$$\sim (\exists x)\,(\exists y)\,(\exists z)\,\big[\,dlz\,(x, y, z)\,.$$
$$.\,x \in U(\{u\}, \alpha_0)\,.\,y \in E(P)\big]$$

will belong to $Gm(P)$.

This consequence can be avoided by a procedure analogous to that applied to $H(Y)$, i.e., by strengthening the condition L and constructing a partial definition with its help. Woodger suggests another solution. He proposes to replace the original definition of $Gm(P)$ by the following one:

$$Gm(P) = (\hat{u})(\exists v)(\exists X)\big[u(u, v) \in X\,.$$
$$.\,(x)(y)(z)(dlz(x, y, z)\,.\,x \in X\,.\,y \in E(P) \supset z \in H(P))\,.$$
$$.\,H(P) \neq \wedge\big],$$

where $u(u, v)$ is a functor denoting the zygote which is formed by the union of gamete u with gamete v. But this proposal does not seem to remove all defects of the original definition. It is not quite clear what is the logical status of the functor $u(u, v)$. If it were a functor in the proper sense[6] the following condition should be satisfied:

For every u and v there is one and only one zygote which is formed by the union of u with v.

But this is obviously false, and would be so even if we decided to restrict the range of variables u and v to the class of gametes which actually fuse to form zygotes. $u(u, v)$ could not, therefore, be introduced as a proper functor. It may then be treated as a relational description in the Russellian sense. This, however, will lead to some undesirable consequences. Suppose that u_1 has not formed any zygote at all. Then:

$$\sim (\exists v)(\exists X)(u(u_1, v) \in X)$$

and, according to our definition:

$$\sim u_1 \in Gm(P).$$

This is hardly an acceptable conclusion. And if u_1 has formed a zygote which has not developed in an environment belonging to $E(P)$, u_1 will belong to $Gm(P)$. In that case, the new definition of $Gm(P)$ leads to the same paradoxical conclusion as the old one. Thus, again, the only solution seems to be a partial definition of $Gm(P)$.

The case examined in the present paper is not an exceptional one. On the contrary, it is typical for all "theoretical concepts". This was first recognized by Carnap, whose theory of reduction sentences was constructed in order to deal with these concepts. Our analysis provides an example illustrating his considerations. Let us state its results in a somewhat metaphorical way. Theoretical terms may be said to denote certain invisible structures. These structures reveal themselves as observable phenomena under certain observable circumstances. Where no such circumstances exist, the structures do not manifest their presence at all. Thus, on the observational level, they acquire the character of dispositions. Genotype may be thought of in the same way. It refers to an unobservable structure – the genetic constitution of organisms – which manifests itself by some kind of observable behaviour – the genetic behaviour of these organisms. If an organism has not produced any offspring, it is impossible to say how it behaves genetically and, consequently, what its genotype is like. That is why the meaning of this concept, like that of all dispositional concepts, can – at the observational level – be specified only partly. Our analysis was intended to demonstrate this in a more formal and detailed way. An analysis of this sort has been made possible by the previous construction of a precise genetical language which has succeeded "in making apparent a real complexity in the subject-matter which the natural language conceals". This seems to be a necessary prerequisite of any methodological analysis.

University of Warsaw, Warsaw, Poland

REFERENCES

1. *Biology and Language,* p. 214.
2. *Biology and Language,* p. 208.
3. R. Carnap, 'Testability and Meaning', *Philosophy of Science* 3 (1936) 419–471 and 4 (1937) 1–40.
4. H. Mehlberg, 'Positivisme et Science', *Studia Philosophica* 3 (1948) 211–294; H. V. Stopes-Roe, 'Some Considerations Concerning "Interpretative Systems"', *Philosophy of Science* 25 (1958) 143–156.
5. E.g. A. Kaplan, 'Definition and Specification of Meaning', *Journal of Philosophy* 43 (1946) 281–288; H. Mehlberg, '*The Reach of Science,* The University of Toronto Press, Toronto, 1958, pp. 258 ff.
6. R. Carnap, *Introduction to Symbolic Logic and Its Applications,* Dover Publications, Inc., New York, 1958, pp. 71 ff.

PAUL G. 'ESPINASSE

GENETICAL SEMANTICS AND EVOLUTIONARY THEORY

We seem to be essentially so constituted that in our thinking we find it easiest to use discrete counters. Indeed the very act of counting itself supposes initially things which can be relied upon to retain their identities for as long as our operations continue, or which, if they change, do so only in certain essentially understandable ways.

Our analyses of structures or of situations are usually directed toward providing ourselves with separable and recognisable sub-structures or factors with the aid of which we hope to be able to do things and to understand things which we could not do and could not understand before we undertook the task of analysis. Confronted with a question beginning 'How?' or 'Why', we generally begin in science by asking further questions very frequently beginning 'How many?' or 'How often?'. These further questions constitute an attempt to analyse the original into pieces which seem more manageable, in the spoken or unspoken hope that the sum of our answers to these question-pieces may turn out to be the answer, or an answer, to the original question. Sometimes our hopes are realised; sometimes they are not. When they are not we commonly suspect that this is because there was something wrong with our original analysis: that is to say we have been seeking answers to the wrong subquestions.

In seeking to account for the structure of contemporary living nature Darwin was led to compare it with that of living nature in the past as he deduced it from the fossil record and the facts of the contemporary distribution over the earth of animals and plants. As a consequence of making this comparison he came to be convinced that changes adding up to an evolutionary process had occurred through secular time and were probably to be supposed still to be occurring in persisting populations. It is perhaps not always remembered that in publishing *The Origin of Species* Darwin finally robbed men of one of their familiar counters. A population – for instances the birds of one kind to be found in the 19th century on one of the islands of the Galápagos – had until Darwin been commonly regarded as an entity given at the moment of creation and continuing immutable and recognisable. Previous attacks on this view, in particular that of Lamarck, had met with very limited success because they did not propound any satisfying mechanism to account for the occurrence of changes. Further, Darwin's predecessors were in no position to support the proposition that such changes had occurred with any really overwhelming mass of detailed evidence. Darwin, however, was able to do just this and it is for this reason that after him the time-honoured and familiar entity, the immutable species, simply vanished.

In its place appeared an entity characterised in ways most seriously different. Men were asked to think of a continuing population as before, but now the population had to be thought of no longer as a given immutable entity but as an entity of quite a different quality. The reader of *The Origin* on its publication had to accustom himself to playing with counters which were likely not only to change colour, which would have been confusing enough, but were liable to divide themselves into two counters, neither of which need have the colour of the original which he had started by thinking about. Our grandfathers were right to be disturbed; the wonder is that so many of them kept their heads at all. Their situation was like that of Alice, who could no doubt think clearly about croquet mallets or about flamingoes, but who was understandably concerned at finding one turning into the other.

The question at once arose in Darwin's mind that has arisen in the minds of all who have tried to think about evolution ever since: What is the mechanism that brings about change in a continuing population? This question raises in our minds as it raised in the mind of Darwin the enantiomorphic one: What is the mechanism which insures that over short, or in some cases over long, periods of time populations often do in fact remain recognisably the same? We are so used to having this last problem with us that it is easy to forget that before Darwin it really scarcely arose. There is little room for it in a world of immutable species, each the result of an act of special creation. The moment, however, that species are seen as entities liable through time to change and to division, species that remain steady and unchanging have to be seen as having somehow resisted this universal liability to change and division, and at once the unchanging species needs as much explaining as a changing one. The two questions above are seen to be truly complementary to each other.

To answer both these questions was early seen to require a considera-

tion of heredity, since it is obvious that for a population to remain unchanged over a period much longer than the lifetime of an individual
member of it requires that descendants shall resemble ancestors, and for
a population to change requires that they shall in some orderly way differ
from them. No adequate or detailed mechanism of heredity either
physical or conceptual was available when Darwin wrote, and certainly
none adequate to play its proper part in the origin of species by natural
selection. It may perhaps be suggested that this is not really surprising.
Given immutable species there is no really very compelling reason why
men should seek a detailed mechanism of heredity adequate to 'explain'
the evolution of species in any way, because the evolution of species was
not, for them, there to be explained.

Without knowledge of any such detailed mechansims plant- and
animal-breeders had had great success, but their intra-specific activities
never put the species in jeopardy. For Darwin the species was in jeopardy,
and he spent a lot of time and trouble trying to find out how plant- and
animal-breeders established the intra-specific differences with which they
were concerned in the hope that there he would light upon the mechanism
of heredity he needed to explain what he was sure was the fact of orderly
natural selection: a fact which he realised eroded the very idea of the
immutable species. He was in the event no more successful than Lamarck
had been in his search for a mechanism of heredity to 'explain' evolution
through the inheritance by the offspring of characters acquired by an
individual. It may be worth considering just why both Lamarck and
Darwin failed to get what they wanted from breeders.

What breeders have always done is to analyse their material in terms
of desirable and undesirable abstract qualities, denoted by abstract nouns
like tallness and shortness. They then breed selectively from animals or
plants possessing or not possessing these qualities and note that the
possession or non-possession is in some way handed down to future
generations. The discrete counters they use in their thinking are thus
discrete abstractions, and not discrete physical objects having dimensions
in space. The 'factors' proposed by the mathematician Mendel were also
discrete abstractions and it was their discreteness that made their distribution susceptible to mathematical treatment. But it was only when they
came later to be related to physically discrete observable chromosomes
that they took on physical dimensions in any way at all. What Darwin

really needed to account in materialistic terms for the physical resemblances and differences between parents and offspring was a mechanism of heredity in terms of entities having dimensions in space, and so subject to the 'laws' of physics and chemistry as well as to those of mathematics, and these were just what the breeders he consulted could not give him. His own attempts to envisage such a physical mechanism failed. Ultimately, however, it was in the event the mathematically inclined breeder Mendel who took the step which made it possible after his death for his successors to think of the elements of heredity as having dimensions in space. It is an odd fact that though Mendel was a practical breeder as well as a theoretician it was the theoretical scientists, interested in evolution, and not the breeders, who recognised at the beginning of this century the importance of Mendel's formulation and who realised that the mechanism he proposed might be used to explain evolution by natural selection, and so the origin of species, in physical terms, and for fifty years evolutionary thinking has been done in a Mendelian climate, and in Mendelian language.

The terms of this language satisfy our requirement for comfortable thinking, in that they provide entities which can be regarded as distinct and which can be relied upon usually to remain recognisable in some sense over the periods of time with which geneticists are commonly concerned. These entities – these units of particulate inheritance – graduated from being conceptual factors in the mind of Mendel to being in some sense material objects regarded as occupying loci on observable chromosomes, and acquired the title of genes. As such they constituted the fundamental units of genetic and of evolutionary thought, and it seems reasonable to enquire closely just what the word gene has meant in the past, means now, and is likely to mean in the future, both in the field of genetics and in that of evolutionary studies. To answer these innocent-seeming questions turns out to be far less easy than might perhaps have been supposed.

The 'factors' of Mendel were thought of as causal agents. The very ease with which this phrase may be written or spoken may be dangerous: it can mean so many things. For instance, pathogenic bacteria having extension in time and space can be spoken of as the causal agent of an illness or a death neither of which can be very easily assigned extension in space. A fall in prices, to which we are scarcely accustomed to assign extension in space, can be regarded as the causal agent of unemployment and of an almost unlimited host of other things regarded as consequential. In science we are accustomed to avoid the infinite and stultifying complications into which the relationships suggested above lead us by adopting, consciously or unconsciously, certain conventional limitations in our uses of the verb 'to cause'. Similar considerations of prudence lead us to be very wary of ever using the noun 'cause' at all. What we actually do can perhaps be most readily made plain by considering some possible questions and some possible answers to them.

When, in real life, we are asked a question concerned with cause *and having the form* 'Why, whenever A then B?' or 'Why, whenever I plant nasturtium seeds do I get nasturtium flowers?' we always have a perfectly clear object in view in framing our answer. Our object is simply to satisfy our questionner in the context of the question so that he will stop asking us questions – to shut, in a word, our interlocutor up. Different questionners are satisfied by different answers, and scientists as a group are, it is suggested, perhaps distinguished by having commonly agreed among themselves to be satisfied by a distinct and recognisable *kind* of answer to the very ordinary question 'Why, whenever A then B?'.

Clearly there are different *kinds* of answer which we can give to the question we are concerned with. Some of these appear to be as follows: (1) 'Because A causes B'. But if the meaning of 'causes' is queried here, then it is said 'Whenever A then B', which is a mere repetition of the original statement underlying the original question. (2) 'Because there is a *power* in A'. But when the meaning of 'power' is queried, then the answer is either to repeat once more the statement underlying the original question or to say that this 'power' is essentially unexplainable in any terms whatever. Answers of this kind are perhaps appropriate in some fields of, for instance, theology. But there is a further possible kind of answer: (3) 'Because A is in fact an XZ, and we all know that whenever an XZ then B.' To the natural further enquiry 'Why whenever an XZ then B?' the answer is 'Because XZ is really an MN, and we all know that whenever MN, then B.' Now, this series of questions and answers started in (3) may, either synthetically or analytically, go on (1) forever; (II) to a definite end that is in time explained in terms of something that has gone before and is therefore a circularity; or (III) to a definite but at the moment unexplained end that *satisfies the questionner*.

In science, it is here suggested, we agree by convention to accept (3) (III) above and we choose as our criterion in judging whether to be satisfied by an answer in terms of XZ or of MN our recognition in one case or the other that the terms make it possible for us to *do* something which we could not have done before adopting them. For instance, it may at once seem clear to us that we can perform an experiment or make an observation to find out whether or not the statement 'A is an MN' is in fact true, while we perhaps cannot think of an experiment or of an observation which will tell us whether the statement 'A is an XZ' is true or not. When we see this difference between XZ and MN we agree to be satisfied for the moment with MN but not with XZ. If we find by experiment that the statement 'A is an MN' is untrue we discard the explanation which depended upon the truth of this statement and seek a further translation in such terms as 'Because A is a PQ', and repeat the process until we find a translation which gives a statement that is supported by an experiment or an observation. When we do find a translation that is so supported by experiment or observation science takes a step forward, and we next ask not 'Why whenever A then B?', but the new question 'Why whenever a PQ then B?' instead. We recognise progress in science by the emergence of new questions each arising in its turn from an experimental or observational confirmation or denial of a previous statement.

An analytic example of the sequence of activities described above would be a consideration of the artificially simplified question 'Why, whenever two X-chromosomes in a Drosophila egg then a female Drosophila?' The 'scientific' answer is 'Because X-chromosomes are physical objects which have at places in them molecular configurations such that in their presence, together with that of all the other molecular configurations also normally present, the metabolic processes of a Drosophila egg in a normal environment will result in a female Drosophila'. To the further question 'Why, whenever these molecular configurations then a female Drosophila?' biochemists are now seeking an answer. But they could not now be doing this unless the original question by a sequence of translations had come to be posed for them in these terms. Before this had been done, the question 'Why whenever two X-chromosomes then a female Drosophila' was simply not a question in biochemistry. Now it is. We are still unsatisfied, but what are we unsatisfied about is now biochemistry.

It seems possible that in the not very distant future the biochemists and the biophysicists will offer us a molecular model which will explain how it comes about that certain molecular configurations, by their chemically necessary actions and interactions, now result in a female Drosophila, or in a Drosophila with peculiarities in its wings for instance. Such a model is only beginning to take shape.[1] Enormous amounts of work remain to be done before it can be used satisfactorily in genetic thought. Let us suppose, however, that this biochemical model is completed and does satisfy the geneticist's needs in the sense that genetics becomes a branch of biochemistry. Any 'gene' of genetics would then, it seems, become the name of a particular molecular sub-configuration. This must surely be the goal of one great branch of biochemistry and biophysics. When, or if, this goal is reached, it will presumably become possible, at least in theory, to write down in chemical terms the complete constitution of an egg and to say that for inescapable chemical and physical reasons this egg, in normal conditions, must now develop into an animal having these peculiarities and not those. It may of course turn out that practical difficulties will put this achievement beyond our reach, but in principle it may be thought of as possible. Genetics would then have been 'explained' in terms of chemistry and physics, rather as chemistry appears to be in process of becoming 'explained 'in terms of physics.

In this so far unrealised situation what will be the situation of genetics? Can it continue to have an existence of its own as a discipline in which are made statements characteristic of it alone? Woodger[2] calls mating descriptions the characteristic statements of genetics. One must suppose that this phrase may be taken by extension to cover the statements made by geneticists studying micro-organisms which lack the more familiar forms of sexual reproduction. Mating descriptions, in the imagined situation of the future, will surely remain possible, but will they not become incomplete in a new way? So long as we cannot know in relevant chemical terms the complete characteristics of an egg or of a sperm we have no duty to try to write them down. If we do know these must we not write them into any mating description as we now write 'genes'? And if we do so write them into future mating descriptions will not mating descriptions become statements in chemistry?[3]

Now, what, in the imaginary future, will happen to evolutionary theory? Evolutionary theory, as has been said, has for half a century or more adopted the language of Mendelian genetics. But this language has, like everything else, had a history. Initially, the units of Mendelian genetics, during its development at the beginning of the century (regarded as causal agents and later assigned physical dimensions and positions on chromosomes) came trailing clouds of one-to-one continuing causality. The individual gene came to be named, for this reason, as it were backwards from what it had been at first supposed uniquely to cause. In the early nineteen-twenties, for instance, an undergraduate reading Botany or Zoology could hardly do other than think of a recessive gene which, present twice, uniquely 'caused' vestigial wings in Drosophila. He was taught to call it in speech and writing 'little v' because little v was the initial letter of what it was supposed uniquely to cause, namely vestigial wings. If a fly had little v twice it had vestigial wings, and if it had not then it had not vestigial wings. This was the proper language of naive Mendelism – the 'scientific Calvinism' of the time of Bateson. This language is still in use, but its appropriateness even in short-term Mendelian genetics as it has developed since Fisher published *The Genetical Theory of Natural Selection* has become less and less. It has been clearly recognised since 1930 that the causal relation to be supposed between gene and quality such as wing-shape is certainly not always a simple one-to-one relation. In the genetics of micro-organisms the 'qualities' considered may be the possession or non-possession of enzymes and here Horowitz and Leupold[4] for instance have argued that there may indeed be a simple one-to-one relationship between the possession of one 'gene' and one enzyme, and the occurrence of the different blood-groups in man may furnish another example of a similar relationship. But enzymes collaborate and genes interact. The 'qualities' of the geneticists concerned with making mating statements about Metazoa and Metaphyta commonly result from this interaction in the context of the metabolism of a developing egg. For this reason the custom, from which we do not seem yet to be able to escape, of naming genes to commemorate the qualities which they were once naively thought uniquely to determine has an inappropriateness in neo- or sophisticated-Mendelism which it did not have in the past days of naive Mendelian model-making. At the very least it is responsible for an awkwardness in neo-Mendelian language, if not for an uncomfortable approach to near nonsense, even when it is used in speaking or writing about far from nonsensical short-term genetic experiment. If, or when, the biochemists and biophysicists provide a unit for short-term genetic thought which is really a definable molecular configuration, can this language possibly continue to be used? If not, then what is to take its place? Unless this is given proper consideration ahead of the event, it seems likely that the labours of biochemists and biophysicists will lead to a situation of great confusion.

Our linguistic difficulties seem to have been made even more acute of recent years by some work in short-term genetics, particularly that of Waddington[5] and Bateman[6], undertaken with the clearly expressed aim of providing long-term evolutionary theory with a genetic model more informative than those previously on offer. This has brought out into higher relief the inappropriateness of our habit of naming genes for the characters for which they were once thought to be simply responsible even in short-term thinking, and still more seems to have rendered it altogether unsatisfactory in a language to be used in long-term evolutionary theory.

What has happened is this: Waddington has now shown in a number of examples, for instance that of the crossveinless and of the bi-thorax condition in Drosophila, that a quality very strongly selected for can in a few generations become 'due' to genetic situations which did not as a rule lead to it before the start of this strong selection for it. Flies have for long been known to lack the crossvein in their wings when they possess a certain recessive gene, little cv, twice. This crossveinless condition was described as being 'due' to the double dose of this gene. It has also for long been known that the crossveinless condition could be brought about in a large proportion of individuals by heating the pupae of wild-type Drosophila which lacked these genes altogether but carried instead the normal allelomorphs to them. Such flies are called phenocopies, using Goldschmidt's term. The crossveinless condition brought about in this way would not of course be expected to be inherited. In the wild-type flies used by Waddington in his work it was known that occasional crossveinless individuals might appear without the heat treatment. Waddington most rigorously selected, within the wild-type population he started with, individuals showing crossveinlessness however they attained it. At the beginning of the experiment the only crossveinless flies he had were those which had attained this condition 'because' they had been heated as pupae, that is to say they were phenocopies. After about a dozen generations of this intense selection in which only cross-veinless flies were allowed to breed he obtained crossveinless flies from pupae which had not themselves been heated. The crossveinless condition in these flies was inherited by their offspring. It was easily shown that they still did not possess the gene little *cv*. They were crossveinless for a new genetic reason. Repetition of the work gave other strains of flies crossveinless for still other genetic reasons. Similar results have been obtained in other cases. Waddington calls this process 'genetic assimilation', and has proposed a most elegant conceptual model to explain it. [5]

Genetic assimilation, thus defined, seems to be well established as a reality. Indeed it may be argued that it should be accepted as a logical development of Fishers' original work on the interaction of genes in which, developing the previous investigations of Timoféeff-Ressovsky, he showed that a gene could in general be expected to have one effect on the metabolism of an egg when acting as a unit within one total genetic situation and another when acting within another genetic totality. The language of genetics, as does the language of any science, struggles along behind its concepts. It just happens that in genetics the lack of correspondence between language and concepts had become exceptionally serious.

If through secular time by genetic assimilation a character selected for can change, so to speak, its genetic basis, then information about the genetic basis of a character now is not information about the genetic basis which that character may have had in the past or may have in the future. If this is once admitted, can the language of genetics, in which the gene is named for what it seems to do now, be a language in which long-term evolutionary theory can be treated at all?

Consider two populations, one of lampshells which have remained practically unchanged (so far as can be judged from their fossilised remains) over immense periods of time, and the other of cephalopods in which obvious changes have occurred during their fossil history. We have no reason whatever to suppose that the mutation rate or the genetic recombination rate was low in the lampshells and high in the cephalopods. Indeed we have reason to think it likely that both populations have always been genetically on the boil with mutation and recombination constantly occurring. Why did one remain apparently unchanged while the other changed?

If genetic assimilation is a fact, as it most certainly seems to be, we have absolutely no reason to suppose that two lampshells looking alike but living at times separated by very long periods indeed had the same characters for the same genetic reasons. All we can say of them is that they look alike and that we therefore suppose that they must each have had some one of what must now be seen as an indefinite number of totally unknown and unknowable genetic complexes each such that its possessor would have the appearance which, because we see it in both, we must take to have had selective value at both of the two widely separated points in time at which they respectively lived.

But is this worth saying? It seems no more than an explanation of what did happen (two lampshells at widely separated points in time which looked alike) in terms of a lampshell-producing genetic system in which nothing need remain constant between the two points in time except the production of lampshells looking alike. In what sense can this genetic system rightly serve as an explanation at all? It is not easy to feel satisfied with an explanation of what did happen to a population of lampshells in terms of an underlying genetic system in which any of an indefinite number of things may have happened, for all we can tell, without disturbing the production of lampshells looking alike. If the nexus between the underlying genetic system and the appearance of the lampshells over very long periods is not a constant one we seem to lose the particulate genetic system with genes as its units as an explanation of evolution by natural selection. It may be that further work and further thought can re-establish an explanatory status in this context of the underlying genetic system, but its status seems at the moment to be in peril. If the explanatory status of the underlying genetic system has really been lost – if we really may suppose that over very long periods qualities of selective value can change their genetic basis – information about the genetic basis of a quality now has no very clear relevance to evolutionary thought. Here we are left, in thinking about lampshells, with Natural Selection, but we seem to have lost genetics. We are left, in fact, with Darwin.

If we turn now to cephalopods which, unlike the lampshells, did change over a long period of time, the underlying genetic system seems of little more help to us. We can only say, once more, that Natural Selection here favoured differences in appearance, and that the underlying cephalopod-producing mechanisms brought about the production of these differences by some one of an indefinite number of possible chemical pathways.

In the two cases, that of the lampshells and that of the cephalopods, all that is really being said is that the mechanism of heredity, as we now know it, is of such a kind as to permit the evolution of species by natural selection. If we believe that we understand the mechanism of heredity, and if we believe that the evolution of species by natural selection has occurred, and is probably still occurring, it would indeed be surprising if the one did not permit the other.

We may be about to receive from the biochemists and the biophysicists a chemical and physical explanatory model of genetical and developmental processes to be seen now. It appears from a consideration of the situations described above that the important word in the previous sentence is 'now'. Information supplied by biochemists and biophysicists about the parts played by molecular configurations within the total genetic make-up of an egg now is not information about the parts these configurations played in the past, or may play in the future, if genetic assimilation is a reality. These parts played in the past, and to be played in the future, are, and must, it now seems, always in principle be, simply unknowable. The original simple gene named to commemorate what it was once naively supposed uniquely to cause has been robbed of all permanent and persisting relation with any particular peculiarity of its possessor by geneticists just when it may be going to acquire in short-term genetics a new status conferred upon it by the biochemists. If it does acquire this new chemical status can it then have any but a purely chemical name and a purely chemical meaning? How are evolutionary theorists going to use molecular configurations which can only be known now, when they are thinking of geological time, if they may not infer particular genetical molecular configurations in the past from fossil records of that past? And this is just what it now seems that they may not, in principle, do.

Department of Zoology, The University, Hull, England

REFERENCES

1. For a discussion of this see, for instance J. A. Butler, *Inside the Living Cell*, Allen and Unwin, London, 1959.
2. J. H. Woodger, *Biology and Language*, Cambridge University Press, 1952, p. 104.
3. M. Calvin, 'Round Trip from Space', *Evolution* **13** (1959) 362.
4. N. H. Horowitz and U. Leupold, 'Some Recent Studies Bearing on the One Gene – One Enzyme Hypothesis', *Cold Spring Harbor Symposia on Quantitative Biology* **14** (1951) 65.
5. C. H. Waddington, 'Evolutionary Adaptation', *Perspectives in Biology and Medicine* **2** (1959) No. 4, University of Chicago Press.
6. K. G. Bateman, 'The Genetic Assimilation of Four Venation Phenocopies', *J. Genet.* **56** (1959) 443.
7. K. G. Bateman, 'The Genetic Assimilation of the Dumpy Phenocopy', *J. Genet.* **56** (1959) 341.

Note added in proof: This essay was written in 1960.

HEINZ HERRMANN

BIOLOGICAL FIELD PHENOMENA: FACTS
AND CONCEPTS*

At the turn of the century biologists faced both unprecedented success in the application of mechanistic concepts to certain parts of biology and the apparent failure in the mechanistic interpretation of other aspects of life. This dilemma was particularly evident in the field of embryology with its mechanistic interpretations of development by Roux and the vitalism of Driesch's concept of entelechy as the two opposite poles. The history of these trends has been interestingly described by J. M. Oppenheimer (1955).

The term "field" may be regarded as an hybrid expression to include both the mechanistic and non-mechanistic aspects of biological phenomena. The content of the term "field" is illustrated by the abstracts in Appendix I which are taken from representative sources dealing with this problem.**

In biological usage "field" indicates the capacity of groups of embryonic cells to become organized into larger multicellular units, each with its own definite form and function. For example, "field" properties were regarded as an attribute of a geometrically simple aggregate of embryonic cells which gives rise to the T-shaped breast bone of the chick. This peculiar structure is formed not only in its natural location but also in explants outside the organism (Fell, 1939).

Initially the term "field" was a designation for phenomena similar to this example. However, it never became clear whether this term implied the existence of new and specific biological "field" forces or "field" agents which differed from known physico-chemical properties of isolated cells and could not be fully understood through analysis with conventional physico-chemical methods (see Appendix I, quotations 1–4). Also, it re-

---

* Contribution number 41 of the Institute of Cellular Biology at the University of Connecticut.
** The "field" concept has found extensive consideration in textbooks and treatises by Huxley and De Beer (1934), P. Weiss (1939), C. M. Child (1941), and L. G. Barth (1949).

343

mained uncertain whether this term suggested some relation to "field" concepts in physics and thus implied a theory of certain biological phenomena (Burr and Northrop, 1935). It is the purpose of this paper to show that the term "field" as used in biology is of little conceptual or theoretical significance and to consider the limitations of biological concepts in general.

THE EMPIRICAL BASIS OF THE BIOLOGICAL TERM "FIELD"

The meaning of the term "field" can be illustrated by several specific examples.

1. *Regeneration of the alga acetabularia*

Although the term "field" has been used predominantly in connection with observations on multicellular organisms, the main aspects of "field" phenomena can be recognized in unicellular organisms as well. The alga acetabularia has large cells having the shape of an umbrella. If the canopy part of the umbrella is cut off, the remaining handle forms a new canopy of approximately the same size and shape as the original (Hammerling, 1934, 1958).

2. *Regeneration in multicellular organisms*

The term "field" is associated with the problem of various forms of regeneration in multicellular organisms, in particular with the regeneration of limbs of amphibian larvae. When parts of extremities are amputated, the removed portion is reformed with a remarkable degree of likeness of the amputated morphological structures such as hands and toes and of the distribution of muscle, bone, and skin tissues. Similar observations were made with head and tail structures of certain flatworms and with the liver of rats. In all these instances, following the removal of large portions of the original organs, rather close reduplication occurs within a short time (Barth, 1955; Nicholas, 1955a).

3. *Development of an extremity*

The most frequent use of the term "field" occurs in the investigations of embryonic development. For example, the development of an avian or amphibian limb begins with the condensation of mesodermal cells in the body wall. Here these cells proliferate more rapidly in the direction of the

344

long axis of the limb. The properties of a "field" are assigned to this embryonic primordium for the following reasons:

(a) If some of the cells of the initial condensation are removed, the remaining cell mass gives rise to a complete limb.

(b) Division of the limb area leads to formation of two well-formed limbs instead of one; fusion of two beginning limb primordia leads to formation of a single extremity. Thus, the original limb area shows a great lability in its organization (Nicholas, 1955b).

(c) The rapid elongation of the limb primordium in one direction and the formation of its typical structures depends upon the interaction between two cell types, the ectoderm and mesoderm, of the limb bud (Saunders, 1948; Zwilling, 1956).

4. *Induction*

The term "field" has been used extensively in amphibian embryology in reference to the chordamesoderm (Holtfreter and Hamburger, 1955). At an early stage of development (blastula), the amphibian embryo has the shape of a hollow sphere. A part of the surface invaginates and forms the mesodermal layer of the embryo. The mesoderm gives rise to the tissues of the central axis of the embryo, the vertebral skeleton and the back musculature. One of the first products of differentiation in this developmental process is a thin axial, cylindrical tissue called the chorda. Hence, the whole central portion of the mesoderm is called the chordamesoderm. The properties of the chordamesoderm leading to the designation of a chordamesodermal "field" are the following:

(a) If a fragment of the early chordamesoderm is excised and grafted elsewhere in the embryo, it develops into the complete pattern of axial structures in this foreign region.

(b) If suitable undifferentiated material is implanted into the chordamesodermal "field" district, it is incorporated and becomes a part of the developing tissues without disturbing the characteristic developmental pattern.

(c) The chordamesoderm also influences the overlying tissue. As a result of this action, known as induction, the neural tube arises from the undifferentiated ectoderm.

5. *Pigment patterns*

Another type of "field" phenomena is the pigmentation patterns, for example, of the amphibian skin. These patterns are brought about by the collective behavior of neural crest cells in their response to each other and to their cellular environment (Twitty and Niu, 1948).

GENERAL CONSIDERATIONS OF THE TERM "FIELD"

From the preceding examples, it is apparent that the term "field" is used to denote one property of frequently observed biological systems. This is the capacity of an embryonic or regenerating tissue to produce a functionally and structurally complete unit, e.g., a limb, a liver, or a head from a group of cells with a labile organization. The designation of chordamesodermal "field" refers, in addition to the immediate observation of a small piece of embryonic tissue, to the postulate that this area of the embryo will form a chorda and subsequently the vertebral skeleton. It is also predicted that chordamesoderm and ectoderm stand in the relation of "the former inducing the transformation of the latter into nervous tissue". Moreover, it can be postulated that this interaction is limited to definite periods of development and that interactions such as between the chordamesodermal "field" and the entoderm do not occur. Therefore, terms like limb "field" appear as generalizations of a very low order (Woodger, 1948), which have the character of simple constructs (Margenau, 1950) or laws (Campbell, 1957).

It must be realized, however, that the degree of generalization in using the term "field" is as low as in the following example:

"The X-ray spectrum of silver is uniformly associated with a density of 10.5, a melting point of 960°, solubility in nitric acid" (Campbell, 1957, l.c. p. 116). This definition of silver is an isolated instance of a uniform association of properties without relation to other similar statements, e.g., comparable definitions of other metals. As such the statement does not show a logically necessary relationship of the enumerated properties and has no explanatory function.

The apparently simple term "field" refers to a great number of unknown processes.* Involved are cell proliferation, changes in the properties of the proliferating cells, and physiological and biochemical processes controlling such changes and mediating the interactions of the cells

* The main types of processes subsumed under the term "field" are given in quotations 5–9 in Appendix I.

in a "field". The use of the term "field" as a general statement of the occurrence of such an interdependence is trivial. Other numerous forms of tissue interactions such as controls by endocrine or nervous systems are not denoted by a general term. Another objection is that a general term "field" is actually misleading since it suggests a much greater uniformity of the mechanisms of interactions than actually does exist and deters from the exploration of the nature of specific forms of interaction.

While the term "field" was used originally as a designation of something unknown, it is apparent from the following quotation that later on a vague explanatory function was attributed to it.

"...once an apical region is produced, it then exerts an influence on other organs and regions within the old tissues of the fragment: this influence is, however, limited in extent. Accordingly, the apical region has been called by Child the 'dominant' region. In terms of the field-concept, the apical region establishes a field of a certain extent, which it dominates so as to control the morphogenetic processes of the other regions of the field. The control is exerted in such a way that the various morphogenetic processes occur in harmonious relation with each other: this is because it exerts its control through the establishment of a field." (Huxley and De Beer, 1934, l.c. p. 285) In this sense the term "field" is tautological since a morphogenetic process of an organ is called a "field" phenomenon and the meaning of a "field" is given as the morphogenetic process of certain organs.

An explanatory function was attributed to the term "field" by relating it to the more familiar explanatory "field" concepts of physics (Burr and Northrop, 1935; see also Appendix II). Actually these two terms belong to quite different classes of logical categories. The basic difference between the biological term "field" and the "field" concept in physics (Margenau, 1950, l.c. p. 194) is in the degree of abstraction and generalization associated with the terms. A "field" of mechanics is defined by a certain magnitude S (a strain or force) which is given as a continuous function of space coordinates x, y, z, at a given time t. In electromagnetic "fields", a force F exerted on an electric charge H is a continuous function of the space coordinates at a given time. The properties of the electromagnetic continua are also expressed by the Maxwell equations which define its wave character.

Inadequate as such a statement about physical fields must be in this context, it shows that the term "field" as used in physics is a state of a system defined by state variables. The changes in the state of such a system are expressed in the form of differential equations. The related state variables are of the most general nature, such as location in space or electrical charge. There is no reference to the specific properties of the object which carries the charge or to the system which generates the electromagnetic forces.

It is expected in developing such a system of equations that definition of specific instances can be deduced from these general expressions. Actually, the practical application of field mathematics consists in just such a deduction. Whenever a specific concrete instance can be derived as a logically necessary consequence from these generalizations, an explanation of the specific situation has been achieved.

This is in agreement with the analysis of the meaning of explanation by Campbell (1957, *l.c.* p. 113) and Hempel and Oppenheim (1948). According to the latter authors, explanation consists in relating certain antecedent conditions (I) by a general law (II), to the observed phenomenon (III). In this scheme I + II is called the explanans and III the explanandum. In the explanation III must be exhibited as a logical consequence of I + II. Thus the immersion of a thermometer into hot water (I) is related by the theories of statistical mechanics (II) to the rise of the mercury column (III).

None of the properties given above for physical fields can be found in statements about biological "fields". The definitions of biological "fields" given in terms of lability, self-reconstitution, and elaboration of functional and structural wholes seem general only because they remain vague. Even though the properties of biological "fields" can be quantitated (see Appendix 1, quotation 6), the meaning of such quantitative parameters is unclear. For these reasons the biological term "field" is not more of a generalization than the term limb forming area and no significant role in an explanatory theory can be attributed to it. One of the main shortcomings of the term "field" as an explanatory category is indeed its failure to give rise to abstractions which are fruitful in the construction of theories and for their explanatory significance as extensive generalizations.* It will be important in the following sections of this paper to elaborate this point.

* In this sense, the concepts of organismic biology as developed by Beckner (1959) or the concepts of directive correlation suggested by Sommerhoff (1950) are similarly restricted in their generality.

APPLICATION OF MOLECULAR CONCEPTS TO THE EXPLANATION OF
BIOLOGICAL "FIELDS" *

During the last decades attempts to explain biological phenomena in terms of molecular events have rapidly increased in number: the numerical relations in the transmission of hereditary characters and the cytological correlates of genes are being reinterpreted in physico-chemical properties of macromolecules of deoxyribonucleic acids; growth is described as biochemical synthesis of proteins and nucleic acids; the mechanical and physiological laws of muscle work are stated in terms of interactions of protein molecules and certain metabolites; the initial phase of vision and the histology of the perceptors of light are related to changes of vitamin molecules. Similarly, attempts have been made to explain the process of morphogenesis and development in molecular terms. Some of the relevant results are indicated under the following headings.

1. *Formation of microscopic subcellular and cellular structures from molecular subunits*

(a) The shape and pattern of microscopically visible units of connective tissue or of muscle can be explained by the interaction of protein molecules according to general theories of the nature of chemical bonds *(Biophysical Science: A Study Program*, J. L. Oncley, ed., 1959).
(b) Molecular mechanisms for the interaction of subcellular particles have been described in connection with the interdependence of structure and function in the production of metabolic energy and in photosynthetic assimilation. Interaction between several types of molecules is required in forming microscopic structures such as the chloroplasts or the mitochondria. The spatial alignment of catalytically active enzyme molecules in such microscopic structures is essential for their effective function in metabolic processes *(Biophysical Science: A Study Program, op. cit.).*

2. *Effect of chemical substances on cell shape*

(a) The formation of round individual cell buds rather than of long filiform undivided yeast cell threads depends upon the depolymerization

---

* The substance of this section is documented in *Biophysical Science: A Study Program,* (J. L. Oncley, ed.) 1959.

of cell wall proteins by metabolic reduction catalyzed by flavin enzymes (Falcone and Nickerson, 1959).

(b) The formation of morphologically distinct gonads in primitive coelenterates (Loomis, 1957) and the differentiation of fungi (Ward, 1959) are dependent upon the content of carbon dioxide and other simple metabolites.

3. Specific reactions of cell growth and tissue organization to chemical substances

(a) A protein, isolated from several natural sources, increases several fold the growth specifically of two types of nervous tissue (Cohen, 1958).
(b) A protein fraction was found which leads to transformation of undifferentiated embryonic ectoderm to neural tissue. The type of neural tissue resulting depends upon the physical state of the applied protein fraction (Yamada, 1958).

4. Cell interaction

The presence at the threshold concentration of a chemical compound leads to the aggregation of individual amoeboid cells of the slime mold *Dictyostelium* into a multicellular mass (Bonner, 1947; Sussman, 1958). These cells do not merely follow a chemical stimulus in their migration, but also interact with their surfaces in forming a stable aggregate which moves as a whole and differentiates into a fully developed slime mold.

Numerous immunological and molecular schemes have been considered as mechanisms for surface interaction between cells, *viz.*, in fertilization (Perlmann, 1959), cell aggregation (Moscona, 1957; Weiss, 1958) and cell cohesion (Schmitt, 1941; Steinberg, 1958), and in the formation of the embryonic neural tube (Brown, *et al.*, 1941). Definite metabolic pathways have been proposed as explanation of metabolic cell interactions which are the equivalent to the metabolic gradients of earlier embryological literature (Flickinger, 1959; Herrmann, 1960).

These examples show that the formation of structural aggregates as well as the initiation and progress of morphogenetic processes can be interpreted in terms of molecular reactions (see Appendix III). Are, then, molecular concepts and the theories of physics and chemistry adequate as a basis for an explanatory hypothesis of developing systems as called for by Woodger (1948) in his lucid and far-sighted "Observations on the

350

Present State of Embryology"? Woodger himself does not give a decisive answer to this question, but two statements in his article indicate a convergence of the position taken in this paper with Woodger's line of thought. "... In the same way, simply to assert that gradients exist in embryos does not help at all. The gradient concept will only be really useful when it enters genuine explanatory hypotheses, such as a hypothesis concerning cilia formation, a hypothesis concerning invagination of endoderm and stomodaeum and so on. Now a hypothesis concerning cilia formation will involve a hypothesis about submicroscopical cell architecture. It will necessitate hypotheses about the structure of cilia, about the state of a cell in which cilia formation is going on, and about the sort of conditions external to such a cell which can bring about or inhibit this state. Similarly with invagination. This will presumably involve hypotheses regarding the mutual tensions and pressures between cells in an epithelial aggregate, and the relation of these tensions and pressures to the tonicity of the surrounding water. Thus the consideration of the sort of further hypotheses required leads us to an analysis of what might be called the *elementary* processes of development, and takes us into the field of general physiology, where a good deal of preliminary work has already been done. Moreover, it seems likely that if such hypotheses could be constructed and tested in relation to echinoderm development, they would at once have a wide range of applicability to other fields of embryology. They would be of relatively high order and would take us a long way towards the kind of key hypotheses of which I have spoken."

"Now it seems clear that we cannot hope to make much progress towards a theory of submicroscopical cell architecture without the collaboration of biochemists and X-ray crystallographers. We shall need the help of their hypotheses in framing our own. But it must be repeated that the statements to be derived from the missing hypotheses are embryological statements containing, therefore, embryological set-designations. But if, from a chemical hypothesis C, containing no embryological set-designations, an embryological statement E is to be derived as a necessary consequence, this will only be possible if C is conjoined with a definition stating that the embryological set-designations of E are abbreviations for combinations of the chemical set-designations of C. Failing such definitions the hypotheses we require must be mixed hypotheses containing both kinds of set-designation. This necessitates a two-way collaboration.

351

Moreover, there are other considerations bearing on this point. It is in the highest degree likely that the required hypotheses, if and when they are discovered, will seem extremely unorthodox. Not only because key hypotheses have usually had this property in the past, but because there are positive indications that they will have it in the present case. . . ."

In elaborating Woodger's position, and observing, in particular, the requirement of a logically necessary consequence in deriving the explanandum from the explanans, the following considerations are presented.

By a reinterpretation of biological "fields" in terms of molecular concepts, it is possible to meet the similar requirements of Woodger (1948) and Hempel and Oppenheim (1948) for an explanation of observed phenomena. Physico-chemical concepts of molecular properties are part of general theories which relate the molecular antecedents to the supra-molecular explanandum. The transition from molecular to supra-molecular levels of organization has been discussed by Hempel (1951). It should be repeated here that, *e.g.*, the transition from molecular concepts of muscle contraction to contraction of fibrils and of the whole muscle does not pose principal difficulties (Herrmann, 1953). The concept of molecular interactions provides conceptual continuity by which the explanandum of specific aspects of cell behavior can be derived as a logically necessary consequence from the molecular explanans.

Therefore, the suggestion is made to abandon the apparent explanation of developmental processes by biological "field" concepts and to adopt instead the concepts of molecular interactions as a basis for explanation. Developmental processes seemed, up to recent times, least amenable to explanations on the molecular level. Does, then, the adoption of molecular concepts for the explanation of developmental processes suggest that biological phenomena in general can be explained only by the use of concepts of physics and chemistry? The answer to this question depends upon the kind of explanation which is expected. If it is satisfactory to replace the less familiar by the more familiar or the seemingly complex by the apparently simple (Campbell, 1957, *l.c.* p. 113 ff.), or to explain in terms of low order generalizations, biological concepts can provide explanations as, for example, the ideas of evolution or the laws of genetics. Explanations of biological events by introduction of theories with far-reaching generalizations can arise from the use of molecular concepts only. The reason for this is plain. By definition, physics and chemistry are those branches of science which have developed concepts and theories for the most general forms of matter and energy. Therefore, these concepts must form the basis for the introduction of general theories for explanation of specific and complex forms of matter and energy. Since biological organisms are among the most highly organized forms of matter, their logically consistent interpretation will have to depend upon the elements of the theories concerning abstract general forms of matter as dealt with in the physical sciences.

APPENDIX I. QUOTATIONS INDICATING THE CONTENT OF THE TERM "FIELD"

1. *Statements about the general meaning of the term "field"*

(1) "In general, the term *field* implies a region throughout which some agency is at work in a co-ordinated way, resulting in the establishment of an equilibrium within the area of the field. A quantitative alteration in the intensity of operations of the agency in any one part of the field will alter the equilibrium as a whole. A field is thus a unitary system, which can be altered or deformed as a whole; it is not a mosaic in which single portions can be removed or substituted by others without exerting any effect on the rest of the system.

"The agencies operative within biological field-systems have not yet been identified with certainty. In many cases, as in the regeneration of hydroids and worms, it has been suggested with a good deal of probability (on the basis largely of experiments on the differential susceptibility of the regions of the system to toxic and narcotic agents) that they concern a gradient in the rate of some fundamental metabolic process. However, the precise nature of the processes in question is irrelevant to the general discussion, and for the time being we shall refer to them under the noncommittal term of *activity-gradients*. In other cases, such as the limb-producing capacities of the Urodele limb-field which concern the morphogenesis of a single restricted region, the simplest assumption is that there exists a graded concentration of the specific chemical substances responsible for limb-production and laid down by chemo-differentiation.

"In all examples so far studied, the agencies in question appear to be graded quantitatively in somewhat simple patterns, frequently (Hydroids, Planarians, many eggs) in the form of a single gradient with high point at

one end and low point at the other, the direction of the gradient coinciding
with the long axis of the organism. It was this aspect of biological field-
systems which first attracted attention, and led Child to formulate his
theory of "axial gradients". It is preferable to combine the two ideas in
a single phrase by speaking of *field-gradient systems*." (Huxley and De
Beer, 1934, *l.c.* pp. 276–278).

(2) "By a biological field-system, then, is meant a system which has
the following characteristics. It is a spatial unity, in respect of certain
properties at least. It is also an interrelated unity, in the sense that it may
be deformed as a whole, and that, in regard to certain essential biological
phenomena, events in one portion of the field have an important influence
on events in other portions. It represents an organized whole with certain
unitary activities, which must be studied as a unit, not merely as summative
resultant of its parts and their activities." (Huxley, 1935, *l.c.* pp. 269–70).

(3) "A district whose activities show field character may be called a
field district. A field district is characterized by the fact that none of its
elements can be identified with any particular component of the field,
although the field as a whole is a definite property of the district as a
whole. Just remember the oversimplified but instructive example of the
magnet in which it is likewise impossible to identify the dynamic characters, "south pole" or "north pole", with any intrinsic differences of the
iron particles at either end." (Weiss, 1939, *l.c.* p. 293).

(4) "A field is primarily an entity and not a mosaic. Its structure has
the capacity of self-conservation, and in this regard it is merely a special
kind of what we have called above ... physical systems. The stable
structure of a field, i.e., its equilibrium, is characterized by a definite and
unique distribution of properties, the field pattern. Following external
disturbances this pattern can be restored, within limits, to its typical
original form. (Weiss, 1939, *l.c.* p. 293).

2. Statements about the organization of a "field"

(5) "From the behavior of fields, such as the animal and the vegetative
field of the sea urchin, it can be seen that each field has a focal point in
which its intensity reaches a maximum; with increasing distance from this
center the field intensity declines. This gradual decrease of the field intensity around the imaginary field center has led to the conception of
field gradients. The field gradient in any given point of the field can be

354

defined as that direction along which the field intensity falls off most rapidly. It must be borne in mind, however, that field gradients are merely convenient symbols to indicate the direction and rapidity of the decline of the resultant field action; as physical entities, they are just as fictitious and non-existent as is the field center." (Weiss, 1939, *l.c.* p. 291).

(6) "Fields, at least in the most specialized forms, are heteroaxial, which means that their structure varies along the three coordinates of space, and heteropolar, which means that their effects differ in the two opposite senses along the same axis." (Weiss, 1939, *l.c.* p. 293).

(7) "The polarity (of fields) is associated with two other characteristic features of such total field-systems. (a) polarity is the expression of a gradient of activity (or of material, or of both), running along the polar axis; (b) the high or apical end of this gradient is in general associated with the appearance of that portion of the center of organization which, as we have seen, inhibits the formation of other centers. This center is usually spoken of as the dominant region.

"The gradient in activity has been demonstrated especially by means of differential susceptibility to poisons, and differential staining with various stains. . . . It appears also to be associated with bioelectric phenomena, the apical end of the gradient being electro-negative (externally) to the basal end. . . . The nature of the activity or activities thus graded is still uncertain, although in some cases they are associated with differences in oxygen uptake or CO_2 production. . .; but even so it cannot be said with certainty whether the metabolic differences are the cause or the effect of the morphogenetic gradient.

"When a gradient in visible material substances is associated with polarity, as for instance with the yolk and/or fat content of various eggs, it appears to be a secondary consequence of the activity-gradient for in several cases such materials can by centrifugation be restratified in a new gradient at an angle to the old without interfering with the egg's inherent polarity. . . . Once present, however, such material gradients may react back into the chain of morphogenetic processes; this is also seen in certain centrifuge experiments, e.g., on frog's eggs where a brain will not be formed if the fat content is made too high anteriorly . . . and in Arbacia where the dorso-ventral (not the primary apicobasal) axis can be altered by altering the distribution of certain materials with the centrifuge. . . ." (Huxley, 1935, *l.c.* pp. 271–72).

3. Statements about the relation of "fields" to the material they are made up of

(8) "Field activity is invariably bound to a material substratum. We cannot accept the view advanced by some authors that fields have an existence of their own independent of the materials upon which they act, because it does not seem that this latter, strictly vitalist, notion can draw any conclusive support from the observed facts. . . . Consequently, a field exists only so long as the material substratum exists with which it is connected, and it dies with the latter." (Weiss, 1939, *l.c.* p. 293).

(9) "The specific qualities of the field are due to the presence of a specific substance whose concentration is highest subcentrally. Such fields would therefore possess, in this respect at least, a more or less circular gradient-system, in contrast to the elongated gradient-system of the primary field. However, as later set forth it is highly probable that this chemical aspect of regional fields merely determine the type of organ to be produced, and in some cases also the ease or intensity of its production; whereas the actual process of morphogenesis of the specific organ is due to a regional individuation field, similar in its nature to total morphogenetic fields. Total extirpation of a regional field area, either before differentiation in the embryo or after it even in the adult, leads to non-appearance or non-regeneration of the organ concerned. . ., this is consonant with the view that the supply of some specific substance has been removed." (Huxley, 1935, *l.c.* p. 277).

4. Statements about conceptual aspects of the term "field"

(10) "The field concept is an abstraction trying to give expression to a group of phenomena observed in living systems. Essentially it is but an abbreviated formulation of what we have observed. Being an abstraction, we cannot expect it to return more than what we have put into it. Its analytical and explanatory value, therefore, is nil. Its utilitarian value, however, is considerable. It permits us to bring a certain, if only temporary, order into the observed facts; moreover, the consistency of the order thus obtained convinces us that the underlying principle cannot, after all, be mere fiction but must have real existence." (Weiss, 1939, *l.c.* p. 292).

(11) "The field concept has its roots in purely empirical grounds; i.e., the properties of fields are reconstructed from the observed facts; in this

manner, such field attributes as individuality, heteropolarity, gradation, have been arrived at. Only secondarily, in a move toward hypothesis, has the attempt been made to inject physical sense into the symbolic term. The fact that practically all developmental phenomena exhibit field-like characters in one or the other respect, is, indeed, a strong indication that the field concept is not only a useful circumlocution, but an expression of physical reality." (Weiss, 1939, *l.c.* p. 292)

(12) "The field concept stresses the dynamic nature of developmental organization, as against the static character accorded to it in the old mosaic concept. Every phase of development is the result of interactions between the material whole with its field properties on the one hand, and the material parts on the other. Every germ must be viewed in this double light; as the originator and carrier of organizing activities, i.e., fields; and, at the same time, as an assembly of individual parts exposed and reacting to those field activities. This duality of viewpoint applies to fractions of the germ as well. A group of cells acts collectively through its field, but, at the same time, its constituent cells react individually to the field influences. For example, remember the case of micromeres transplanted into an animal half. . . : first, collectively, they project a vegetative field into the surrounding animal material, and individually they react to this field by forming skeleton, as called for in the vegetative field pattern." (Weiss, 1939, *l.c.* p. 295).

APPENDIX II

In one instance the observation of gradients of electric potentials was mistakenly regarded as a basis for an electro-dynamic theory of developmental processes. This approach is illustrated by an interpretation of experiments by Burr, *et. al.*, in Northrop's *Science and Humanities* (1948, *l.c.* p. 166):

"On a fertilized egg of amblystoma in one of its early stages of embryonic development, when no directly inspectable differentiations of the natural history type were observable, H. S. Burr found, with his experimental apparatus, a definite organized pattern of electron potential differences. Using dyes, he designated an inspectable pattern on the surface of the organism corresponding to the postulated potential differences designated by his apparatus. When the organism was allowed to grow, later, inspectable, natural history differentiations appeared at precisely the locations which his dyes, as guided by his electro-metric readings, had indicated. This suggests quite definitely that electrical differentiation is at the basis of the developing differentiation of the natural history data. It indicates, also, that the requirements of sound, scientific method for solving the problem of biological organization are satisfied. That is, epistemic correlations* have been established between the postulated electrical distinctions of physical theory and the directly observable data of the natural historians' inductive observation of biological organization.

"It would appear, therefore, that in the relationship between the fundamental concepts of chemical theory, thermo-dynamical theory, and the electro-magnetic theory of field physics, a scientific theory for the solution of the problem of organization in biology is to be found. Chemical theory provides the postulated entities at the basis of the material constituents of living organisms; thermo-dynamics provides an understanding of their dependence on energy factors from without; and the electro-dynamic theory provides the irreducible relatedness necessary for an understanding of the organization of the constituents as worked upon by the energy".

It can be seen from this quotation that, in the view of Northrop and Burr, statements about "substance" give information only about the seperate molecular entities and not about their dynamic relations by which they form complex living organisms. This is refuted in the last section of this paper.

By substituting the electric "field" as an explanatory principle, they attempted to establish a magnitude representing supramolecular organization. But, in doing so, they have to disregard the specific properties of the material substrate of the electric "field". This is a necessary prerequisite for physics. The particular properties of the condenser (its metallic composition, color, or shape) which delimit the electric "field" do not enter into the laws of "field" theory.

This portion of physics deals primarily with the distribution of free electric charges, and the system of physical theories does not possess properties of significance except those which are ascribed to it in the postulates. Therefore, the electromagnetic "field" is adequately described by the Maxwell equations.

---

* Correlations of concepts and observations.

358

On the other hand, if a biological system is described in terms of electrical potentials, the number of unknown factors is very great even if the correlation between the observed electrical data and the biological phenomenon is high. It must be realized that, for example, the oxidation-reduction potentials which are observed in certain secreting tissues can be correlated with great reliability with the secretory activity of the respective tissue. However, in this instance the relation of the secretory process and the electric phenomenon is not given like the relation between electrons and field properties in physics. In secretory processes the electrons are transported by a larger series of enzymatic systems, each of which is distinguished by highly specific spatial configurations and specific relations to the other systems. Further, each system can be influenced individually by various means. The systems are often contained in different types of cells, and the electron transport occurs across cell boundaries. In such an instance, electron transport appears as a supracellular mechanism. However, the system is not completely defined unless the entire series of reacting systems is known. This whole body of knowledge would be required in order to give the description of the system the same logical coherence as in the case of physical systems.

APPENDIX III. OBJECTIONS TO THE MOLECULAR EXPLANATION OF DEVELOPMENTAL PROCESSES*

In objecting to this approach, it has been pointed out that because active substances (lithium, metabolic reduction oxidation systems, nucleic acids) are considered to be distributed in a random fashion, these substances cannot effect the formation of a highly organized (non-random) structure from an unorganized aggregate of parts. This difficulty refers to the experimental finding that by the addition of certain substances to the developing organism "field" structures can be elicited or existing "field" structures can be markedly altered.

It was observed, for example, that the addition of lithium ions to sea urchin embryos leads to a predominance of the development of the organs (gut) which develop from the vegetative "field". The fallacy of this is borne out by the following consideration. The "active" substances never

* From the unpublished Proceedings of a Symposium on Concepts in Embryology, 1950.

act on unorganized systems of randomly distributed constituents. Even in the first phase of embryonic development in the fertilized egg, there is a highly organized system in which the distribution of the added substance will not be at random in the moment it is bound to the structures of the cells. Therefore, the substances will exert their action in a non-random fashion and will not create non-randomness out of randomness. They simply modify the existing non-random structure of the embryonic cells. In reacting with the complex structure the active substance is forced into very definite spatial patterns of reaction. Once random substances react with simple forms of structural aggregates, the formation of new and more complex aggregates seems possible without difficulty. A case in which the relation of a random substance to a non-random system in biology can be well exemplified is the process of muscle contraction. There the addition of a "random substance", called adenosine triphosphate, to the highly organized non-random system of the muscle fiber is responsible for the change in the state of the muscle fiber from relaxed to the contracted state. This shows a possibility of how a random substance can very readily and specifically influence the state of a highly organized system. Other examples are listed in section one of this paper.

University of Connecticut, Storrs, Connecticut, U.S.A.

REFERENCES

Barth, L. G., 1949, *Embryology,* rev. ed., The Dryden Press, New York.
Barth, L. G., 1955, 'Invertebrates', in *Analysis of Development* (edited by B. H. Willier, P. A. Weiss and V. Hamburger), W. B. Saunders Co., Philadelphia, p. 664.
Beckner, M., 1959, *The Biological Way of Thought,* Columbia University Press, New York.
Bonner, J. T., 1947, 'Evidence for the Formation of Cell Aggregates by Chemotaxis in the Development of the Slime Mold Dictyostelium discoideum', *J. Exp. Zool.,* 106, 1
Brown, M. G., *et al.,* 1941, 'Density Studies on Amphibian Embryos with Special Reference to the Mechanism of Organizer Action', *J. Exp. Zool.,* 88, 353.
Burr, H. S. and F. S. C. Northrop, 1935, 'The Electro-Dynamic Theory of Life', *Quart. Rev. of Biol.,* 10, 322.
Campbell, N. R., 1957, *Foundations of Science: The Philosophy of Theory and Experiment,* Dover Publications, New York.
Child, C. M., 1941, *Patterns and Problems of Development,* University of Chicago Press, Chicago, Ill.
Cohen, S., 1958, 'A Nerve Growth-Promoting Protein', in *The Chemical Basis of*

Development (edited by W. D. McElroy and B. Glass), Johns Hopkins Press, Baltimore, p. 665.

Falcone, G. and W. H. Nickerson, 1959, 'Enzymatic Reactions Involved in Cellular Division of Microorganisms', in *Biochemistry of Morphogenesis*, vol. VI, IVth Int. Cong. of Biochem., Vienna, 1958, Pergamon Press, New York, p. 65.

Fell, H. B., 1939, 'The Origin and Developmental Mechanics of the Avian Sternum', *Phil. Trans. Roy. Soc. London, Ser. B.,* **229**, 407.

Flickinger, R. A., 1959, 'A Gradient of Protein Synthesis in Planaria and Reversal of Axial Polarity of Regenerates', *Growth,* **23**, 251.

Hammerling, J., 1934, 'Über formbildende Substanzen bei Acetabularia mediterranea, ihre raumliche und zeitliche Verteilung und ihre Herkunft', *Wilhelm Roux' Arch. Entwicklungsmechanik,* **131**, 1.

Hammerling, J., *et al.*, 1958, 'Growth and Protein Synthesis in Nucleated and Enucleated Cells', *Exp. Cell Res., Suppl.* **6**, 210.

Hempel, C. G., 1951, 'General System Theory and the Unity of Science', *Human Biology,* **23**, 313.

Hempel, C. G. and P. Oppenheim, 1948, 'Studies in the Logic of Explanation', *Phil. Sci.,* **15**, 135.

Herrmann, H., 1953, 'An Account of Recent Biological Methodology: Causal Law and Transplanar Analysis', *Phil. Sci.,* **20**, 149.

Herrmann, H., 1960, 'Direct Metabolic Interactions Between Animal Cells', *Sci.* **132**, 529.

Holtfreter, J. and V. Hamburger, 1955, 'Amphibians', in *Analysis of Development, op. cit.,* p. 230.

Huxley, J. S., 1935, 'The Field Concept in Biology', *Leningrad Inst. Zh.,* **10**, 269.

Huxley, J. S. and G. R. De Beer, 1934, *The Elements of Experimental Embryology,* The University Press, Cambridge.

Loomis, W. F., 1957, 'Sexual Differentiation in Hydra', *Sci.,* **126**, 735.

Margenau, H., 1950, *The Nature of Physical Reality,* McGraw Hill, New York, pp. 54–101.

Moscona, A., 1957, 'Development in vitro of Chimeric Aggregates of Dissociated Embryonic Chick and Mouse Cells', *Proc. Nat. Acad. Sci.,* **43**, 184.

Nicholas, J. S., 1955a, 'Vertebrates', in *Analysis of Development, op. cit.,* p. 674.

Nicholas, J. S., 1955b, 'Limb and Girdle', *ibid.,* p. 429.

Northrop, F. S. C., 1948, *The Logic of the Sciences and the Humanities,* The MacMillan Co., New York, pp. 133, 219.

Oncley, J. L., ed., 1959, *Biophysical Sciences: A Study Program,* John Wiley and Sons, New York.

Oppenheimer, J. M., 1955, 'Problems, Concepts and Their History', in *Analysis of Development, op. cit.,* p. 1.

Perlmann, P., 1959, 'Immunochemical Analysis of the Surface of the Sea Urchin Egg – An Approach to the Study of Fertilization', *Experientia,* **15**, 41.

Saunders, J. W., Jr., 1948, 'The Proximo-Distal Sequence of Origin of the Parts of the Chick Wing and the Role of the Ectoderm', *J. Exp. Zool.,* **108**, 363.

Schmitt, F. O., 1941, 'Some Protein Patterns in Cells', *Growth,* **5**, 1.

Sommerhoff, G., 1950, *Analytical Biology,* Oxford University Press, London.

Steinberg, M. S., 1958, 'On the Chemical Bonds Between Animal Cells. A Mechanism for Type-Specific Association', *Amer. Nat.,* **92**, 65.

Sussmann, M., 1958, 'A Developmental Analysis of Cellular Slime Mold Aggregation', in *The Chemical Basis of Development, op. cit.,* p. 264.

361

Twitty, V. C. and M. C. Niu, 1948, 'Causal Analysis of Chromatophore Migration', *J. Exp. Zool.*, **108**, 405.
Ward, J. M., 1959, 'Biochemical Systems Involved in Differentiation of the Fungi', in *Biochemistry of Morphogenesis, op. cit.*, p. 33.
Weiss, P., 1939, *Principles of Development*, University of Chicago Press, Chicago, Ill.
Weiss, P., 1958, 'Cell Contact', *International Rev. Cytol.*, **7**, 391.
Woodger, J. H., 1948, 'Observations on the Present State of Embryology', in *Growth*, Vol. II, Symposia of the Soc. Exp. Biol., Academic Press, Inc., New York, p. 351.
Yamada, T., 1958, 'Induction of Specific Differentiation by Samples of Proteins and Nucleoproteins in the Isolated Ectoderm of Triturus-gastrulae', *Experientia*, **14**, 81.
Zwilling, E., 1956. 'Interaction Between Limb Bud Ectoderm and Mesoderm in the Chick Embryo', *J. Exp. Zool.*, **132**, 157.

JOHN DAVISON

ANIMAL ORGANIZATION AS A PROBLEM
IN CELL FORM*

Early exponents of the cell theory considered the cell to be more than the basic unit of reproduction or synthetic activity. The words of Virchow clearly suggest an atomistic role for the cell (to compare cells with the building blocks of classical chemistry): "Every animal appears as a sum of vital units, each of which bears in itself the complete characteristics of life" (Virchow quoted in Wilson, 1928). This fundamental or unitary interpretation of the cell was attacked by Whitman in his delightful essay, 'The inadequacy of the cell theory of development' (Whitman, 1893). He pointed out, among other things, that structures which are multicellular in certain animals may be formed from fewer or even a single cell in other species. His comments on the then current views are pertinent even today: "We are so captured with the personality of the cell that we habitually draw a boundary line around it, and question the testimony of our microscopes when we fail to find such an indication of isolation". Perhaps the most dramatic extremes of what one might call sub- or non-cellular differentiation are to be found among the Protozoa, especially the ciliates, which may contain whole "organ systems" within the limits of a single reproductive unit, leading some to regard these creatures as acellular rather than unicellular animals. Another observation disturbing to a building block concept of the cell is differentiation without cleavage as studied by Lillie (1906) in the marine annelid *Chaetopterus*. Though imperfect, larval structures may be formed, under appropriate conditions, in the complete absence of cell (but not nuclear) division.

Recognizing that the cell is a somewhat questionable organizational entity, one might ask the question; what, if not cells, are the units of animal organization? This proves to be a rather embarrassing question since there seem to be no other convenient structures to consider. A more realistic question might be; what *kind* of an organizational entity is the cell? In obtaining an answer to this question, we may still be able to

* This work was supported in part by funds from USPH grant H–4527 and by the Research Council of Florida State University grant #60 (27).

JOHN DAVISON

interpret animal organization in the familiar cellular fashion to which we have grown accustomed.

It is the purpose of this essay to examine the properties of the cell as a unit of animal organization. Many points of view might be taken. Mine will be through a consideration of polyploidy as an experimental variable affecting cell number, cell size and cell form, primarily in amphibians. It is not the purpose to attempt to discuss all the many interesting aspects of polyploidy, but only those directly bearing on the problem. Excellent reviews are available summarizing polyploidy in plants and animals and these may be consulted by those interested in the general aspects of this field (Briggs and King, 1959; Fankhauser, 1945, 1954, 1955).

One difficulty in Whitman's analysis depends on the comparison of homologous structures in taxonomically distinct animals. Since we know little concerning the mechanism of gene action in determining the organizational properties of such animals, the comparisons do not readily lend themselves to rigorous analysis. An ideal situation might be to vary simple parameters, as for example, the size and number of cells comprising the animal without simultaneously introducing genetic variables of unknown dimensions.

Such a means is available by producing polyploid animals in which the fertilized egg contains more than the usual two sets of chromosomes. Nuclear and cell volumes are roughly if not exactly directly proportional to the degree of polyploidy, at least in the cases which have been carefully examined (Fankhauser, 1955; Davison, 1959). Haploid animals with a single set of chromosomes are abnormal, nearly without exception, due not only to homozygosity but to other factors as yet unknown (Moore, 1955). One of the simplest means for inducing polyploidy is by the inhibition of the second meiotic division in anuran or urodele eggs by heat or cold treatment immediately or shortly following fertilization. The unreduced diploid egg nucleus combines with the usual haploid sperm nucleus to produce a triploid fertilization nucleus, and after differentiation, a triploid animal. Such animals (and I will restrict the discussion to amphibians) are normal in size and gross structure and proceed with organogenesis and growth in the usual manner (Fankhauser, 1952).

Since diploid and triploid amphibians are of the same size and since the triploid cells are $\frac{3}{2}$ the size of the homologous diploid cells, it follows that the triploid has $\frac{2}{3}$ as many cells as the normal diploid. Similarly a

364

tetraploid ideally would have $\frac{1}{2}$, and a pentaploid $\frac{2}{5}$ the normal number of cells, etc.

Such modifications of cell volume and number present some interesting problems in the control of normal structure. To a geometrician, the simplest solution might be that the homologous cells in polyploid animals all would remain the same shape (geometrically similar). Comparing diploid and triploid cells for example, whatever the shape of the cells may be, they will remain of the same shape if the corresponding linear dimensions of the cells are in the ratio of the cube roots of 2 and 3 or approximately in the ratio of 1.14 for the triploid to 1 for the diploid. Such a solution implies certain basic alterations in the physiological properties of the cellular system. Within any collection of similar solid figures, volume is proportional to any corresponding linear dimension cubed (L^3), while surface is proportional to any corresponding dimension squared (L^2). Therefore, for dimensional reasons, the surface to volume ratio (proportional to L^2/L^3 or $1/L$) would vary inversely as the cube roots of the cell volumes or as 1.14 to 1 for diploids and triploids respectively.

Not only would the surface/volume properties of the cells be changed, (and concomitantly any physiological properties dependent on such considerations) but of necessity any tissue consisting of a single or definite number of cell layers would also be altered. Polyploid kidney tubules would have thicker walls, as would the intestinal epithelium or the epidermis, to choose a few examples. Such deviations from the normal structure would increase in direct proportion to the cube roots of the polyploid/diploid ratio.

Apparently, the cell knows nothing of geometry or the principle of similitude. The actual situation observed in two polyploid tissues is represented in Figure 1. It is clear that cell form is not autonomous, but is altered in such a way that the larger organizational unit (kidney tubule or epidermis) remains constant in overall form. It is as if the cells were molded by the larger structure to fit into the structure irrespective of differences in cell volume or number. Such adjustments are typical of a general class of phenomena described by embryologists as regulations. The problem remains, however, to explain how these regulations occur, a point recently emphasized by Briggs and King (1959).

One approach to the problem is to ask; is there any feature common to the various polyploid series? There do seem to be such constant features. In the epidermis series for example, the thickness of the constituent cells remains independent of the polyploidy. In addition to constant thickness, the kidney tubule illustrates another interesting feature. The radius

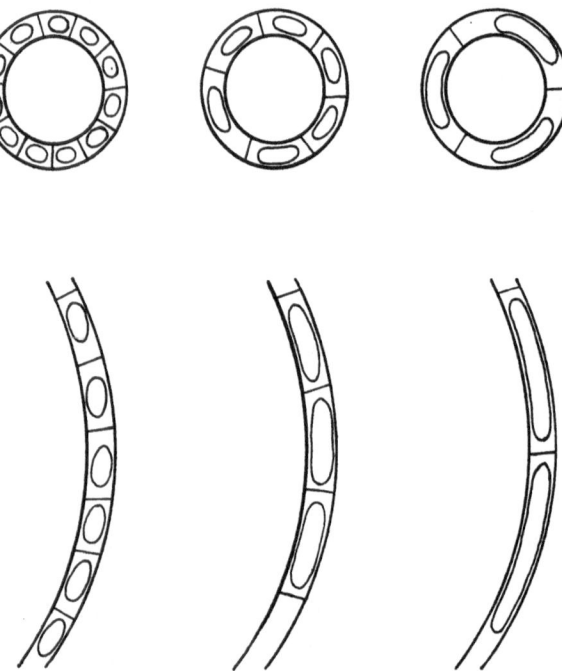

Fig. 1.   Diagrams of cross sections through a pronephric tubule (above) and the lens epithelium (below) in a haploid (left), diploid (center) and pentaploid (right) *Triturus viridescens*. Normal size and structure are maintained with cells of different sizes by adjustment of number and shape of individual cells. (From Fankhauser, 1955).

of curvature remains constant for the kidney cells, with the net result that not only the wall thickness but the diameter and lumen size are unchanged, in this case from haploid to pentaploid. It is significant to note that in each case the dimension which remains unchanged is the one most intimately related to the physiological properties of the cell system. For example, the thickness of the epidermis determines the extent to which the tissue functions as a simple diffusion barrier, as well as the path distance from outside to inside the animal for materials entering or exiting by surface combination or other means of active transport. Similar considerations apply to the kidney tubule with respect to resorption of substances from the tubule or excretion of materials into the tubule from the blood. One might even question whether cell to cell junctions are of any physiological significance in tissues comprising continuous structures such as epithelia, tubules or cylinders. In one case at least, the skeletal muscle fiber, such junctions have been eliminated entirely leaving a syncytial solid cylinder. Also it is well known that rotifers and other aschelminths have eliminated cell boundaries in several of their organ systems.

It is clear that the number of cell to cell boundaries will decrease as the degree of polyploidy increases. If it is true that cell junctions of the type described are physiologically inert in such tissue systems, one may generalize the situation as follows: In homologous polyploid tissues, *cell dimensions perpendicular to physiological surfaces remain unaltered.* It then becomes reasonable that surface-dependent activities can be carried out with equal efficiency irrespective (within reasonable limits, of course) of the degree of polyploidy, and accordingly such functions become independent of both the size and number of cells making up the tissue system. These interpretations placed on polyploid tissues are in complete agreement with Whitman's views stated almost seventy years ago. He commented on nephridial structure in certain invertebrates as follows: "The nephrostome is a nephrostome all the same, whether it consists of one cell, two cells or many cells. Its form and function are both independent of the number of component cells. Cells multiply, but the organ remains the same." And to make his point unmistakably clear he concludes the paragraph: "So far as homology is concerned, the existence of cells may be ignored" (Whitman, 1893). Such a frontal assault on the organizational properties of cells find clear support in more recent studies based on polyploid tissue comparisons.

Certain special cases present some interesting problems. One such case is that of Mauthner's cells, a pair of giant ganglion cells in the medulla at the point of entrance of the seventh and eight nerve roots. A tetraploid has a single pair as does a diploid, while a haploid may have two pairs of Mauthner's cells (Fankhauser, 1955). It is clear that perfect regulation is impossible simply because one cannot create fractional numbers of entire cells. Imperfect regulation of gross structure also characterizes the retina which is profoundly affected in all layers in tetraploid animals (Fankhauser, 1955). The nervous system is especially interesting since it is

367

composed in large part of discrete neurones making specific communication with other neurones, sense organs and effectors. Unfortunately, very little work has been done in an attempt to clarify the effects of reduced cell number in this organ system.

There still remains the problem of the mechanism of regulation of cell form. When the change in shape of a homologous cell in a polyploid series is described, one in no way explains why it had that shape in the first place, not to mention the reasons why the shape changes as a function of volume. The real problem thus becomes the determination of cell form itself. The whole issue is undoubtedly complicated by the fact that cells comprising tissue sheets and layers are in intimate contact with each other and subject to chemical and physical influences from each other and the environment. These influences are all too frequently of unknown character and extent. It might therefore be wise to select a cell which is not involved in a tissue system but rather exists as a free cell. The egg cell is an example of such a cell; a sphere for all practical purposes, which is of course the figure of minimal surface. Similarly one can account for the tetrakaidecahedral form (a fourteen sided figure with eight hexagonal and six quadrilateral faces) of close packed isodiametric cells, simply as the form with minimal surface again, a fact first deduced by Lord Kelvin and discussed at some length by Thompson (1944). The fourteen sided condition is actually approximated in plant pith cells (Lewis, 1946) and more recently a similar configuration has been demonstrated in the annular ligament of the goldfish (Brown, 1958). I mention these cases because they remind us that the form of the cell is frequently a consequence of physical factors imposed from outside the cell (or in the case of the sphere, of the absence of such deforming forces.) Another feature of cellular morphology is the striking similarity cells may bear to fluid drops, emulsions and films, again indicating that cellular, like inert systems, must often conform to physical forces acting to reduce the surface energy of the system to a minimum. Of course, cells are not simply fluid drops or films, yet the many similarities with fluid systems make such comparisons a fruitful avenue for the study of cell form. D'Arcy Thompson stated the view succinctly in his treatise on growth and form. "If the sphere be the one surface of complete symmetry and therefore of independent equilibrium, it follows that in every cell which is otherwise conformed there must be some definite cause of its departure from sphericity; and if this cause be the obvious one of resistance offered by a solidified envelope, such as an egg shell or firm cell wall, we must still seek for the deforming force which was in action to bring about the given shape prior to the assumption of rigidity" (Thompson, 1944).

The amphibian red cell is an excellent example of both a free cell, not involved in permanent contact with other cells, and at the same time, a cell with easily determined form characteristics. The erythrocyte of frogs and salamanders, like that of most lower vertebrates, approximates a thin elliptical disc in form, with an elliptical bulge in the center due to the nucleus. That the nuclear bulge is not requisite for the elliptical form is clearly demonstrated by the plethodontid salamander *Batrachoseps*. The majority of the red cells in this animal are anucleate, presumably generated by cytoplasmic division form circulating nucleated red cells. The anucleate cells, called "erythroplastids", are clearly of elliptical disc form differing only in the absence of the bulge characteristics of the nucleate cells (Emmel, 1924).

An analysis of red cell form in diploid and triploid *Triturus* larvae disclosed that the cells are clearly of different form (Davison, 1957). Figure 2 shows the reconstructed average forms of diploid and triploid red cells calculated from the measured areas and eccentricities of the cells. The ratio of the major to minor axes (a/b) may be taken as an index of the eccentricity of the cell (a/b is not the same as the mathematician's eccen-

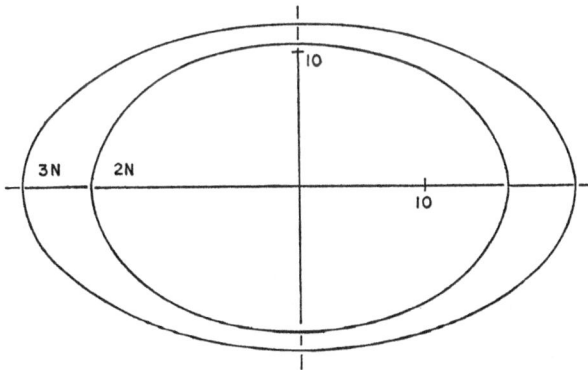

Fig. 2. The mean forms of triploid (outer) and diploid (inner) blood cells of *Triturus viridescens*. The ellipses were constructed on coordinate paper from the measured areas and eccentricities employing the basic property of ellipses that: $x^2/a^2 + y^2/b^2 = 1$
(From Davison, 1957).

tricity but is a valid index of form nevertheless). Knowing the eccentricity and the area of the ellipse ($\pi\ ab$) one can calculate the form of the red cell employing the basic property of ellipses that: $x^2/a^2 + y^2/b^2 = 1$, in which a and b are the major and minor semi-axes of the ellipse respectively. The observed cell areas in Figure 2 are almost exactly in the ratio of 3 to 2 for triploid to diploid, or in the ratio of the cell volumes, since there are no detectable differences in cell thickness. Also the triploid cell is clearly more eccentric than the diploid cell (a/b is 1.82 in the triploid and 1.55 in the diploid cell). It is interesting to calculate the expected variations in the surface to volume ratio (S/V) for the diploid and triploid red cells. As we already pointed out, if the cells remained geometrically similar, S/V would vary inversely as the cube roots of 2 and 3 or as 1.14 for the diploid to 1 for the triploid, a difference of 14%. However, due to the increased eccentricity and constant thickness (the thickness is about $\frac{1}{5}$ the width of the diploid cell), one calculates that the triploid S/V is only about 1% lower than that for the diploid cell. It is reasonable that the observed shape changes leave relatively unaffected the physiological properties of the cell related to the exchange of oxygen between the cell and the tissue environment. Once again however, the observation of an *adaptive* form modification fails to provide us with the *mechanism* by which the form change was effected.

One approach to the problem is by means of a series of questions. The first might be; why should the triploid cell be more eccentric than the diploid? It is clear that one cannot deduce the elliptical disc form from considerations of figures with minimal surface, since it is not a sphere. Armed with Thompson's views, the next step is to inquire into the nature of the forces which have altered the cell to produce the present form. To answer this question, one must ask another. What is the physical environment of the red cell? The most obvious answer is the circulatory system, which one may consider ideally as a system of cylindrical vessels varying in diameter from the heart to the capillary. Accepting an essentially cylindrical environment, and recalling the similarities between fluid and protoplasmic forms, the next step becomes the examination of the form assumed by a fluid drop in contact with a cylindrical surface. Such a step is supported further by the observation that the red cells slide through the capillaries (but not the larger vessels) with one elliptical surface in contact with the capillary wall.

By placing a very large cylindrical vessel on its side (the cylindrical axis horizontal), one finds that a pool of mercury poured on the inner wall will assume the form of a thin elliptical disc. By choosing a sufficiently large cylinder, the mercury pool can approach the red cell in its relative dimensions, excluding the nuclear bulge of course. As more mercury is added to the pool, it becomes a more eccentric elliptical disc but undergoes only slight increases in thickness. Comparing the cylinder to the capillary and the mercury to the red cell, the model mimics the form differences observed in the red cells of polyploid *Triturus* (Davison, 1957). It is also clear that with a constant volume of mercury, the pool will be more eccentric the smaller the cylinder. The model thus suggests some definite causal relationships between capillary diameter, red cell size and red cell form.

One requirement for increased eccentricity in the triploid cell is that the capillary must not increase in diameter with increased cell size. Measurements of capillary diameter disclosed no differences in diploid and triploid *Triturus* as one might expect from the observed regulation of another tubular element, the kidney tubule, already discussed. By choosing larger numbers of diploid and triploid red cells it is possible to break the pooled population into size classes and thus determine how cell eccentricity varies as a continuous function of cell size, and at the same time determine the relationship between cell size, cell shape and capillary diameter. Such an analysis was carried out in diploid and triploid Spanish newts (*Pleurodeles*) (Davison, 1959). Figure 3 demonstrates the relationship between eccentricity and area in the red cells of this animal. It is clear that the smallest cells are circular discs, $(a/b = 1)$ provided that their area does not exceed the average capillary diameter in the tail fin (the arrow on the abcissa). Larger cells increase in eccentricity, first in a curvilinear fashion and then in a linear fashion. The linear portion of the curve satisfies the equation: $(a/b) - 1 = kA_{cell}$. It is immediately clear what the deforming force might be which confers the elliptical disc configuration on the red cell. Since the red cells are larger than the bore of the capillary, they must be deformed as they pass through. One way to demonstrate the mode of passage of the cells (which agrees with observations in the living tail fin) is to cut out paper or cardboard outlines of the cells and attempt to put them through a glass tube of proportional diameter. The smallest circular cutouts can of course freely tumble through the tube since they are no larger than its inner bore. Cutouts with areas and eccentricities corresponding to those of the curvilinear portion of Figure 3 are bent as they are pushed through the glass tube but they do not slide on one elliptical surface. That is, they assume an oblique configuration such that the major ellip-

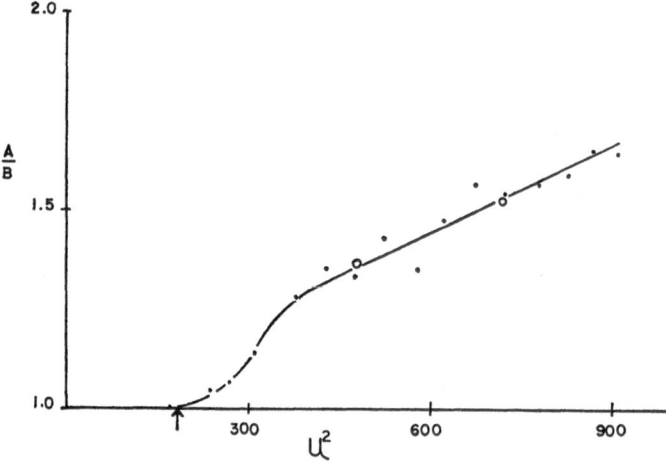

Fig. 3. The relationship between cell area (u^2) and cell eccentricity (a/b) in diploid and triploid *Pleurodeles*. The open circles represent the mean values for diploid and triploid red cells. Other points were obtained by breaking the total sample into size classes of 50 u^2 and plotting the mean values for area and eccentricity in each class. The arrow indicates the average value for the cross-sectional area of the capillary of the tail fin. (From Davison, 1959).

tical axis is not parallel to the capillary axis. The largest paper cells corresponding in relative form and area to the linear portion of Figure 3 can only pass through the tube with one elliptical surface in contact with the wall of the tube and with the major axis of the ellipse parallel to the axis of the tube. These observations are fully consistent with observations made on the passage of the red cells through capillary loops in the living tail fin. Figure 4 indicates in diagrammatic fashion the flowing configurations of: *A*, the circular red cells which can pass through without deformation; *B*, red cells with areas and eccentricities corresponding to the curvilinear portion of Figure 3; and *C*, the largest red cells corresponding to the linear portion of Figure 3. The larger cells are symmetrical elliptical discs as observed in the living condition outside the

372

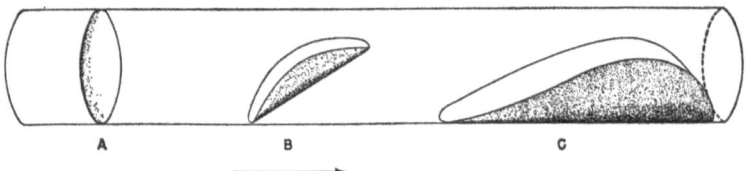

Fig. 4. Diagramatic representation of the flow configurations of: A, circular red cells which may pass through the capillary without deformation: B, red cells with areas and eccentricities corresponding with the curvilinear portion of Figure 3: C. the largest red cells corresponding to the linear portion of Figure 3.

circulation. However, as they are passing through the capillary, they assume a pear or tear-drop shape with the more pointed end trailing the rounded end, (C in Figure 4). It is also clear that the trailing end of the disc is thinner than the advancing end. This configuration can be nicely duplicated with the mercury model by tipping up one end of the cylinder and letting the mercury pool slowly flow along the cylindrical surface in response to gravity. The symmetry of the majority of red cells may be explained most simply as due to random entry into capillaries on successive passages. If the same end repeatedly entered the capillary the ultimate cell shape would probably be a pear-shaped rather than an elliptical disc. As a matter of fact, a small percentage of the red cells actually show a non-symmetrical form approximating a pear. Such a small percentage of cells would be expected to pass through the capillary many times in succession with the same end first, on the basic of chance alone. Unlike the model then, the red cell form is determined by dynamic rather than static forces, but the basic similarity of the systems is apparent.

A second consideration of some interest is the relative effect of capillary size on red cell form. Conditions affecting capillary diameter result in corresponding changes in cell shape, as would be expected from the model system. The larger the capillary, the less eccentric the red cell, under conditions of constant cell volume. The inverse proportionality observed between a/b and capillary area satisfies the mathematical relationship: $(a/b) - 1 = k/A_{cap}$. We have already seen that: $(a/b) - 1 = kA_{cell}$. These two functions can be combined to give a general equation relating cell form to cell area and capillary area: $(a/b) - 1 = K(A_{cell}/A_{cap})$. This equation satisfies the linear portion of the curve in Figure 3 for a large number of different amphibian species varying widely in cell form, cell size and capillary size. A plot of a/b versus the ratio of cell to capillary area results in a curve essentially identical in form to Figure 3 based on the single species *Pleurodeles* (Davison, unpublished).

One special case permits an analysis of the nature of the form determination. *Triturus* maintained in the refrigerator for an extended time are found to have very eccentric red cells and small capillaries, while those maintained at room temperature have larger capillaries and less eccentric cells. The obvious experiment is the examination of the time course of capillary change and cell form change when the cold adapted animal is placed at room temperature. One might anticipate two different results. If cell form is determined at some specific time in the history of the red cell and remains fixed for the balance of the life of the cell, one should expect a very slow alteration in cell form after capillary adaptation with the total time required determined by the life span of the red cell. However, such is not the case. When the animals are taken from the cold and placed at room temperature, capillary diameter rapidly increases to the warm-adapted value, and with a short lag, red cell eccentricity rapidly decreases, the whole process being completed in five or six days. Such a rapid equilibration of cell form clearly favors the view that cell form is plastic and subject to continuous or phasic alteration (Davison, 1959). In fact, Seifriz, many years ago, described the red cell with the terms plastic and elastic, based primarily on the response of the cell to deformations produced by micromanipulation (Seifriz, 1926). That the cell is elastic is clear from the manner in which it can assume the flattened tear drop form during its passage through the capillary, while its plasticity is dramatically demonstrated by the rapidity with which its form is altered when the deforming environment of the capillary undergoes a change in diameter. These observations make it clear that not only is red cell form determined initially by the physical environment of the capillary, but that the form is constantly subject to these deforming forces. The form of the cell is never fully determined but is a dynamic consequence of circulatory morphology.

The limiting form of the elliptical disc is the circular disc ($a/b = 1$) which in the cases cited is characteristic only of the smallest cells which do not exceed the capillary in area. An interesting example is the observation that in rapidly growing *Rana pipiens* larvae, nearly all the red cells are circular or very slightly eccentric discs. The rapidly growing tadpole has very large capillaries, thus permitting many red cells to tumble through and provide a rich blood supply to the tissues. When the animals are fasted, capillary diameter rapidly decreases and red cell eccentricity increases in the manner already described for adult *Triturus* (Davison, unpublished). The two forms, circular and elliptical disc, suggest an adaptive significance associated with these two configurations. When the elliptical disc slides through the capillary, the number of red cells per unit length of capillary (and accordingly the supply of hemoglobin) is limited by the long axis of the cell, since they always pass through in single file. However, under conditions of rapid tissue growth, large capillaries permit a much richer supply of red cells since the number of circular discs per unit length of capillary is limited by the thickness of the cells rather than the length, as is the case for the elliptical form. So the rapidly growing tadpole has circular red cells like the mammal, an animal with similar high tissue oxygen requirements. As tissue demands are decreased, as in fasting the larvae, the system undergoes an adaptive change in morphology resulting in a lower rate of supply of oxygen to the tissues of the animal. Similar relationships between capillary and red cell size may pertain to the situation in mammals. That the biconcave form of the mammalian red cell is independent of the disc and elliptical forms discussed here is suggested by two observations. First, the circular red cells of amphibians are not biconcave, even the anucleate ones described by Emmel (1924). Secondly, the camel and relatives present a most confusing picture, since their cells are not only elliptical, but biconcave as well.

The original problem was the analysis of the nature of the cell as a unit of animal organization. However, in order to proceed with that problem we must first establish the nature of form determination, since form and structure are the essence of organization. We really know very little about the mechanics of form determination. To attempt to understand organization without a knowledge of these basic mechanics is in my view impossible. The red cell offers a unique opportunity to reduce the mechanism of form determination to quantitative and predictable terms.

The study of the red cell as well as earlier investigations on the nature of form determination have served to emphasize certain basic properties of cell structure which may guide future work on problems of this sort.

(1) Cell form is not autonomous but is imposed upon the cell by dynamic structural factors in the cell environment. I do not mean to suggest that intrinsic factors never play a part in form determination, but certainly the expression of endogenous chemical and physical factors is subject to profound environmental influence. (2) Cell form is not determined or final but is plastic and changeable. There is good reason to believe that cell form is in dynamic equilibrium with the cell environment, in much the same way as we regard the metabolism of the cell. Alterations in the physical environment lead to a new equilibrium form which will be maintained until the environment is again changed. (3) The analysis of cell form requires a new approach. Ordinary analytical methods will not suffice. One cannot look at a capillary or kidney tubule for example, and deduce the shape of the structure from first principles. Neither can one understand the shape simply from chemical analysis of the cells making up the structure. To surmount such difficulties requires a synthetic approach in which the first step might be to mimic the configuration with a model system of some sort. Once the key structural and force environment is reconstructed in the model, the *in vivo* structure acquires meaning and the experimental means are then available to test the model hypothesis in the living system.

Unfortunately, we still tend to think in terms of the classical dilemma; does form determine function or does function determine form? We favor the second of these alternatives when we understand the form as soon as we see that it is adaptive. However, this in no way explains the structure any more than does the proof that the particular configuration is under genetic control. Actually the dilemma is not a dilemma at all. Using the red cell as a single example, it is clear that the form of the capillary determines the form of the cell. The cylinder determines the elliptical disc. It is far more in keeping with the epigenetic view, or what some have called the principle of progressive differentiation, to choose a third alternative and escape the dilemma with the conclusion, *form determines form.*

Louisiana State University, Baton Rouge, Louisiana, U.S.A.

REFERENCES

Briggs, R. and King, T. J., 'Nucleocytoplasmic interactions in eggs and embryos', *The Cell* 1 (1959) 537–617. Academic Press, New York.

Brown, H. D., 'Cellular morphology of the annular ligament of the goldfish, *Carassius auratus*', *J. Morph.* 103 (1957) 255–280.

Davison, J., 'A fluid drop model of the elliptical red blood cell', *Experientia* 13 (1957) 472.

Davison, J., 'Studies on the form of the amphibian red blood cell', *Biol. Bull.* 116 (1959) 397–405.

Emmel, V. E., 'Studies on the non-nucleated elements of the blood', *Amer. J. Anat.* 33 (1924) 347–405.

Fankhauser, G., 'The effect of changes in chromosome number on amphibian development', *Quart. Rev. Biol.* 20 (1945) 20–78.

Fankhauser, G., 'Nucleo-cytoplasmic relations in amphibian development', *Internat. Rev. Cyt.* 1 (1952) 165–193.

Fankhauser, G., 'Interactions of nucleus and cytoplasm in cell growth', *Dynamics of Growth Processes* (11 Growth Symposium), Princeton University Press, Princeton, N. J., pp. 68–94.

Fankhauser, G., 'The role of nucleus and cytoplasm', *Analysis of Development,* Saunders Co., Philadelphia, pp. 126–150.

Lewis, F. T., 'The shapes of cells as a mathematical problem', *Amer. Scientist* 34 (1946) 359–369.

Lillie, F. R., 'Observations and experiments concerning the elementary phenomena of embryonic development in *Chaetopterus*', *J. Exp. Zool.* 3 (1906) 153–269.

Moore, J. A., 'Abnormal combinations of nuclear and cytoplasmic systems in frogs and toads', *Advances in Genetics* 7 (1955) 139–182.

Seifriz, W., 'The physical properties of erythrocytes', *Protoplasma* 1 (1926) 345–375.

Thompson, D'Arcy, *Growth and Form*, Macmillan Co., New York, 1942.

Whitman, C. O., 'The inadequacy of the cell theory of development', *J. Morph.* 8 (1893) 639–658.

Wilson, E. B., *The Cell in Development and Heredity*, Macmillan Co., New York, 1924.

R. F. J. WITHERS

MORPHOLOGICAL CORRESPONDENCE AND THE
CONCEPT OF HOMOLOGY*

In a stimulating essay called 'On Biological Transformation'[1] published in 1945, Woodger attempted to make precise the concept of morphological correspondence which, as he showed, is basic to the concept of homology. He indicated how the concept of morphological correspondence might need modifying if it were to cover the same ground as the concept of homology covers in contemporary biological thought. If one considers the ways in which the concept of homology is applied it turns out that it covers a number of related but different ideas. We talk of a wing of a bird being homologous with a leg of a mammal. We talk of the fore and hind limbs of a vertebrate being serially homologous. We find parts being designated as homologous because they develop from embryonic parts which are homologous, and that the adult homology is based on this. We find two parts are homologous because they have common ancestors. We even find chromosomes are said to be homologous as well as genes, and recently the concept of homologous sera has been put forward.

I do not intend to discuss in detail the meaning of all these uses of the term 'homology'. However, an amplification of the concept from Woodger's original formulation and an examination, even if brief, of the application of this concept to some of the wider uses of the term in various meanings of homology will, I hope, show the extent to which use of Woodger's analytical methods can clarify concepts in the biological sciences.

THE APPLICATION OF MORPHOLOGICAL RELATIONS
TO AN ORGANISM

In his paper[2] Woodger pointed out that "When we compare two things we set up a one-one relation or *correspondence* between the parts of one and the parts of the other and proceed to state how the corresponding

---

* This paper is based on part of a dissertation submitted for the degree of M. Sc. in the University of London, 1948.

parts resemble or differ from one another with respect to certain sets of properties".

Woodger pointed out that when the comparison is morphological it must be made with respect to a set of morphological properties derived from certain morphological relations. Morphological relations hold between *parts* of organisms when these parts are considered from a spatial point of view. The relationship between a nerve and muscle when they are *functioning* is not merely a spatial relation, but for purposes of morphological comparison we only consider the spatial relations involved.

Of the many specific spatial relations that could be taken as primitive notions in any axiomatic account of morphology, I propose to take three. (1) *Anterior to* $[A]$ (We shall say that the converse of this relation is *posterior to*). (2) *Dorsal to* $[D]$ (We shall say that the converse of this relation is *ventral to*). (3) *To the right hand of* $[R]$. (We shall say that *to the left hand of* is the converse relation).

In the practical application of any calculus based on these three relations, there must be a set of semantic rules which enables the practical man to use the calculus. In describing these some difficulties arise and must be discussed. The basic problem is to decide on a procedure by which an organism could be orientated.

Let us consider how to orientate an organism with respect to the anterio-posterior axis. It is not sufficient to describe as the anterior end that end which possesses a head, for it can be said that a Paramoecium has an anterior end, but not a head. Moreover, not all parts named by the word "head" have their parts in morphological correspondence with one another and this we might require as a consequence. What is it that they share that has earned them the same name?

When parts are named in organisms there appear to be two sorts of criteria which are used in the process. The first is based on the similarity of *functioning* of the parts named by the same name. The legs of an insect and the legs of a horse are so named on the basis of their similarity of function, and it may or may not be (in this case not) that the two objects sharing the same name would turn out to have parts in morphological correspondence. The second method of naming parts is a morphological one. The names 'anterior vena cava', 'chondrocranium' are examples of such names.

It turns out to be very difficult to decide which of these criteria should be used to name the "head". The head of a cockroach and the head of a man, looked at in isolation, would certainly present very few parts which could be put into morphological correspondence.

However, it is possible, for many animals, to define "head" with respect to one of the other systems of the body, i.e. the nervous system. This means that not merely must we have the morphological relations as primitive in the system, but also certain classes of parts characterised by their histological structure, e.g. *Nerve, Muscle, Bone*. These words name sets of parts which have their usual histological description as their semantic interpretation.

From the use of such a set, *Nerve*, we can define a *non-circular* nervous system as one in which it is possible to distinguish two ends, in contrast to those cases where there is a nerve-net. We find, in practice, that all non-circular nervous systems have a swelling at one end. Moreover, this swollen part of the nervous system is at the same end of the organism as is the mouth – the opening through which food enters the body. We can therefore say that the part of the body which possesses these two parts, or at least one of them, is to be called "the head".

We are now therefore in a position to classify the animal kingdom in terms of the type of nervous system that we find, and from this to categorize the head end of the animal.

(A) Those animals which either have no nervous system, or which have a circular one, e.g. Amoeba, Hydra, an adult Asteroidean.

(B) Those animals which have a non-circular nervous system, of which one end is differentiated more than the other. This group of organisms can be further subdivided.

(i) Those organisms whose nervous system at one end encircles the gut so that on one side of the gut, the nervous elements continue in the opposite direction to that of the main body of nervous system on the other, e.g. Annelids.

(ii) Those whose nervous system does not encircle the gut at one end, and whose nervous system is confined in bulk to one side of the body only. This last feature is not strictly characteristic of the group but serves to make clearer the opposite feature of group B(i).

We are now in a position to orientate the body of organisms of either class with respect to our original relations.

(1) We shall call that end of the body possessing the nervous system in its enlarged state, the "head end" or the "anterior end". If any two parts x and y are such that x is *Anterior to* y, then x is nearer the head end of the body than y. We can note that another correlation, this time with reference to movement, is that usually it is the head end which moves foremost in progression. This breaks down in animals such as Crabs which are described as moving sideways. However, it must be admitted that definition of anterior end in terms of the direction of progression is the only way in which we can determine this end in animals with a circular nervous system, or with none at all, such as Paramoecium.

(2) The other end of the body will therefore be called the posterior end.

(3) In class B(i), the side of the body on which the main body of the nervous system is situated is said to be the *ventral* side, and that which is opposite to this, and on which are the anterior parts of the nervous system, is said to be the dorsal side.

(4) In class B(ii), the side of the body possessing the main body of the nervous system is said to be the *dorsal* side, and therefore the opposite side is called the ventral side. It must be noticed that this makes the dorso-ventrality of an organism something which can be determined relative to the organism, and in particular to the adult organism.[3]

The orientation might have been carried out by considering an earlier time-slice of the organism. This was not done because I wished to be able to compare simply two adult moments of two different organisms with one another. Sir Gavin de Beer[4] has suggested that use might be made of Child's concept of axial gradients in this connection. However, this concept has now been found to have too limited a generality for application in this context.

(5) In class B we find that it is possible to observe a plane which lies dorso-ventral, and passes from the anterior to the posterior end of the organism, in such a way that this plane divides the parts of the organism into two sets, between the members of which there is one, and only one, correspondence determined by the relation of mirror imaging. This plane (there being one, and only one, plane satisfying this condition), is called the plane of *bilateral symmetry* of the organism.

(6) Consider a three dimensional system of rectangular co-ordinates with the X and Y axes horizontal and the Z axis vertical. Let the X and Z

axes lie in the plane of symmetry of a time slice of a bilaterally symmetrical animal a. The origin of the co-ordinate system is O and may be anywhere in the plane of bilateral symmetry within the organism. Further let the X co-ordinate of all points anterior to the point O be positive and let the Z co-ordinates of all points dorsal to the point be *positive*. Then we shall call part of the animal a *right hand* part if, and only if, the Y co-ordinate of every point of that part is negative; and we shall call any part of the animal a *left hand* part if, and only if, every point of that part is positive.

We are now in a position to define the relations we shall use.

(7) We shall say that a part x is *anterior to* a part y if there is some point of x for which the X co-ordinate is greater than that for any point of y.

(8) We shall say that a part x is *dorsal to* a part y if there is some point of x for which the Z co-ordinate is greater than for any point y.

(9) We shall say that a part x is *to the right of* a part y if there is some point of x for which the Y co-ordinate is less than that for any point y.

It must be noticed that these relations are transitive and asymmetrical, and that they are not mutually exclusive. They have a further useful property – that of stratifying the parts which they relate.[5] Each of the classes produced by these relations can be correlated with an inductive cardinal so that we can number the strata in such a way that the nth stratum corresponds to the nth power of one of the relations. Since there are three stratum-generating relations involved, any part in the body will be a member of a stratum of each of the three relations. In principle, therefore, each part can be designated by a three figure term referring to the number of the strata for each relation, and this three figure term will designate, uniquely, each part.

So far we have considered the morphological relations without considering how we shall determine what their field shall consist of. The difficulty which now has to be faced is to delimit the sort of *parts* which can be used as the relata in this co-ordinate system. It must be admitted at the outset, that the delimitation will be arbitrary. If we could use the relation *part of*, we can also use its hierarchical nature to give us further numerical correlations with the levels[6] of the part relation for descriptive purpose. The problem, however, is to decide on the *nature* of the parts which can serve as major, or first level parts in the hierarchy. Shall we take head, thorax and abdomen as the major parts? Another method could be to determine the major parts with reference to the physiological division of the body. Thus the major parts could be the nervous system, the skeleton, the vascular system and so forth. Both of these systems lack generality and would have to be abandoned.

Another way of subdividing the parts would be to consider the organisms in their time extent. Woodger has pointed out[7] that "development proceeds by the production of parts in existing parts, followed by the appearance of differences between those parts".

But he also says "If we consider . . . a life as a whole, we see that the lower the level to which part belongs the longer is its time extent, because it appears earlier than do parts of higher levels. (You have the beginning of the head before a brain, the beginning of a brain before an eye, the beginning of an eye before an iris, and so on)." On the basis of this it would appear that all that has to be done is to look for the parts which have existed the longest and call these major parts. Woodger says it is the head, and hence uses the head as one of the major parts. This would be unsatisfactory for our system, for we have defined the head in terms of the nervous system. We might decide to study even earlier time slices of the organism. For example, we may put grey crescents or the dorsal lip of the blastophore in various embryos into morphological correspondence. The difficulty we are then faced with is the extension of the concept of correspondence between two parts in two time slices to include the relation of development. Since this frequently implies that we have to decide the adult parts first and then compare earlier stages containing the anlagen of the adult parts, this may lead to confusion.

However we struggle with the concept of major parts in embryological time-slices of an organism using the germ layer theory or any other arrangement we are faced with the fact that not all our major parts come into being at one time. This means that if the parts were really parts in a spatial hierarchy of a momentary time slice of an organism, there would *always* be major parts, for throughout its ontogeny the animal must be divisible into parts. We have therefore to realise that we are specifically only considering as major parts those which can be described as such in the adult. These parts will differentiate at different times in ontogeny, but once differentiated they can generate parts at higher levels in their own spatial hierarchy.

If we use the other primitives we have described – that of a class of parts delimited in the basis of histological structure we can distinguish between two types of collection of parts. We can call a collection of parts *homogeneous*, if all its members belong to the same set with respect to their histological structure. Thus, the muscular tissue of the body will be a *homogeneous* collection of parts. We can then call a collection of parts which are members of different classes of histological structure a *hetero-geneous* collection of parts. For example, a head or an arm is a hetero-geneous collection of parts. So far, we have assumed that our sets of parts are homogeneous. If we are considering parts as functional units, then we must consider them as heterogeneous collections of parts. This means that there are mutually exclusive sets of histological properties which can distinguish the members of the sets, within a part as a functional unit.

Using the sets *Nerve, Muscle*, etc., we can for each of the major systems of the body, proscribe the limits of its parts in terms of their spatial boundaries. This is what we do intuitively when we call a nerve the fifth cranial nerve. Frequently we find that one such set is used as the arbiter for the designation of another. It is interesting to notice that once again the set used to determine other sets is frequently the set of nerves. When we say that the nervous system is very constant, this is what we mean. The correspondence of some of the bones and muscles in the body, depends upon their position relative to the parts of the nervous system.[8]

To summarise, we have seen that sets of parts may have members which are homogeneous, or which are heterogeneous. If the latter we may use the property of class membership of a set based on histological properties to determine the limits of the members, for we usually first order a set of parts which are homogeneous and then include some of its members in a set with parts from other histological sets, using the first sort of parts as arbiter for the others, or as a means of defining the spatial boundaries involved.

Finally, we must notice that parts related by the morphological re-lations should be parts of the same level in the spatial hierarchy. It is not that a different procedure cannot be used, but it is a rule of the procedure for comparison that we must compare parts from similar levels in their respective hierarchies of the part relation within the particular histo-logical set of parts. We do not wish to compare bones with osteoblasts, or the sciatic nerve with a Schwann cell.

MORPHOLOGICAL CORRESPONDENCE

We are now in a position to describe how to arrive at correspondence between the members of sets of parts. Because a set of parts can be ordered in the way we have described by the three morphological relations (*A*, *D* and *R*) each part which is a member of the field of all three of the relations will be a member of a class which is the logical product of three strata of each of these relations. Thus a part *p* will be designated by (*An, Dm, Rq*) if it is a member of the stratum *An*, when the set of parts of which *p* is a member is ordered by the *anterior* relation and if it is a member of the stratum *Dm* when the set is ordered by the relation *dorsal to*, and if it is a member of the stratum *Rq* when the set of parts is ordered by the relation *to the right of* 'n', 'm' and 'q' will be inductive cardinals. We can abbreviate this by retaining the inductive cardinals, but omitting the relation prefix, *provided* we always remember to keep the product in the same order. Thus each part will be designated by a triple, and for any set of parts, there will be a set of triples determined by the morphological relations such that there will be a one-one correspondence between every part of the set and every member of the set of triples.

There is a difficulty however, which must be overcome. On reflection we can see that on the one hand we might have a set of bones, and on the other hand we might have a set of cells the members of which could put into *identity* correspondence with respect to some set of triples. We have already shown that we must insist that the members of both sets must be members of the same levels in their respective hierarchies which are determined within a set of histologically similar parts. However, there is another way of ensuring that we shall be comparing satisfactory sets of parts. We must only consider those parts to be in correspondence which have as many *sub-parts* in correspondence as possible at the various levels of the sub-parts. Then, if when we put members of a set of parts into correspondence, each part is in morphological identity correspondence with a member of the second set but none of the parts at other levels fall into correspondence, we shall discard the pairing and seek for one which, over all the organisms we wish to consider, puts as many of the higher level parts into correspondence as possible.

Woodger[9] suggested that when we compare a head with a head we do so intuitively, because we know that this will bring noses, mouths and brains and all the other parts of the heads also into correspondence, so that our comparison can go right to the highest level of organisation. It may be asked what happens if we cannot put all the parts into correspondence, do we have one set with twelve members and the other with eight, four parts thus having no corresponding parts, or what do we do? Woodger suggested that we choose those relations which hold over the greatest number of forms. This is not really satisfactory for the relations will hold for *every* form in space provided we ignore the histological sets we have mentioned. What is true is that we chose the set of triples for which there are correlates for the greatest number of forms within a given set of histologically determined parts. This set of triples is then reflected back in the set of parts in that we only choose and compare those parts in two given individuals which correlate with this set of triples.

We can now summarise this method of establishing correspondences.

(1) The morphological relations and the histologically determined sets of parts give rise to strata of parts which are correlated with inductive cardinals.

(2) There will be a set of triples of inductive cardinals for every part in a set of parts which has been ordered by the morphological relation, and with respect to their histological set membership.

(3) We can therefore establish a one-one relation between a set of parts and a set of triples if the set of parts has been ordered by the set of morphological relations.

(4) The set of parts that we choose for comparison is that set which correlates with a set of triples for which there is correlation over the greatest number of forms.

(5) If a part x is in correspondence with a part y because there is a triple to which they are both correlated, then the parts are in morphological *identity* correspondence, provided that if x is a member of a set[10] characterised by its histology, then y is also a member of.[10]

(6) This is not sufficient, for we want to be sure that not only parts of the level n in different organisms are in correspondence but also parts of level $n + m$, $n + p$ and so on. When this is done we can say that the parts of level n are in *maximum* identity correspondence.

The definition of maximum identity correspondence given by Woodger suggested that the parts might be considered to be in correspondence if they were themselves parts of parts already in correspondence. He felt that this was necessary on account of "the pairing of parts like the bone of the pentadactyle limbs which are based partly on the fact that they are parts of parts which are in morphological correspondence, and partly on their own morphological relations, but are not in maximum identity correspondence in the above sense because we do not require that parts of the bones should be in correspondence".

This is reasonable, for in the situation I have described we must be careful not to take the requirement about the other levels of part too literally. But we may nevertheless wish to bring some parts of bones into correspondence, and to bring as many levels of parts into correspondence *as possible* seems to me to do what is required.

Our modification of Woodger's definition of maximum identity correspondence is therefore as follows: S is a *morphological maximum identity correspondence* between a set γ of parts of some life and a set δ of another life, with respect to a set κ of morphological relational properties and sets $\lambda_1, \lambda_2, \ldots \lambda_i$ of histologically determined parts, if and only if S is a one-one pairing of the members of γ with those of δ and

(1) S brings a maximum number of parts into correspondence so that if x is a member of γ paired by S with a member y of δ then there will be parts x' and y' also paired by D and belonging to γ and δ respectively and such that x' is a part of x and y' is a part of y; and

(2) if any member of γ has any property belonging to κ then the member of δ paired by S with it also has that property; and finally,

(3) if any part x is paired by S to part y, and x is a member of a set λ_i characterised by histological character then y is also a member of λ_i.

The discussion above on the method of correlation of the sets of parts with sets of triples makes explicit what is involved in the set κ of morphological relational properties. We can therefore make the correspondence more explicit by saying that:

If any member of γ is correlated with a member of a set of triples, and a member of δ is also correlated with the same member of the set of triples, then the relation S stands between them.

From these statements two definitions can be obtained.

(1) *Morphological correspondence* is the relation between two parts x and y if, and only if, x is a member of a set of parts γ and y is a member of a set of parts δ, and x is paired by S with y, and S, γ and δ satisfy the

above conditions with respect to a set κ of triples obtained from the numbering relation and the set of morphological relations, and there is a set λ characterised by histological characters such that x and y are both members of γ.

(2) A set of parts γ is *isomorphic* with a set of parts δ if, and only if, there is a pairing of the two sets which satisfies the above conditions with respect to the correlation of each set with a set of triples.

It must be pointed out that there is a difference between a set of parts, and the thing which is the sum of that set. A set of bones A may be isomorphic with a set B of bones, but the sum of the set A may be the skeleton of the forelimb of a mammal, while the sum of the set B may be the skeleton of the hind limb of a mammal. As sets of parts they are isomorphic, but as corresponding limb skeletons they are not in morphological correspondence, for they have very different relations with the rest of the body. In other words the sum of each set of parts must be regarded as themselves members of sets of parts which are *not* isomorphic. Thus there is no significance in describing as morphologically corresponding such things as feathers, hairs etc. on the same animal as some of the older morphologists would have us do.

However, we must examine this distinction a little further, for morphologists would certainly say that the skeletons of the hind limb and forelimb of a mammal, are both examples of the "pentadactyle limb". How does this fit into the system? As Woodger[12] pointed out this is the name given to the abstractive class generated by the symmetry and transitivity of the relation of morphological correspondence and hence of isomorphy. He used the term *Bauplan* for such an abstractive class, and this term replaces the older idea of the Naturphilosophen of Platonic Archetype.

MORPHOLOGICAL CORRESPONDENCE AND HOMOLOGY

We are now in a position to examine some of the ways in which morphological correspondence has been, and still is being used, by comparative anatomists to determine what they call "homologies". We can distinguish some uses of the term homology which involve morphological correspondence.

1. *Morphological correspondence and the theory of 'Archetypes'*

This is the way in which the older morphologists, including Owen[13], used the term homology. They applied the concept of morphological correspondence (more intuitively than has been done above, it is true) to mean homology. In case it is felt that we do not do this at the present time let us consider briefly the use of the term 'caridoid facies' for describing a set of parts found in the Crustacea. This term names a Bauplan, and the Crustacea differ from this Bauplan by "excess or defect". We still seem obliged to keep to some sort of "Type" theory, except that we can define a Bauplan without resorting to Greek philosophy or without feeling that we *must* fit our concept of science into such a scheme of philosophy as did the early German anatomists. The mention of "excess or defect" raises important questions concerned with the maximum group of organisms exhibiting the Bauplan. This involves questions of taxonomy and is dealt with elsewhere.[14]

2. *Morphological correspondence between two parts using a Bauplan of another system of parts as an arbiter*

We find that the morphological correspondence of some parts of a vertebrate animal, for example, the chondrocranium, is always established with reference to parts of the nervous system of the group. This is because the relations of the nervous system are sufficently constant over the vertebrates, whereas the cartilages which make up the chondrocranium are not found to have constant relationships to one another. Therefore if A_x orders the set of cranial nerves of the animal x and A_y orders the set of cranial nerves of the animal y, then the part of the chondrocranium which stands immediately anterior to A_x5 is in morphological correspondence with the part of the chondrocranium of y which is immediately anterior to A_y5.

If A_x5 is the fifth cranial nerve in one of these organisms, then we are describing the pila antotica which is immediately anterior to it. There is really nothing being added to the concept of morphological correspondence by this usage, for we are merely increasing the number of parts in the sets which form the Bauplan covering these parts. The increase in number is because of the addition of a non-nervous part in a definite relation to one of the members of the original set, thus, the one-one relation between any two members of the enlarged set still holds and therefore the definition of morphological correspondence is satisfied. Moreover, we have nowhere required that a set of parts need be homogeneous

389

with respect to any set characterised by histological character. The set of parts we are now considering is in fact heterogeneous.

3. *Relations between isomorphic sets of parts*

It has already been pointed out that because two sets of parts are isomorphic the sum of those sets are not necessarily in morphological correspondence. Of course, it may be the case that the sums of the sets of parts may be members of isomorphic sets of parts, the latter being of a different level from the original set of parts. However, with this in mind it is possible to arrange as many parts in correspondence as there are parts of the organism. Because of size differences this correspondence will ultimately break down in any case at the cellular level. With this limitation in mind, it is possible to say that, for example, two men would exhibit a *maximum* Bauplan which we could define as the set of sets of parts of a level which is not too high in the spatial hierarchy but which when considered as the sum of each one of these sets of parts is the whole organism. From this it is obvious that two organisms which exhibit the same maximum Bauplan must be members of the same species. In fact, as a Bauplan tends towards maximum, we could say that the organisms exhibiting it were closely related. Therefore, although the relation between two sets of parts in isomorphy is not necessarily to be included in homology, the relationship may perhaps nevertheless be useful in determining particularly close relationships between animal forms.

4. *Morphological correspondence in one organism*

Occasionally we find the concept of serial homology used. This would be precluded by our definition of morphological correspondence. This definition can be altered by the modification of the phrase "of another life" in our definition (p. 387), so that the definition would read: "S is a morphological maximum identity correspondence between the parts of two lives or parts of one life.

Then we can say that in one organism, if a part x is in morphological correspondence with a part y, and the part x is a member of a set of parts α and the part y is a member of a set of parts β, then x is serially homologous with y if, and only if, the sets α and β are both included in the set of left hand parts or the set of right hand parts, or the set of median parts, and the set α is anterior to the set β or the set β is anterior to the set α.

390

5. *Homology of parts in two organisms based on serial homology*

A variation on the theme which has just been discussed is used in the establishment of the homology of the hyomandibula cartilage in various organisms. The homologies are based on the serial homology of these cartilages with the epibranchial cartilages in any one fish. Thus the homology of the hyomandibula cartilage in two fish is an expression of the ordinal similarity of the two cartilages with epi-branchial cartilages which are in morphological correspondence, the correlating relation being that of *being the serial homologue of*. The exact expression of the relation depends on the use of the number relation explained above, for we must know the exact number of steps we need to take from any one epibranchial cartilage to the hyomandibula cartilage.

6. *The developmental use of homology*

Homologies have always been believed to necessitate the similarity of spatial relations throughout development. Expressing this another way we can say that if a part x in a momentary organism a is in morphological correspondence with a part y in a momentary organism b, then the part x is *homologous* with the part y if, and only if there is an organism a' which will develop into the organism b, x' is part of a' and will develop into x, and y' is part of b' and will develop into y and x' is in *morphological correspondence* with y'. Moreover, this should hold for any x' and y' in any organism a' or b' which will develop into a or b.

Now, if the parts x and y are in morphological correspondence they are *homologous* according to the views of Owen. If the parts are *not* in morphological correspondence, then there cannot have been a similarity of development. This concept can be expressed in two propositions for which we may be able to find counter-examples.

(1) For any parts x, x', y and y' in lives a and b (considering the lives as time extended) in which x' will develop into x in life and a and y' will develop into y in life b, then if x is in morphological correspondence with y, x' *will have been* in morphological correspondence with y'.

(2) Under the same conditions, if x' is in morphological correspondence with y', then x will be *homologous* with y.

Most morphologists could find counter-examples of the first proposition, but most would agree with the second. "The similarity in relations between the processes ascendens and the pila lateralis of Amia led Allis to conclude that they were homologous, but as De Beer showed, this cannot be, for the pila lateralis is a neurocranial and not a visceral arch structure." [15]

This is a rejection of proposition 1. Yet the case of the homology of the thyroid and endostyle would be an example of the truth of proposition 2. Therefore we find that this sort of homology is also ordinally similar to morphological correspondence but with *develops into* as the correlating relation. Even so the matter is not so simple as this and is discussed elsewhere.

7. *The use of homology with an evolutionary bias*

The last use of the concept of homology in the classical period of comparative anatomy is that which not merely believes that morphological correspondence between parts of two organisms is indicative of a common ancestor of the two organisms, but which believes that even if there is morphological correspondence between the two parts, the parts are not homologous *unless* there has been a common ancestor. This view requires us to believe that a correspondence without evidence of a common ancestor is indicative of convergence, and yet at the same time exhorts us to use the morphological correspondence to *determine* the presence of the common ancestor. This piece of reasoning is fallacious and can only be made on the basis of some such *postulate* as the evolutionary postulate mentioned by Woodger in his paper. [16]

8. *Homologous chromosomes and homologous genes*

We cannot define a set of parts which can be ordered by the morphological relations so that two chromosomes will be correlated with the same triple in a set of triples. However, it is possible to use another morphological relation on the basis of which the chromosomes themselves might be subdivided into a set of parts, and with which parts of different chromosomes could be put into morphological correspondence. These parts of chromosomes may be correlated with the appearance of parts in adults lives and may be called genes. Therefore, there is a way of describing homologous genes by reference to the parts in the adults which are in an extremely complicated way correlated with them. Once again this needs a further extended treatment.

MORPHOLOGICAL CORRESPONDENCE

9. *Homologous serum*

The history of the introduction of this usage has not been traced, but it appears to come from a totally different system of terminology. We find it defined as[17]: "The antiserum that produced by the inoculation into a suitable animal of a particular bacterium is frequently referred to as a *homologous* serum. Used in this sense the term is useful and logical and may be applied either to an anti-serum in relation to a bacterium, or to one another, implying in either case complete correspondence between actively combining groups."

It appears therefore that there is no connection between morphological correspondence and this use of the term homology.

10. *Homologous responses and homologous variation*

These two uses of the adjective are related to morphological correspondence in that they refer to phenomena exhibited by some parts which are in morphological correspondence. The term "homologous response" is used by Weiss to describe the responses given by structures in morphological correspondence to various stimuli which act upon them in the course of development. The law of homologous series in variation described by Vavilov applies to the similarity of variation of forms over closely related genera. Both of these usages need further analysis to find the way in which morphological correspondence is sued in them, but since so much else is obviously involved, I feel that the term "homologous" would better be avoided in these connections.

We can conclude that the pioneering work of Woodger in clarifying the concept of morphological correspondence has helped biologists to see what else is involved in their use of the notion of homology. I would like to thank Professor J. H. Woodger for his encouragement and helpful criticism in this study.

Department of Biology Applied to Medicine, Middlesex Hospital Medical School, London

REFERENCES

1. J. H. Woodger, *On Biological Transformations* in *Essays on Growth and Form* (edited by W. E. Le Gros Clark and P. B. Medawar), Oxford, 1945, pp. 95–125. This will be subsequently referred to as *B.T.*
2. *B.T.* p. 98.
3. This is in distinction to a system whereby dorso-ventrality is related to the position of the animal relative to the earth's surface, or any other similar system.
4. Private communication.
5. The relations which have this stratifying property have been called hierarchy generating relations. They have been discussed in detail by J. H. Woodger, *The Axiomatic Method in Biology*, Cambridge, 1937, pp. 42–47.
6. The time 'level' used in connection with the *part* relation is synonymous with the term 'stratum' used previously with the morphological relations. It appears to be worthwhile to use the separate terms to distinguish between the hierarchy generating relations involved.
7. *B.T.* p. 100.
8. The central nervous system will be the sum of the set *Nerve*, – the sum of a set being that defined in J. H. Woodger, *Axiomatic Method in Biology*, Cambridge, 1937, p. 29, Definition 0.15.
9. *B.T.* p. 101.
10. Will be one of the sets *Nerve, Muscle*, etc.
11. It must be noticed that the maximization used here refers to the bringing of as many levels of part into correspondence as possible and *not* to the taxonomic distribution of these parts as in *B.T.* p. 103.
12. *B.T.* p. 104.
13. R. Owen, *On the Archetype and homologies of the vertebrate skeleton*, 1848.
14. R. F. J. Withers, *Morphological Correspondence in Taxonomy and Evolutionary Theory*. In preparation.
15. G. R. De Beer, *The Development of the Vertebrate Skull*, Oxford, 1938, p. 420.
16. *B.T.* p. 109.
17. A. Boyden, *American Midland Naturalist*, 37 (1947) 648.

ON CATEGORY OVERLAPPING IN TAXONOMY *

In his book *The Language of Taxonomy*[1], J. R. Gregg, following the lead of J. H. Woodger[2], has applied the techniques of modern logic and set theory to examine a problem of importances in biology. In Gregg's case the problem is that of elucidating the structure of the taxonomic systems used in biological classification. His achievements in this direction are considerable, but the model he sets up still suffers from certain short-comings. The one serious defect of the model is, as he himself takes pains to point out, that it leads to "category overlapping". This arises because taxonomic groups[3] are defined as sets of organisms, and sets with the same membership must be identified. Thus, to take one of Gregg's examples, the groups called "*Apteryx*" (a genus), "Apterygidae" (a family), and "Apterygiformes" (an order), all contain the same organisms (the flightless kiwi birds of New Zealand) and must accordingly be considered identical. Therefore in this case[4] the categories "Genus", "Family", and "Order" cannot be distinguished. This, then, is category overlapping: the inability to distinguish between taxonomic groups which, though belonging to different categories, contain the same organisms, and the consequent inability to distinguish the categories themselves, at least "locally". Another way of putting this is that taxonomic categories, as such, cannot be defined *within* Gregg's model.

Various remedies for category overlapping have been proposed. A suggestion that has been made is that "higher" groups be defined, not as sets of organisms, but as sets of "lower" groups: thus a "genus" would be a set of species, a "family" a set of genera, and so on. Since it is custom-ary in set theory to distinguish a set from any of its members, even extending this distinction to the case of a set with only a single member[5], it becomes impossible to confuse taxonomic categories. This suggested cure, however, has the air of being worse than the disease. Were the

* I should like to thank Professor Gregg for several stimulating discussions, both in person and by correspondence, of the contents of this paper, albeit these contents, and the opinions that are part of them remain my personal responsibilities. (A.S.)

suggestion adopted, it would be simply false to say "A human being is a mammal [i.e., a member of the set 'Mammalia']", since "Mammalia" would not be a set of organisms, including human beings, but, say, a set of sets of sets of sets of sets of organisms[6]. This is clearly unsatisfactory.

Another viewpoint has been expressed by Murray Eden. In a review[7] of Gregg's book, he says: "Category overlapping is a consequence of the particular logic implicit in the formulation of set theory employed by the author. This logic is denotative, i.e., if two names in a language denote the same object, for example, 'John R. Gregg' and 'author of *The Language of Taxonomy*', then one name can always substitute for the other in that language. . . . This is certainly not the intention of the taxonomist. There is, after all, a very good reason for placing the kiwi in an order all by itself. In some ill-defined sense the kiwi is very different from all other birds; its ancestors branched off the evolutionary tree rather early in history . . .

"To provide a model that would retain the evolutionary aspect of taxonomy requires a logic capable of preserving the *sense* as well as the *denotation*. Such logics, so far as they have been studied, are exceedingly intricate, but they may very well be unavoidable if the structures of scientific disciplines that concern themselves with properties of natural objects are to be formalized".

To which one can only add "amen". Unfortunately, the sense-preserving logics invoked by Eden are, for any present practical purposes, nonexistent; and even if they were available, their use would seem to require some agreement on the "sense" of the objects under consideration. Now taxonomists evidently do not agree on the "sense" of their taxonomic groups and categories; for if they did agree they should all come up with the same classification schemes and they do not[8].

What seems to be needed therefore, if it is desired to extend Gregg's results, is something rather primitive and purely formal, something taxonomists can use without reference to the meanings they attribute to their constructs. Such primitive, formal devices are available, and the purpose of this note is to indicate one such device. In doing this we shall take for granted the elementary notions of set theory: membership in a set, set identity, subsets of a set, union and intersection of sets, relations on a set, etc. Other notions will be defined, and in particular, some of the definitions of Woodger and Gregg will be reviewed and restated.

Firstly, a set S is (*partially*) *ordered* if there is a relation (which we shall denote by $\subseteq$) on it, such that:

O1   For any member m of S, $m \subseteq m$.

O2   If m_1 and m_2 are members of S and we have both $m_1 \subseteq m_2$ and $m_2 \subseteq m_1$, then $m_1 = m_2$.

O3   If m_1, m_2, and m_3 are members of S, and we have $m_1 \subseteq m_2$ and $m_2 \subseteq m_3$, then $m_1 \subseteq m_3$.

Notice that it is *not* part of the definition that we must have:

O4   If m_1 and m_2 are members of S, then m_1 and m_2 are *comparable*, i.e., either $m_1 \subseteq m_2$ or $m_2 \subseteq m_1$.

If O4 does hold on S in addition to O1–3, then we say that the ordering is *total* or *linear* and we refer to S as a *chain*.

Next, if m_1, m_2 are *distinct* members of an ordered set, and if $m_1 \subseteq m_2$, then we shall say that m_2 *outranks* m_1.

A *hierarchy*[9] is an ordered set S satisfying the following conditions:

H1   There is a (necessarily unique) member m_0 of S that outranks every other member of S. We refer to m_0 as the *initial member* of the hierarchy.

H2   If, of three members m_1, m_2, m_3 of S, we have $m_3 \subseteq m_1$ and $m_3 \subseteq m_2$, then m_1 and m_2 are comparable.

H3   For any member m of S, there are only a finite number of distinct members of S that outrank m. The exact number of such members is the *rank* of m; the rank of m_0 is 0, while any other member of the hierarchy has rank exceeding 0.

The nth *level*[10] of a hierarchy is the set of all members of the hierarchy whose rank is n. The zeroth level of a hierarchy contains only the initial member of the hierarchy; but any other level may have any number of members, even infinitely many. It is easy to see, though, that a hierarchy is also a chain if and only if every level has precisely *one* member. A hierarchy that is at the same time a chain will be referred to as a *chain-hierarchy*.

Now the members of a hierarchy may themselves be sets, and the order relation $\subseteq$ may be the ordinary relation of set inclusion, i.e., $m_1 \subseteq m_2$ may simply mean that m_1 is a subset of m_2. This is the case with the hierarchies called *taxonomic classifactory systems* by Gregg[11], and which we will call *Gregg hierarchies*. Specifically, a Gregg hierarchy is a hierarchy which, in addition to H–13, satisfies the following conditions:

G1   The members of the hierarchy are themselves non-empty sets (of the objects to be classified; their exact nature is irrelevant here), and the order relation in the hierarchy is the relation of set inclusion.

G2   If two members of the hierarchy are incomparable (i.e., neither is included in the other), they are *disjoint* (have no members in common).

This is the schema that Gregg applies to taxonomic classification, and this is the schema that will now be modified to get away from category overlapping The modification will consist essentially in systematically enlarging Gregg's sets to include not only the objects to be classified, but also certain extraneous "index"-objects. This can be done in such a way that taxonomic groups belonging to the same taxonomic category contain the same set of index-objects, while groups belonging to different categories contain different sets of index-objects. Toward this end, let G be a Gregg hierarchy with initial member g_0 and let C be a chain-hierarchy (the "indexchain") whose members, like those of G, are nonempty sets ordered by the relation of set-inclusion. We shall assume that C has at least as many levels as G, and that c_0 (the initial member of C) and g_0 are disjoint[12]. Note that, because g_0 and c_0 are disjoint, the following is true: if g is a member of G, c a member of C, and t the union of g and c, then g and c can be recovered from t by intersecting t by g_0 and c_0, respectively. In the usual notation of set algebra, this reads: if g_0 and c_0 are disjoint, $g \subseteq g_0$, $c \subseteq c_0$, and $t = g \cup c$, then $g = t \cap g_0$ and $c = t \cap c_0$. We call g the *G-component of t*, and c the *C-component of t*.

By forming unions of members of G with members of C we can construct a new hierarchy T having the following properties:

T1   Every member of T is a union of a member of G and a member of C.

T2   Every member of G is a G-component of at least one member of T.

T3   The hierarchy T, like G and C, is ordered by the relation of set-inclusion.

T4   If t_1 and t_2 are members of T, and t_1 outranks t_2, then the C-component of t_1 outranks the C-component of t_2.

A trivial way of forming such a hierarchy T is to take every member g of G and form the union of g with that member of C whose rank is the same as the rank of g. In general, there will be many other ways of forming T.

We are now ready to define taxonomic systems, groups, and categories.

A *taxonomic system* is any hierarchy satisfying T1–4. (Note that this implies the existence of a Gregg hierarchy G and a corresponding index-chain C.)

A *taxonomic group* is any member of a taxonomic system.

Two groups of a taxonomic system T are *taxonomically equivalent* if they have the same C-components.

A *taxonomic category* is a set containing all groups taxonomically equivalent to a given group.

To see how such a system could work out in practice, we return to the problem of classifying kiwis among birds. Let, therefore, the objects to be classified consist of all birds. G will be taken to be a rather rudimentary division of birds into one class, and various orders, families and genera, so that C may be taken to be a chain-hierarchy with only four members: c_0, c_1, c_2, c_3. The initial member c_0 may consist of anything *except* birds: let us take it to consist of the four *words*: "class", "order", "family", "genus". Then c_1 can be the set {"order", "family", "genus"}, c_2 the set {"family", "genus"}, and c_3 the set whose only member is the word "genus". A taxonomic system T can now be formed in such a way that its initial member, which we refer to by the conventional designation "Aves", is the union of the set of all birds and the set c_0, i.e., contains all birds and also four additional items that are not birds. Among other members of T are the set "Apterygiformes", which is the union of the set K of all kiwis and the set c_1; "Apterygidae", which is the union of K and c_2; and "*Apteryx*", which is the union of K and c_3. Every kiwi is a member of all three sets; no other organism belongs to any of the sets; yet the sets are distinct. "*Apteryx*" is a subset of, but is not identical with "Apterygidae" which in turn is a subset of, but is not identical with "Apterygiformes". "*Apteryx*" is a member of the category "Genus" which consists of all sets of T containing the word "genus" and no other word. Similarly, "Apterygidae" is a member of the category "Family" which consists of all members of T containing the words "genus" and "family" and no other words; while "Apterygiformes", containing the words "genus",

399

"family" and "order", is a member of the category "Order". "Aves" is a member (in this instance, the only member) of the category "Class". No two categories overlap.

From this example, it should be clear how the same method could be applied in any other classification scheme.

It will be objected that the method outlined above is artificial and *ad hoc*. But so is, to a large extent, biological classification itself. True, the broad outlines of taxonomy are "natural"; but the finer details almost always reveal complications that can be fitted into a classification scheme only by the fiat of the taxonomist. Accordingly, there would seem to be no reason for the taxonomist who wants to straighten out the formal structure of his systems to reject an additional small degree of artifice, particularly when this artifice is noncommital and does not affect the logical "type" of taxonomic groups and categories, nor, to a very large extent, their mutual interrelationships as sets containing organisms.

Illinois Institute of Technology, Chicago, Illinois, U.S.A.

REFERENCES

1. J. R. Gregg, *The Language of Taxonomy*, Columbia University Press, 1954.
2. J. H. Woodger, *The Axiomatic Method in Biology*, Cambridge University Press, 1937; 'From Biology to Mathematics', *The British Journal for the Philosophy of Science*, 3 (1952) 1–21.
3. J. H. Woodger, in *Biology and Language* (Cambridge University Press, 1952, p. 235] introduces the term "taxonomic set" for a taxonomically recognized set of related organisms. For the purposes of this note, it has seemed better to stick to Gregg's term "taxonomic group".
4. Which is, of course, far from being unique. Gregg discusses this and other examples on pp. 63–68 of *The Language of Taxonomy*.
5. Thus "the object *A*" is distinguished from "the set whose only member is the object *A*".
6. Using only the "obligatory" categories "Class", "Order", "Family", "Genus", and "Species". If finer distinctions are made, then the "type" of "Mammalia" [in the sense of the theory of types] goes still higher.
7. M. Eden, *American Scientist* **43** (1955) 128A–130A.
8. Such agreement may, of course, become possible if and when methods become available for determining the exact degree of relationship of two given organisms (expressed perhaps in number of years or generations back to a common ancestor, if any).
9. Cf. Woodger, *Axiomatic Method, loc. cit.*, pp. 42–47; *From Biology to Mathematics, loc. cit.*, p. 11; and Gregg, *Language of Taxonomy, loc. cit.*, p. 26.

10. Cf. Woodger, *Axiomatic Method, loc. cit.,* p. 44; Gregg *Language of Taxonomy, loc. cit.,* p. 29.
11. Gregg, *ibid.,* p. 47.
12. These requirements can always be met, even if the required number of levels is infinite (and this possibility is explicitly not excluded: see Gregg, *ibid.,* remark on p. 51). For example, if g_0 contains no positive integers, we can take C to be the set whose members are $c_0, c_1, c_2, \ldots$, where c_0 is the set of all positive integers, c_1 the set of all positive integers except 1, c_2 the set of all positive integers except 1 and 2, and so on.

LEIGH VAN VALEN

AN ANALYSIS OF SOME TAXONOMIC CONCEPTS*

There have been a number of recent attempts to formalize various aspects of taxonomy, of which the most successful are those of Gregg (1954) and Woodger (1952a). Beckner's analysis (1959), in nearly the only book I have ever seen written by a philosopher who is biologically competent, is also noteworthy and complementary. The concepts of species and taxonomic system, as delineated in the metabiological literature, seem somewhat inadequate biologically. The present paper is an attempt to improve on these. For any interest this may have, I am a zoologist with background about equally in paleontology and genetics. I have discussed some aspects of this paper with John Gregg, to whom I am indebted for many improvements. For easy reference, a glossary of the set-theoretical symbols repeatedly used follows the text.

I. THE CRITERIA FOR THE CLASSIFICATION OF ORGANISMS

There appear to be about four partly conflicting goals in most modern attempts to classify some large or small set of organisms. The first two are the most important.

(1) To reflect phylogeny as closely as practicable. Ideally this would involve construction of a phylogenetic diagram (Figure 1), and the earlier the fork between two phyletic lines the higher the taxonomic level of separation. This has obvious disadvantages by itself in that such a diagram is usually speculative in whole or in part, that a phyletic line such as A would receive the same rank as B, and that rates of evolution differ between and within taxa.

(2) To reflect the diversification and, conversely, the similarity, within taxa. The mammals should certainly not be considered a suborder of the reptilian order Pelycosauria (although von Huene, 1948, has done es-

* Written at intervals during the tenure of three National Science Foundation predoctoral fellowships, a Columbia University fellowship, and a NATO postdoctoral fellowship.

402

sentially this), nor the tetrapods a suborder of the order Rhipidistia (early lobe-finned fishes), although each was undoubtedly derived from these respective groups. This extreme "vertical" procedure could ultimately lead, where strict monophyly is present and the phylogeny is known, to the presence of only a single taxon higher than subgenus!

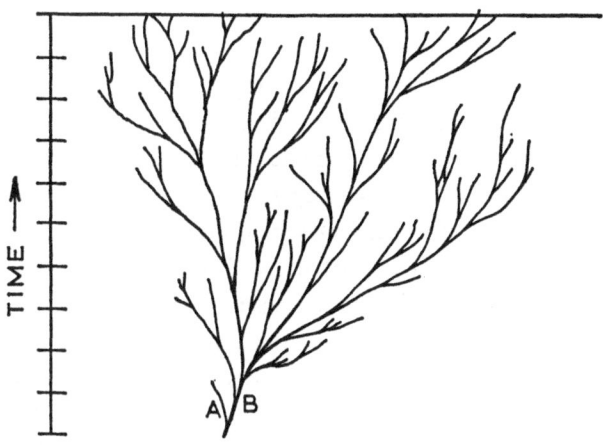

Fig. 1. Complete phylogenetic diagram for one monophyletic taxon. Each strand represents a species.

(3) To separate taxa of equal rank only where more or less discrete gaps occur in the array of organisms classified, generally and quite roughly greater the higher the level of classification. This is often impracticable where more or less continuous series of intermediates are known, as in various fossil lineages. In these situations, however, the intermediates are usually relatively few in comparison to the more typical parts of the taxa. A division is usually made even if the gap does not exist, but this is felt to be less satisfactory than if one did, and disputes are likely to arise as to where the boundary is to be placed. Routine placement of boundaries within continuous series would partly obviate such disputes and would be theoretically largely desirable in groups with a good fossil record, but it would make identification of specimens belonging to known taxa more difficult.

403

This third criterion is related to the facts that the *adaptive zones* of organisms (the ranges of environmental conditions to which the members of taxa are best adapted; the mappings on the total environment of their ways of life) are usually and generally more similar the more closely the organisms are related, and that the empty potential adaptive zones between those filled by two groups usually and generally decrease in size as the taxonomic difference between these groups decreases (Simpson, 1953).

(4) To be usable or convenient. For instance, the family of mice, rats, voles, lemmings, and hamsters (Muridae) is often split into two or three families along rather well-defined lines simply because it is so much larger (although not correspondingly more diversified) than any other mammalian family. However, certain families in other groups, notably among insects and angiosperms, are even larger without being split to this extent.

In other cases the phylogeny may be quite uncertain. A situation related to this common difficulty occurs with parallel or convergent evolution, with the result that it is much easier to group successive stages of different phyletic lines than to group the members of each phyletic line. An example of this is the graptolite genus *Didymograptus* of current classification, which is a stage reached by a half dozen or more different phyletic lines (Bulman, 1955, and references therein). Such taxa are based more on morphological than phylogenetic concepts, and purely morphological taxa are still defended, but both kinds of taxa are readily fitted into a taxonomic hierarchy. Both criteria are at least implicit in the establishment of most taxa. Emphasis on one or the other leads to a more *horizontal* classification (based more on similarity) or a more *vertical* one (based more on phylogeny). See Simpson (1945 and 1961) for a discussion of this problem.

The interspecific diversity present at any one time seems to be brought about entirely by the following factors (cf. Figure 1):

(1) The previous rates of phyletic (non-branching) change in the total organism.

(2) The previous rates of splitting into separate species.

(3) The length of time each rate in each of the above categories has continued, and their order.

(4) The survivorship of the phyletic lines resulting from splitting.

(5) Our knowledge of the phyletic lines resulting from splitting (for taxonomic purposes only).

(6) The degree of parallelism and convergence, which may lead to the polyphyletic origin of certain taxa.

(7) The degree of *introgression, allopolyploidy,* and possibly other mechanisms for the origin of a single taxon from two or even more taxa (*reticulate evolution*), virtually or entirely restricted to taxa at or below the level of species.

The total diversification in a given taxon can be considered roughly as an integral of the diversification over the entire time interval the taxon has existed.

The foregoing account is too abbreviated to be adequate as more than an outline of the problems; most of the points mentioned have been discussed at length by Simpson (1961), in a book written at the same time as the present discussion.

II. SET THEORY AND THE SPECIES CONCEPT

The *parental relation P* may be taken as "The set of all couples (x, y) such that x is parent of y" (Woodger, 1952a, p. 232). This relation is denoted by xPy. I will take it to be asymmetric (i.e. if xPy, then it is not true that yPx) and irreflexive (i.e. it is not true that xPx). Some subsets of P are transitive, for if xPy and yPz, it is possible that xPz as a result of in-cestuous mating or the equivalent. The set of all predecessors of P (i.e. parents) is the domain of P, and the set of all successors of P (i.e. offspring) is the counter domain of P. The union of these is the field of P.

Consider a well-ordered subset P_w of P, i.e. a subset which can be put into orderly one-to-one correspondence with a subset of the non-negative integers. Each couple in P is ordered by the relation P; P_w is formed by choosing couples such that the predecessor in all but one couple is a successor in some other couple in P_w, and therefore that the successor in all but one couple is a predecessor in some other couple in P_w. The couple whose predecessor is not also a successor will be taken as the first couple. The index i_c of any couple c in P_w is defined as Card c, i.e. the integer to which it is associated, and similarly the index i_x of any individual x is the index of its couple. Since x may appear in more than one couple, its index is not uniquely determined.

The *descendant relation D* may now be defined as the set of all ordered couples (x, y) such that x is an element of the domain of some P_w, y is an element of the counter-domain of this P_w, and there exists some i_x and i_y such that $i_x \geqslant i_y$. D is thereby the transitive extension of the converse of P. Gene flow occurs between individuals related by D.

The *mating relation M* is the set of all couples (x, y) for which there exists z such that xPz and yPz, provided that x and y are not identical. M is of course symmetric. M is therefore defined only for cross-fertilizing and parasexual organisms; I will ignore such complications of restricted application as the phenomena in bacteria of transduction and lysogeny, although expansion of the model to include them is not difficult.

It is necessary for the sequel to introduce a new variety of set, *stochastic sets*. To each element of a stochastic set is associated a probability value, and each element is weighted by this probability. A probability of 0 can be associated to as many elements as desired, so long as the sum of the probabilities is 1. Arbitrarily order the elements. A random subset of a stochastic set is the set of elements associated with a random sample from the segment whose upper bound is 1.

A *mating pool W_y* is a stochastic set, where x and y are organisms and the probability $P(x)$ for each x is defined as the probability that x stands in M to y, i.e. that xMy. The union of y with the set of all x for which $P(x) > 0$, i.e. $\{y\} \cup \{x: P(x) > 0\}$, is then W_y. The *potential mating relation MP* is the relation between x and y in any W_y. M is obviously a proper subset of MP.

Consider a subset $V_{y,a}$ of W_y, defined as the union of y with the set of all x for which $P(x) \geqslant a$, i.e. $\{y\} \cup \{x: P(x) \geqslant a\}$, where a is any value less than 1. Now form the mating pools W_x of each x, and take a subset $V_{x,a}$ of each of these, where a has the same value as previously. A *Mendelian population $UV_{x,a}$* is formed by taking the union of any one $V_{x,a}$ with all $V_{x,a}$ with which it has a non-empty intersection, then taking the union of the set so formed with all $V_{x,a}$ having a non-empty intersection with this set, and so on until all intersections are empty or there are no more $V_{x,a}$.

A *species* is the union of a stochastic set X_1 with a stochastic set X_2: X_1 is a Mendelian population $UV_{x,a}$ whose intersection with all other Mendelian populations of the same a is negligible, and X_2 is the union of all Mendelian populations which, if they were sympatric and synchronous with $UV_{x,a}$ would have a non-negligible intersection with $UV_{x,a}$. Beckner (1959) has proposed a roughly similar scheme, and ours are meant merely to formalize the most widespread biological usage and concept of species.

Any adequate definition must be vague in the appropriate areas, unless there is no ambiguity as to the identity and composition of the elements included in and excluded from the set referred to by the definiendum. The word "species" is necessarily vague in several areas, and the present definition allows for this. There are

(1) the exact number of individuals included in a single species
(2) intermediate cases, the potential origin of new species:
    (a) two partially reproductively isolated races
    (b) introgression of one species into another, and similar events
    (c) ancestral-descendant species in phyletic evolution.

These species will be the conceptual units of the taxonomic system presented below, but the taxonomic system is not formally constructed from this level. Analogous classificatory "species" are often provided for groups of organisms whose predominant mode of reproduction (asexual or parthenogenetic) does not permit Mendelian populations to be defined for them. These analogies are on the basis of morphological and physiological differentiation of the organisms, and to a lesser extent on more or less discrete environmental niches that divide them to some extent into groups by selecting out those that are not sufficiently well adapted to particular niches. For example, it is a defensible thesis (because of the enormous numbers of individuals involved) that bacterial "species" are mainly determined not by the potentialities of the bacteria but by the available environments and their greatly differing carrying capacities. At any rate, such species analogues can be formally treated as species, and this is usually done.

III. THE STRUCTURE OF TAXONOMIC SYSTEMS

Before considering hierarchies, a few preliminary concepts should be understood. A relation Y is *one-many* if and only if not more than one element of its field has the relation Y to any other element, i.e.
$(\forall w)(\forall x)(\forall z)((wYx) \cup (zYx)$ implies $(w=z))$. If wYx and xYz, w is said to *stand in a power* (the second in this case) of Y to z. This relation has been defined by Woodger (1937).

The definition of hierarchies can be approached from two ways, depending on whether Y is transitive or intransitive. If Y is transitive, it cannot be one-many. However, most (not quite all, e.g. equality) transitive relations have an intransitive restriction; probably all hierarchy-forming transitive relations do. For example, P is the intransitive restriction of D. Conversely, most or all intransitive relations have transitive extensions.

Woodger's definition (1952b) of a *hierarchy* applies to intransitive relations only, but probably to all hierarchies defined from either approach. His definition is as follows, with the substitution of Y for R: "Y is a hierarchy if and only if Y is one-many and if the converse domain of Y is identical to the set of all terms to which the first term of Y stands in some power of Y." This type of hierarchy may be called an *intransitive hierarchy*; the parental relation P is an example of it in asexually or parthenogenetically reproducing organisms. The hierarchies of Parker-Rhodes (1957) constitute a very special case of this definition.

Sklar (this volume) has come to the concept of hierarchy from the viewpoint of transitive relations. Since "standing in some power of an (intransitive) relation" is itself a (transitive) relation, such an approach is necessarily valid for all hierarchies. Y is a *transitive hierarchy* if and only if

(1) Y is transitive.

(2) If any two different elements of the field of Y have the relation Y to any third element, then one and only one of the former elements has the relation Y to the other, i.e. $(\forall w)(\forall x)(\forall z)((wYx) \cup (zYx)$ implies $((wYz)$ or $(zYw)))$.

(3) There exists a unique object B in Y having the relation Y to every other element of Y. B is known as the *beginner*. The descendant relation D normally forms transitive hierarchies. Note that this definition is rather closely analogous to the definition of one-many. Criterion (2) implies that Y is an ordered relation.

I am not sure that the definitions of Woodger and Sklar are equivalent, i.e. that all the hierarchies of one are also hierarchies for the other. I cannot prove this to be the case, but I can find no contrary examples and do not see exactly where, if at all, they differ. For convenience I will use in this paper Sklar's definition as formulated above. An *inclusion hierarchy*, e.g. a taxonomic system, is one in which Y is inclusion, and the objects are therefore sets.

Consider two finite sets of organisms:

R: the set of all organisms that have ever existed (if ordered in terms of time, it has an indefinite lower bound, while its upper bound shifts with the passing of the present).

S: the set of all organisms ever seen by taxonomists or otherwise available for classification.

Taxonomy is an attempt to classify *R*, using *S* as the basis for this. Obviously, *S* is a proper subset of *R*.

I will define a *taxon*, which is a taxonomic group of organisms, as any element of a set *G*. *G* has the following two properties:

(1) $(\forall E \epsilon G)(E \subset S)$, i.e. all taxa are composed of organisms in the set *S* defined above.

(2) $(\forall E \epsilon G)(\forall F \epsilon G)(\text{either}(E \cap F = \emptyset)$ or $(F \subset E)$ or $(E \subset F))$, i.e. all taxa are either disjoint or one is included in the other. Note that a taxon may be empty. Let Card *F* be the non-negative integer expressing the number of elements of *F*.

Taxa are arranged in *categories* to form the taxonomic system; examples of such categories are "order", "genus", "superfamily". A category *K* is a set of elements of a hierarchy of sets, and has the following properties:

(1) All the elements of *K* are disjoint.

(2) Each element of *B* is an element of some element of *K*.

(3) At least one element of *K* is nonempty.

In addition:

(4) If *H* and *K* are categories, $E \epsilon H$, $F \epsilon H$, $E' \epsilon K$, $F' \epsilon K$, and *EYE*, then it is not true that *FYF* unless $F = F$.

These properties are essential because it is implicit in the use of a taxonomic system that every organism belongs to one and only one taxon in each category, and that these categories do not cross each other. If the taxa of a given category were a partition of *B* (which they would be except that empty taxa are not excluded), then the taxa of each category lower than this one would be a refinement of this partition. Category *H* is *below* category *K* if and only if each element of *H* is included in some element of *K*, and $H \neq K$. This is equivalent to saying that Card $H >$ Card *K*.

The most important variety of hierarchy may be called a *canonical hierarchy*. This is one in which no two categories have the same number of elements, i.e. if *H* and *K* are categories, then Card $H \neq$ Card *K*.

Gregg (1954) defined a taxonomic system in such a manner as to exclude the empty set $\emptyset$ from G, the set of all taxa. His definition implies, as he pointed out, (1) that taxa with the same elements are identical, even if one is a genus and another a family, and (2) that therefore categories in this system do not entirely correspond to the usage by taxonomists of such terms as "genus" or "family". Both these results are taxonomically objectionable. The first is objectionable because other taxa of the lower category may later be discovered or split off, and the second because then the set-theoretical system is not an accurate description of taxonomic systems as they are used. These difficulties occur only in non-canonical hierarchies and hierarchies including a non-canonical hierarchy.

Three alternatives have been proposed to meet objections of this type. Eden (1955) believed an intensional (connotative) definition must be used. But an extensional system is much simpler when possible, and one can in fact be salvaged, although the present attempt has overtones of the intensional. Sklar (this volume) has given a rather complicated system, associating terms with objects, e.g. "species" with all species, "genus" and "species" with all genera, etc. This is formally satisfactory, but it does not lead to what a taxonomist considers a taxonomic system, where the taxa are sets of organisms only. A third possibility, which has been proposed by Parker-Rhodes (1957) is to define a category so that its elements are the taxa in the next lower category, not organisms. This alternative is open to the same sort of objection as is Sklar's; see Sklar's paper for further discussion.

A more direct means of overcoming the difficulties of Gregg's model is to associate with each taxon containing known members a second taxon, which is empty with respect to S. To do this we must define another new type of set, *conditionally empty sets*, symbolized by $\emptyset_X$ (see Figure 2). $\emptyset_X$ is a necessarily proper subset of some set X and disjoint from each of a set of sets, the intersection of each of which with X is non-empty.

A further definition must be given, that of category K being *immediately below* category H. This occurs when H is the smallest category larger than K, i.e. the next category on the opposite side of K from the category containing B. This definition applies only to canonical hierarchies. K may then be written as $(H - 1)$. A more general definition, applicable to all finite hierarchies, is that K is below H and there is no category that is below H and that is above K.

410

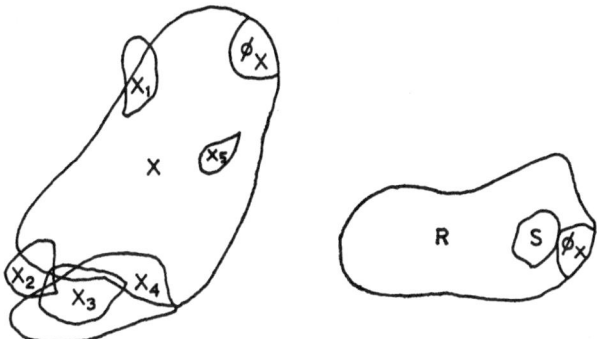

Fig. 2. Diagrams to illustrate the concepts of a conditionally empty set and taxon. See text.

Now I will define a *conditionally empty taxon* $\emptyset_X$ (Figures 2 and 3). Let H and K be categories, K immediately below H. Let $X \epsilon H$, $F \epsilon K$, $D \epsilon K$, $F \subset S$, $X \subset S$, $F \subset X$, $D \subset X$. Let $D \cap S = \emptyset$. Then $\emptyset_X = D$. In other words, $\emptyset_X$ is a set of unknown organisms (or of none at all), associated with a known taxon and included with it in the next higher taxon X. A known taxon is any non-empty taxon.

A taxonomically and methodologically adequate definition for a *taxonomic system T* appears to be the following:

T is an inclusion hierarchy with the following three properties:

(1) $(\forall t \epsilon T)(t \epsilon G)$, i.e. a taxonomic system is a set of taxa.

(2) $(\forall s \epsilon S)(\forall H \epsilon Q)(\exists F \epsilon H)(s \epsilon F)$, i.e. every known organism is a member of some taxon in every category. For less inclusive taxonomic systems, and more generally, $b \epsilon B$ may be substituted for $s \epsilon S$.

(3) $(\forall (H - 1) \epsilon Q)(\forall X \epsilon H)(X \neq \emptyset)(\exists \emptyset_X \epsilon (H - 1))(\emptyset_X \subset X)$, with reference to the definition of $\emptyset_X$ above. This means that with every known taxon above the lowest category, there is associated a conditionally empty taxon on the category immediately below it. Note that there is only one empty set for any universe of objects and that $\emptyset_X$ is empty with respect to S, so any category really contains only one $\emptyset_X$. This $\emptyset_X$ is associated,

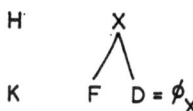

Fig. 3. Diagram to illustrate the use of a conditionally empty taxon. See text.

411

however, with many X's, and many $\emptyset_X$'s are presumably non-identical in R. Category H thus differs from category $(H - 1)$ partly, and importantly, in its containing unknown taxa that are not members of $(H - 1)$.

This definition gives a place for future discoveries, in which event $\emptyset_X$ will of course at that time exclude them also. It also provides a means of separating taxa of different categories that are composed of the same individuals. If as a matter of fact $\emptyset_X$ is empty, i.e. $\emptyset_X \cap R = \emptyset$, then there are no possible future discoveries except those that may be the result of future diversification in nature, and in this case (quite possibly nonexistent and at any rate unprovable) I would have no objection to interchanging taxa between categories. Under these very special and unreal conditions, a specific genus might actually be a family, as maintained more generally by Gregg (1954).

We may conceptually erect a taxonomic system for R parallel to the one for S, and this is to some extent done by taxonomists when they consider incompletely known phylogenies. This "complete" taxonomic system is conceptually basic to the actual one by the preceeding argument.

The objection has been made that the taxonomic system thus defined is never complete (except in the trivial case where $\emptyset_X \cap R = \emptyset$), but this is also true of any real taxonomic system. Thousands of taxonomists spend their lives fitting undescribed material into taxonomic systems and less importantly, splitting some old taxa into new ones in the same category.

It is irrelevant that polyphyletic taxa fit as nicely as monophyletic ones, since the hierarchy is merely a description of taxonomic systems and not a basis for judgment of their validity, which must be based on all the criteria given in the first section. A possible difficulty is the provision that each organism have a taxon in each category, since this is usually not done explicitly for monotypic taxa at some intermediate, non-obligatory category (suborder, etc.). The omission, however, is merely for brevity's sake, could easily be filled, usually is filled implicity, and must be filled if another taxon of that intermediate category is discovered.

An alternative and almost as acceptable a solution would be to define a taxonomic system as a canonical hierarchy, define the categories for this, and let any non-canonical hierarchies that appear in such a taxonomic system remain undefined as taxonomic systems. They will, however, still be part of the more inclusive taxonomic system. There will be

412

no confounding of categories, the model is methodologically sound, and it largely satisfies the requirements of a taxonomist.

Classification is a successive comparison of groups; where there is only one group there is no need for it to appear in more than one category. Thus there is no taxonomic category higher than the seldom-used "empire", in which all organisms are contained in a single taxon. For the same reason, there is no real need to use in one taxonomic system all the categories that appear in a taxonomic system that includes it. This is often done for aesthetic or legalistic reasons even where unnecessary in terms of our knowledge of the structure of S, and offers the only apparent drawback to this second proposal. For example, a subfamily may contain two genera, each with a single species. Nomenclatorial requirements, as well as a desire for symmetry with other parts of a more inclusive taxonomic system, demand the use of both categories. This is quite apart from the possibility of future discoveries. This latter possibility is irrelevant in a model concerned only with S as it exists at a given moment, but becomes important when this model is used in a world where S is expanding.

APPENDIX

Numerous theorems can be proved about taxonomic systems. Gregg (1954) has presented a number of these, most of which are applicable also to the present concept; I have selected two further ones.

THEOREM I. The maximum number of discrete categories is one more than the number of individuals, i.e. if Q' is the set of all non-intersecting categories in the taxonomic system, Card $Q' \leqslant$ Card $B + 1$. S may be substituted for B if desired; a similar theorem can be proved using species instead of individuals.

PROOF: Let $K \epsilon Q'$.

$(\forall K)$(Card $K \leqslant$ Card $B + 1$), i.e. every element in the segment determined by Card Q' is at most equal to Card $B + 1$, by the definition of a taxon.

$\therefore$ Card $Q' \leqslant$ Card $B + 2$, by the definition of an ordinal.

$\therefore$ Card $Q' \leqslant$ Card $B + 1$, since there is no empty category.

Since most of the taxa in even the lowest categories contain many known individuals, except in small geographic areas and short paleontological time intervals, and since (with the same caveats) all or nearly all cate-

413

gories contain more than one more taxon than the next lower category, the number of categories in most real taxonomic systems will be enormously smaller than this.

THEOREM II. The maximum number of taxa of a canonical hierarchy is the sum of the cardinals through Card $B + 1$, i.e. if G is the set of all taxa in the taxonomic system,

$$\text{Card } G \leqslant \sum_{i=1}^{\overline{\text{Card } B+1}} i$$

Again, S may be substituted for B.

PROOF: $(\forall H)(\exists E)(E \epsilon H)$, i.e. each category has at least one taxon, by the definition of a category.

Card $H <$ Card $(H - 1)$, by the definitions of canonical hierarchy and immediately lower category.

Card $H \leqslant$ Card $B + 1$, since there is only one empty set.

∴ Card $(H + 1) \leqslant$ Card $B \ldots$; Card $(H + B) \leqslant 1$.

$$\therefore \text{Card } G \leqslant \sum_{i=1}^{\overline{\text{Card } B+1}} i$$

For the reasons mentioned under Theorem I, this limit is not at all closely approached.

The American Museum of Natural History, New York

GLOSSARY OF SYMBOLS REPEATEDLY USED

B – the beginner of a hierarchy

E, F – taxa

G – the set of all taxa

H, K – categories

Q – the set of all categories

R – the set of all organisms that ever existed

S – the set of all organisms available for classification

X – arbitrary set

x, y – arbitrary elements

Y – arbitrary relation

$\emptyset$ – the empty set

$\emptyset_X$ – conditionally empty set associated with set X

ϵ – is an element of

$\subset$ – is a subset of (may be equal to)

$\forall$ – for all

$\exists$ – there exists

$\cup$ – the union of

$\cap$ – the intersection of

$\therefore$ – therefore

$\{x\}$ – the set whose only element is x

REFERENCES

Beckner, M., 1959: *The biological way of thought*. Columbia Univ. Press, New York. 200 pp.

Bulman, 1955: 'Graptolothina', in *Treatise on invertebrate paleontology* (by R. C. Moore, ed.), Part V. Geological Society of America and Univ. of Kansas Press, Lawrence. 101 pp.

Eden, M., 1955: (review of Gregg 1954). *American Scientist* **43**, 128a–130a.

Gregg, J. R., 1954: *The language of taxonomy*. Columbia Univ. Press, New York. 70 pp.

von Huene, F. R., 1948. 'Short review of the lower tetrapods', *Robert Broom Commemorative Volume* (Spec. Publ., Royal Soc. of South Africa), pp. 65–106.

Parker-Rhodes, 1957: (review of Gregg 1954). *Philosophical Review* **66**, 124–125.

Simpson, G. G., 1945: 'The principles of classification and a classification of mammals', *Bull. Amer. Museum Nat. Hist.* **85**, 1–350.

Simpson, G. G., 1953: *The major features of evolution*. Columbia Univ. Press, New York. 434 pp.

Simpson, G. G., 1961: (A book on the principles of systematic biology) Columbia Univ. Press, New York.

Woodger, J. H., 1937: *The axiomatic method in biology*. Cambridge Univ. Press. 174 pp.

Woodger, J. H., 1952a. *Biology and language*. Cambridge Univ. Press. 364 pp.

Woodger, J. H., 1952b. 'From biology to mathematics', *The British Journal for the Philosophy of Science* **2**, 193–216.

ARISTID LINDENMAYER

LIFE CYCLES AS HIERARCHICAL RELATIONS

> "...language is just as indispensable a tool for the pursuit of biology as microscopes, kymographs and other instruments. If observations are impossible without the one, their recording and the construction of hypotheses are no less impossible without the other". (J. H. Woodger, *Biology and Language*, 1952)

In this paper the development of a theory is attempted, in which the important features of life cycles can be described and the relationships between the various life cycles can be exhibited.

Theories are considered in modern logic as languages. A language, in logical terms, must have syntactic rules, or postulates, governing the use of the primitive notions of the language. The primitive notions remain otherwise undefined. Other terms may be introduced into the language by explicit definitions, with the help of the primitive ones. The primitive and defined notions may be said to constitute the vocabulary of the language, while the syntactic rules to represent its grammar. To have a meaning, the language must also have semantic rules, which establish connections between observations and the theoretical concepts. (A language without semantic rules is called an uninterpreted calculus.) The fewer rules a theory has and the more observations it can account for, the more powerful we would consider it to be.

The life cycle theory to be presented here is construed as a logical language. Only three primitive notions are allowed, in addition to those of the basic logic, those of 'mitosis', 'meiosis', and 'gametic fusion'. Life cycles are considered here as sequences of these three cellular processes, the last two of which are sometimes referred to as the cardinal events of life cycles. Ten postulates are given concerning the three primitive concepts. Further terms, like 'division', 'haploid', 'diploid', and others standing for subclasses of zygotes, gametes and meiocytes, are introduced by definitions. The semantic rules of this theory are those constituting the cytological meaning of mitosis, meiosis, and gametic fusion, with certain

416

reservations as discussed below. Since with present day techniques it is almost always possible to tell whether a cell or a nucleus undergoes mitosis or meiosis or fusion, or not, the interpretation of this theory, i.e., the confirmation or rejection of the theorems by observations, should not present any difficulty.

All the theorems listed can be derived from the postulates and definitions of this theory together with those of the basic logic. The logical language of Whitehead and Russell's *Principia Mathematica* is used throughout, the definitions and symbolism of this system being described in the next section. The logical concept of hierarchical relations, as defined by Woodger, is a particularly useful one, and a considerable part of this paper is devoted to the application of this concept to life cycles.

The postulates and the use of the concept of hierarchies in this theory grew out of Woodger's formal system in *The Axiomatic Method in Biology* (1937). The theory presented here was worked out to a large extent while the author enjoyed the guidance and hospitality of Professor Woodger in London during the tenure of a National Science Foundation fellowship. This year of association with Professor Woodger has been a lasting experience for me and a source of continuing stimulation for work in this area of mathematical biology to which he has contributed so much. The original idea for a theory of life cycles came from many discussions with Professor Ralph O. Erickson in Philadelphia.

The writing of this paper was aided by a Special Summer Research Grant of the University of Pennsylvania and by a U.S. Public Health Service training grant in biomathematics (2G-678) at North Carolina State College.

LOGICAL CALCULUS AND SYMBOLISM

The basic logical language to be used is the sentential and functional calculus of *Principia Mathematica* (abbreviated as PM), with its customary truth functional interpretation. The postulates, definitions and semantic rules of this language are not given here, except for a short explanation of the symbols used.

The symbols may represent either constants or variables. The constants may, in turn, be names or logical constants (sentential connectives, quantifiers, sign of identity, etc.) The logical constants used in this system

are given below together with their approximate equivalents in English.
Sentential connectives:

$\sim \ldots$ not . . .
$\ldots \vee \text{- - -}$ . . . or - - -
$\ldots . \text{- - -}$ . . . and - - -
$\ldots \supset \text{- - -}$ if . . . then - - -
$\ldots \equiv \text{- - -}$ . . . if and only if - - -

In place of the three dots or dashes in each case a sentence or a sentential
variable may be substituted. With these concepts the entire sentential
calculus can be constructed.

The functional calculus deals with statements containing variables.
A formula such as ϕx becomes a statement if a constant is substituted for
the variable 'x', or if the variable is quantified. A variable may be universally quantified, when we wish to state that something is true for *all*
values of the variable, or it may be existentially quantified, when we want
to state that something is true for *some* values of the variable. Quantified
statements are symbolized in the following manner:

$(x) \cdot \phi x$ for all x, ϕx
$(\exists x) \cdot \phi x$ there is an x such that ϕx

In addition, the concept of identity is also needed for the functional
calculus. It represents strict mathematical identity, and can only be used
between names of the same thing. Identity is symbolized by '$=$', and its
negation by '$\neq$'.

The various kinds of variables are distinguished by using letters from
different alphabets or from different parts of an alphabet. Thus 'p', 'q', 'r'
are sentential variables; 'ϕ', 'ψ', 'χ' are functional variables; and 'x', 'y',
'z', 'u', 'v', 'w' are individual variables.

Following the convention of PM, varying numbers of dots are placed
before or after the sentential connectives and quantifiers to indicate the
scope of that particular connective or quantifier in a complicated sentence.
By this method we can avoid the use of multiple parentheses. The rule
about the use of these dots is that a higher number of dots can always
override a smaller number of dots. Where dots are found without accompanying connectives or quantifiers they are interpreted as connectives
standing for 'and'. For instance, '$(p \supset q) \equiv [p \equiv (p \cdot q)]$' can be written
in this notation as '$p \supset q \cdot \equiv \: : p \cdot \equiv \cdot p \cdot q$'.

An extension of this language in PM is achieved by introducing the

418

concepts of classes (or sets) and relations, all of which may be termed 'predicates'. Classes are one-termed predicates, while relations may be two-termed, three-termed or many-termed predicates. Membership in a class, designated by 'ϵ', must be added to our list of logical concepts, its use being governed by the theory of types. According to the postulates of PM concerning types, no class may be a member of itself, classes can be members only of classes of classes. The individuals which have no members are of the lowest type, the classes of these individuals are of the next higher type, and so on. In this paper, 'α', 'β', 'γ', 'δ' are class variables, 'κ' and 'λ' are variables for classes of classes, while 'R', 'P', 'Q', 'S', 'T' are variables for relations.

In the appendix the relevant statements from PM are listed under their original numbers preceded by asterisks; these may be construed as our definitions (the actual definitions of PM are not given here).

In addition to the statements of PM some of my theorems are also listed there; these can be distinguished from the former by the letter 'T' placed before the number. The numbers for these theorems were chosen so as to place them near to related theorems of PM, but not to duplicate any of the numbers in PM. These theorems are followed in parentheses by the numbers of those PM theorems on which their derivation depends. Because of the large number of the PM theorems needed in this paper, they cannot all be reproduced here.

A few of Carnap's (SLA) definitions and theorems are also needed for our exposition, and these are listed under number 99., with the letter 'C' placed in front of the numbers.

It should be mentioned here that there are a few minor changes in the symbols used from those of PM. These deviations are: the use of '$\subseteq$' rather than of '$\subset$' for class inclusion (the former is preferable for mnemonic reasons, in my opinion), the use of '$\bar{\alpha}$' instead of '$-\alpha$' for the complement of a class, and finally the use of '$\cup$', '$\cap$', '$\subseteq$', '$\wedge$', 'V' without dots for both classes and relations. An abbreviation of PM, which is followed here, is to write '$x, y \in \alpha$' instead of '$x \in \alpha \cdot y \in \alpha$', and to write '$(x, y)$' and '$(\exists x, y)$' instead of '$(x)(y)$' and '$(\exists x)(\exists y)$'. As in PM, '$p \equiv q \equiv r$' is written for '$p \equiv q \cdot p \equiv r$', and '$\alpha = \beta = \gamma$', is used instead of '$\alpha = \beta \cdot \alpha = \gamma$'. Following another PM convention, universal quantification of entire sentences is never indicated. Formulas with free variables should be considered as if universally quantified.

With the help of the concept of membership in a class it becomes possible to incorporate the Boolean algebra into this language. Accordingly $\alpha \cup \beta$ is the Boolean sum or join of the two classes α and β (defined by *22.34), $\alpha \cap \beta$ is Boolean product or common part of the two classes (*22.33), and $\bar{\alpha}$ is the complement of α or not $-\alpha$ (*22.35). The Boolean sums, products and complements of classes are classes themselves. The expression '$\alpha \subseteq \beta$', on the other hand, stands for 'class α is included in class β', and it does not refer to a class but is a sentential formula (it becomes a sentence when the variables 'α' and 'β' are quantified or are substituted by constants.) A class is said to be included in another class just in case all the members of the first class are also members of the second class (*22.1). Similarly, '$\alpha = \beta$' is a sentential formula which is equivalent to 'α is included in β and β is included in α' (*22.41). Finally, 'V' stands for the universal class, of which every individual of a given type is a member (*24.01, .104), and '$\wedge$' stands for the empty class, i.e., the class that has no members (*24.02, .105). Similar notions apply to relations (in treatments of logic other than PM relations are considered as classes of ordered pairs). The definitions of the Boolean operation son relations are given in statements *23.1, .33, .34, .35, .41, and those of the universal and empty relations in *25.01, .104, .105.

We will have occasion to use only two-termed relations of the general form 'xRy', where 'x' and 'y' represent individuals and 'R' is the relation in which they stand to each other. Two-termed relations can obtain, however, not only between individuals, but also between an individual and a class, or between classes, or between an individual and a relation, or between relations, or between a class and a relation. We thus will have relations of the forms '$xR\alpha$', '$\alpha R\beta$', xRQ', 'PRQ', 'αRQ', etc. Colloquially we may mention R steps whenever we want to refer to a pair of terms, one standing in relation R to the other. Several relation concepts are defined in the following, the first of which is that of relational descriptions, of the general form '$R'y$', standing for 'the term that stands in relation R to y'. For instance, 'the Queen of England' or 'the father of John' can be considered as relational descriptions, where "being queen of" and "being father of" are the relations involved. One must be careful in using a relational description, since mentioning in any context an expression like '$R'y$' has as a consequence not only that an individual exists which stands in R to y, but also that only a single such individual exists (cf. T30.38 and *30.21). The statement that such a single term exists is written as '$E!R'y$', the formal explanation being given by theorems *30.21 and T30.301.

The converse of a relation is obtained by reversing the order of the terms of that relation, thus x stands in the converse of R to y whenever y stands in R to x. The converse of R may designated by '$\breve{R}$' or by '$Cnv'R$' (*31.11, 31.12). The reason for having two symbols for the same concept is that the first symbol is convenient to use with single letters, while the latter one is convenient with more complex formulas. The converse of the relation "being a parent of" is "being a child of".*

Next, the class of individuals which stand in relation R to a certain individual y is designated by '$\overrightarrow{R}'y$', which may be read as "the R's of y" (*32.18). $\overrightarrow{R}$ is actually a new relation between a class and an individual forming expressions like '$\alpha\overrightarrow{R}y$', and '$\overrightarrow{R}'y$' is its relational description. Similarly, '$\overleftarrow{R}'x$' designates the class of individuals to which x stands in relation R (*32.181), where $\overleftarrow{R}$ is again a new relation forming expressions like '$\beta\overleftarrow{R}x$'. If, for example, R is the parenthood relation, then $\overrightarrow{R}'y$ is the class of parents of y, and $\overleftarrow{R}'x$ is the class of the children of x.

By similar definitions are the concepts of the domains, converse domains, and fields of relations introduced. The domain of a relation R ($D'R$) is the class of all individuals which stand in R to some individual (*33.13). The converse domain of a relation R ($\sigma'R$) is the class of all individuals to which some individual stands in R (*33.131). Finally, the field of a relation R ($C'R$) is composed of all the individuals which either stand in R to some individual or to which some individual stands in R (*33.132), in other words the field of R is the Boolean sum of the domain and converse domain of R (*33.16). Thus the domain of the fatherhood relation is the class of all fathers, its converse domain is the class of all children, while its field is the class of all fathers and children. The expres-

---

* Single quotation marks are used only when reference is made to the *name* of something rather than to the thing itself; while double quotation marks are employed as in common usage, namely, to delimit and separate a phrase from the rest of the sentence, or to indicate doubt concerning the meaning or applicability of a term.

sions '$D'R$', '$\mathit{Q}'R$', and '$C'R$' are also relational descriptions formed with the relation signs 'D', 'Q', and 'C', each of which stands for a relation between a class and a relation (e.g., αDR, $\beta\mathit{Q}R$, γCR).

The relative product of two relations R and P is itself a relation, designated by '$R \mid P$', and this relation holds between two individuals x and z whenever there is an individual y to which x stands in R, and which in turn stands in P to z (*34.1). In other words '$R \mid P$' stands for 'an R step being followed by a P step'. For instance, the relation "being an uncle of" can be considered as the relative product of "being a brother of" and "being a parent of" relations. The relative product of a relation R with itself, i.e., $R \mid R$, may be written as R^2 (*34.02), and the relative product of R^2 with R as R^3 (*34.03), and so on. "Being a grandparent of" is therefore the same relation as "(being a parent of)^2".

We can speak of relations with limited domains or converse domains whenever the members of the domain or of the converse domain of a relation all belong to some class. The formal designations are '$\alpha \upharpoonright R$' for the former, and '$R \upharpoonright \beta$' for the latter (*35.1, .101). We can also describe a relation the field of which is limited to a class, the symbol for it being '$R \restriction \alpha$' (*36.13).

On the other hand, if we want to designate the class every member of which stands in a certain relation R to some member of a certain class β, then we write '$R''\beta$' (*37.1). These descriptions are called plural descriptive functions, and are very useful in our theory. 'The children of color-blind fathers' is such an expression, formed of father-to-child relation and the class of color-blind persons.

For the sake of convenience, identity and its negation (called 'diversity' in PM) are given relation symbols as well, according to *50.1 and *50.11. Thus we can write 'xIy' instead of '$x = y$', and 'xJy' instead of '$x \neq y$'. A further symbol is introduced by definition *51.01, substituting 'ι' for '$\vec{I}$'. This symbol is mainly used in designating unit classes, i.e., classes with a single member. The unit class of x is written as '$\iota'x$', and some individual can be a member of this unit class only if it is identical with x (*51.15).

With the help of unit classes cardinal numbers can be defined in a simple manner, as statements *52.1, .11, *54.101, .28, T54.71 show. Accordingly, '1' designates the class of all classes which have a single member, and '2' designates the class of all classes with two members and similarly for '4'.

These numbers are defined here because we need them for the characterization of relations which hold between a single individual and one or more individuals, or between one or more individuals and a single one, or between specified numbers of individuals. The general definition of such classes of relations is shown under *70.11. The definition for the class of relations between a single and one or more individuals, called 'one-many relations' or '1 → Cls', is given by statements T71.104 and *71.17, while that for relations between one or more individuals and a single individual, called 'many-one relations' or 'Cls → 1', is shown in T71.105 and *71.171. Both of these classes of relations are very frequently referred to in this work.

In order to be able to make even more specific statements with regard to the number of individuals involved, we also use classes of relations which are $1 \to 2$, $1 \to 4$ and $2 \to 1$. The first of these is applicable to mitosis, the second to meiosis, and the last to gametic fusion.

It will be necessary for us to indicate that certain classes are mutually exclusive. According to PM notation, 'Cls²excl' designates the class of all those classes the members of which are mutually exclusive classes (*84.1).

Under the 'power of a relation' we understand, roughly speaking, some finite number of successive steps of that relation. Thus R, R^2, R^3, etc. are all powers of R, and can be defined recursively by stating that $R^1 = R$ and $R^{n+1} = R|R^n$ (where n is any finite positive integer). The symbol for some unspecified power of R is R_{po}. Another symbol to be defined is 'R_*', which simply designates the relation of either a power of R or of identity in the field of R (*91.54). Therefore, if xR_*y, then x is either identical to y and is a member of the field of R, or x stands in some power of R to y.

The actual definitions given for 'R_*' and 'R_{po}' in PM are quite subtle, two statements equivalent to those definitions having been reproduced under *90.1 and *91.62. These definitions involve the concept of hereditary classes. A class α is said to be an hereditary class with respect to a relation R if and only if all the individuals to which the members of α stand in R are also members of α (formally: $\breve{R}''\alpha \subseteq \alpha$). Thus the class of men named Smith is an hereditary class with respect to the father-to-son relation. In biology we usually assume that the genotypes of cells are hereditary classes with respect to mitosis, barring mutations of course. This concept will be important in the development of the biological theory to be presented.

Coming back to the definition of powers of relations, theorem *91.62 states that x stands in some power of R to y if and only if y is a member of every class which is an hereditary class with respect to R and of which all the individuals are members to which x stands in R. The great advantage of this definition of 'R_{po}' over the recursive one of 'R^n' mentioned above is that it does not need the concept of finite integers. In fact, Whitehead and Russell derived the concepts of finite (inductive) cardinals from that of R_*.

Several theorems which are to be used concerning R_{po} and R_* are listed under numbers 90., 91., 92., 94., and 96.

Two more concepts that we need are those of the beginners and of the links of a relation. An individual x may be called a beginner of R if it is a member of the domain of R and not of the converse domain of R (designated as 'xBR') (*93.1). For instance, zero is the beginner of the "less than" relation among positive numbers, while the "greater than" relation has no beginner among these individuals. Some relations may also have many beginners. An individual x is called a link of a relation R, on the other hand, if it is a member of both the domain and the converse domain of R (designated by '$xLkR$', T93.7). We can also designate the terminals of a relation R by the term 'the beginners of the converse of R' (T93.601).

The fact that there is a single beginner of a relation is expressed by '$E!B'R$', and various theorems are given for the case that this is true (T93.61 to .615).

If some of the beginners of a relation R are members of a class α, this can be stated either by '$\overrightarrow{B'R} \cap \alpha \neq \wedge$' or by '$R \in \breve{B}''\alpha$', according to T93.65, the second expression being sometimes more convenient.

Finally, some further classes of relations must be defined using Carnap's (SLA) definitions.

A relation R is called symmetric if and only if R is included in the converse of R (C99.101), or in other words, whenever xRy also yRx. "Being married to" is a symmetric relation. On the other hand, a relation

R is asymmetric if it is included in the complement of its converse (C99.102), i.e., if it is impossible to have xRy and yRx. "Being a father of" is such a relation.

A relation R is transitive if R^2 is included in R (C99.103), i.e., whenever xRy and yRz then also xRz. "Longer than" or "heavier than" are such relations. A relation R is intransitive if it is never the case that xRy and yRz and also xRz (C99.104). The fatherhood relation is intransitive, while the ancestral relation is transitive. The parenthood relation is intransitive in some human societies, but not in general.

A relation R is said to be reflexive if individuals in the field of R always stand in R to themselves (C99.105). A relation R is irreflexive, on the other hand, if an individual can never stand in that relation to itself (by C99.106 and *50.24).

We must note that none of the three pairs of classes defined are mutually exclusive. The contradiction or negation of $R^2 \subseteq R$, for instance, is not $R^2 \subseteq \bar{R}$, but $\sim(R^2 \subseteq R)$. It is therefore entirely possible to have relations which are both symmetric and asymmetric, or both transitive and intransitive, or both reflexive and irreflexive (but either they or their second power will be empty; cf. T99.24).

Definitions and Theorems of Logic

* 22.1	$\alpha \subseteq \beta . \equiv . (x) : x \epsilon \alpha . \supset . x \epsilon \beta$	
* 22.33	$x \epsilon \alpha \cap \beta . \equiv . x \epsilon \alpha . x \epsilon \beta$	
* 22.34	$x \epsilon \alpha \cup \beta . \equiv : x \epsilon \alpha . \vee . x \epsilon \beta$	
* 22.35	$x \epsilon \bar{\alpha} . \equiv . \sim (x \epsilon \alpha)$	
* 22.41	$\alpha \subseteq \beta . \beta \subseteq \alpha . \equiv . \alpha = \beta$	
T 22.75	$\alpha \cap \beta = \alpha \cap \gamma . \equiv . \alpha \cap \beta =$	(*22.35)
	$= \alpha \cap \bar{\gamma}$	
T 22.76	$\alpha = \beta . \equiv . \alpha \cup \beta \subseteq \alpha \cap \beta$	(*22.41, .42, .45, .59)
* 23.1	$R \subseteq S . \equiv : (x, y) : xRy . \supset . xSy$	
* 23.33	$x(\bar{R} \cap S)y . \equiv . xRy . xSy$	
* 23.34	$x(R \cup S)y . \equiv : xRy . \vee . xSy$	
* 23.35	$xRy . \equiv . \sim (xRy)$	
* 23.41	$R \subseteq S . S \subseteq R . \equiv . R = S$	
T 23.812	$R \subseteq S . R \neq S . \equiv . R \subseteq S .$	(*23.41)
	$. \sim (S \subseteq R)$	
* 24.01	$x \epsilon V . \equiv . x = x$	

* 24.02	$\wedge = \breve{V}$	
* 24.104	$(x) . x \,\epsilon\, V$	
* 24.105	$(x) . \sim (x \,\epsilon\, \wedge)$	
T 24.581	$\alpha \subseteq \beta . \supset : \alpha \neq \wedge .$	(*22.621)
	$\equiv . \alpha \cap \beta \neq \wedge$	
* 25.01	$x \vee y . \equiv . x = x . y = y$	
* 25.104	$(x, y) . x V y$	
* 25.105	$(x, y) . \sim (x \wedge y)$	
* 30.21	$E \,!\, R'y . \equiv \therefore (\exists x) . xRy : (x, z) :$	
	$: xRy . zRy . \supset . x = z$	
T 30.301	$E \,!\, R'y . \equiv . (\exists x) . x = R'y$	(*30.2, .3)
* 30.31	$x = R'y . \equiv : xRy : (z) :$	
	$: zRy . \supset . z = x$	
T 30.38	$\phi (R'y) . \supset . E \,!\, R'y$	(*14.21, 30.01)
* 31.11	$x \breve{R} y . \equiv . yRx$	
* 31.12	$Cnv'R = \breve{R}$	
* 32.18	$x \,\epsilon\, \overrightarrow{R}'y . \equiv . xRy$	
* 32.181	$y \,\epsilon\, \overleftarrow{R}'x . \equiv . xRy$	
* 33.13	$x \,\epsilon\, D'R . \equiv . (\exists y) . xRy$	
* 33.131	$y \,\epsilon\, \mathcal{Q}'R . \equiv . (\exists x) . xRy$	
* 33.132	$x \,\epsilon\, C'R . \equiv : (\exists y) : xRy . \vee . yRx$	
* 33.16	$C'R = D'R \cup \mathcal{Q}'R$	
T 33.361	$R \subseteq S . \supset : D'S \cap D'T = \wedge .$	(*33.263, 24.58)
	$. \supset . D'R \cap D'T = \wedge$	
T 33.362	$R \subseteq S . \supset : \mathcal{Q}'S \cap \mathcal{Q}'T = \wedge .$	(*33.264, 24.58)
	$. \supset . \mathcal{Q}'R \cap \mathcal{Q}'T = \wedge$	
* 34.02	$R^2 = R|R$	
* 34.03	$R^3 = R^2|R$	
* 34.1	$x(R|S)z . \equiv . (\exists y) . xRy . yRz$	
T 34.362	$R|Q|P \neq \wedge . \supset .$	(*24.561, .58, 33.24,
	$. R|Q \neq \wedge . Q|P \neq \wedge$	34.3, .36)
* 35.1	$x(\alpha \uparrow R)y . \equiv . x \,\epsilon\, \alpha . xRy$	
* 35.101	$x(R \upharpoonright \beta)y . \equiv . xRy . y \,\epsilon\, \beta$	
T 35.69	$\overrightarrow{(\alpha \uparrow R)}'y = \overrightarrow{R}'y \cap \alpha$	(*32.18, 35.1, 22.33)
T 35.691	$\overleftarrow{(R \upharpoonright \beta)}'x = \overleftarrow{R}'x \cap \beta$	(*32.181, 35.101, 22.33)

* 36.13	$x(R \restriction \alpha)y . \equiv . x, y \in \alpha . xRy$	
T 36.44	$R \restriction \alpha \subseteq R$	(*35.442, 36.11)
* 37.1	$x \in R''\beta . \equiv . (\exists y) . y \in \beta . xRy$	
* 50.1	$xIy . \equiv . x = y$	
* 50.11	$xJy . \equiv . x \neq y$	
* 50.24	$R \subseteq J . \equiv . (x) . \sim (xRx)$	
* 51.01	$\imath = \overrightarrow{I}$	
* 51.15	$y \in \imath' x . \equiv . y = x$	
* 52.1	$\alpha \in 1 . \equiv . (\exists x) . \alpha = \imath'x$	
* 52.11	$\alpha \in 1 . \equiv : (\exists x) : (y) : y \in \alpha . \equiv .$ $. y = x$	
* 54.101	$\alpha \in 2 . \equiv . (\exists x,y) . x \neq y . \alpha =$ $= \imath'x \cup \imath'y$	
T 54.28	$\alpha \in 2 . \equiv \therefore (\exists x,y) \therefore x \neq y : (z) :$ $: z \in \alpha . \equiv . z = x . \vee . z = y$	(*51.232, 54.101)
T 54.71	$\alpha \in 4 . \equiv . (\exists x, y, z, w) . x \neq y .$ $. x \neq z . x \neq w . y \neq z . y \neq w .$ $. z \neq w . \alpha = (\imath'x \cup \imath'y \cup \imath'z \cup \imath'w)$	(Definition)
* 70.11	$R \in \kappa \to \lambda . \equiv \therefore (y) : y \in \mathrm{\Omega}'R .$ $. \supset . \overrightarrow{R}'y \in \kappa : (x) : x \in D'R . \supset .$ $. \overleftarrow{R}'x \in \lambda$	
T 71.104	$R \in 1 \to Cls . \equiv : (y) :$ $: y \in \mathrm{\Omega}'R . \supset . \overrightarrow{R}'y \in 1$	(*33.41, 71.14)
T 71.105	$R \in Cls \to 1 . \equiv : (x) :$ $: x \in D'R . \supset . \overleftarrow{R}'x \in 1$	(*33.4, 71.141)
* 71.17	$R \in 1 \to Cls . \equiv : (x, y, z) :$ $: xRz . yRz . \supset . x = y$	
* 71.171	$R \in Cls \to 1 . \equiv : (x, y, z) :$ $: xRy . xRz . \supset . y = z$	
T 72.39	$S \in 1 \to Cls . R \subseteq S . P \subseteq S .$ $. \supset . \mathrm{\Omega}'R \cap \mathrm{\Omega}'P = \mathrm{\Omega}'(R \cap P)$	(*22.33, 23.33, 71.17)
T 72.391	$S \in Cls \to 1 . R \subseteq S . P \subseteq S .$ $. \supset . D'R \cap D'P = D'(R \cap P)$	(T72.39, *31.14, .4, 33.21, 71.211)
T 72.493	$P \in 1 \to Cls . \mathrm{\Omega}'R \cap \mathrm{\Omega}'S = \wedge .$	(*22.35, 37.32, .462,

$$. \supset . \, \mathcal{C}'R|P \cap \mathcal{C}'S|P = \wedge \qquad \text{51.31, 53.31, 71.163, .37)}$$

T 72.494 $\quad R \, \epsilon \, Cls \to 1 . D'P \cap D'S = \wedge .$ (*22.35, 37.32, .461,

$$. \supset . D'R|P \cap D'R|S = \wedge \qquad \text{51.31, 53.31, 71.164, .371)}$$

T 72.584 $\quad R \, \epsilon \, 1 \to Cls . Q|R = P|R .$ (*37.32, 72.5)

$$. \supset . \, \mathcal{C}'Q \cap D'R =$$
$$= \mathcal{C}'P \cap D'R$$

T 72.585 $\quad Q \, \epsilon \, Cls \to 1 . Q|R = Q|P .$ (*37.32, 72.501)

$$. \supset . \, \mathcal{C}'Q \cap D'R =$$
$$= \mathcal{C}'Q \cap D'P$$

* 84.1 $\quad \kappa \, \epsilon \, Cls^2 \, excl . \equiv : (\alpha, \beta) : \alpha, \beta \, \epsilon \, \kappa .$

$$. \, \alpha \neq \beta . \supset . \, \alpha \cap \beta = \wedge$$

* 90.1 $\quad xR_*y . \equiv : x \, \epsilon \, C'R : (\alpha) :$

$$: \breve{R}''\alpha \subseteq \alpha . x \, \epsilon \, \alpha . \supset . y \, \epsilon \, \alpha$$

T 90.241 $\quad \breve{R}''\alpha \subseteq \alpha . x \, \epsilon \, \alpha . \supset . \overleftarrow{R_*}' x \subseteq \alpha$ (*51.2, 90.24, 53.301)

T 91.505 $\quad D'S_{po}|R \subseteq D'S$ (*34.36, 91.504)

T 91.506 $\quad \mathcal{C}'R|S_{po} \subseteq \mathcal{C}'S$ (*34.36, 91.504)

T 91.507 $\quad D'R \cap D'Q = \wedge$ (*33.32, 91.504)

$$. \vee . \, \mathcal{C}'R \cap \mathcal{C}'Q = \wedge :$$
$$: \supset . \, R_{po} \cap Q_{po} = \wedge$$

* 91.54 $\quad R_* = I \lceil C'R \cup R_{po}$

T 91.591 $\quad R_{po} \subseteq (R \cup S)_{po}$ (*22.58, 91.59)

T 91.592 $\quad (R_{po}|S_{po})_{po} \subseteq (R \cup S)_{po}$ (*34.34, 91.56, T91.591)

T 91.596 $\quad R^2 \subseteq R . \supset . (S|R)_{po}|R \subseteq (S|R)_{po}$ (*22.72, 34.21, .22, .26,

$$\text{.34, 91.57)}$$

T 91.597 $\quad R^2 \subseteq R . \supset . R|(R|S)_{po} \subseteq (R|S)_{po}$ (*34.25, .34, 91.57)

T 91.61 $\quad R^2 = \wedge . \supset . R_{po} = R$ (*34.32, 91.57, .575)

* 91.62 $\quad xR_{po} y . \equiv : (\alpha) : \breve{R}'' \alpha \subseteq \alpha .$

$$. \overleftarrow{R'}x \subseteq \alpha . \supset . y \, \epsilon \, \alpha$$

T 91.63 $\quad (R \cup R|P) \subseteq P . \vee . (P|R \cup R)$ (*23.58, 31.33, 32.19,

$$\subseteq P : \supset . R_{po} \subseteq P \qquad \text{.241, 37.3, .301, 91.53, .62)}$$

T 91.763 $\quad R = R \lceil \overleftarrow{R_*}'x . \equiv . C'R = \overleftarrow{R_*}'x$ (*33.151, .16, 36.25,

$$\text{90.14)}$$

T 92.313 $\quad R \, \epsilon \, 1 \to Cls . \supset . R_{po}|\breve{R}_{po} \subseteq$ (*31.4, 34.34, 91.54, .75,

$$\subseteq R_{po} \cup \breve{R}_{po} \cup I \lceil C'R \qquad \text{92.31)}$$

T 92.41 $\quad R \, \epsilon \, 1 \to Cls . \alpha \cap \mathcal{C}'R = \wedge$ (*32.18, 33.132, 35.1,

	$. \supset . (\alpha \uparrow R_{po}) \, \epsilon \, 1 \to Cls$	.44, 50.5, 71.17, 90.12,
		91.504, .542, .601)
T 92.411	$R \, \epsilon \, 1 \to Cls . \overline{\alpha' R} \cap \overline{\alpha' S} = \wedge .$	(*35.48, 71.25, T92.41)
	$. \supset . S \vert R_{po} \, \epsilon \, 1 \to Cls$	
* 93.1	$xBR . \equiv . x \, \epsilon \, (D'R \cap \overline{\alpha' R})$	
* 93.101	$\overrightarrow{B'R} = D'R \cap \overline{\alpha' R}$	
T 93.105	$xBR . \equiv . x \, \epsilon \, D'R . (y) .$	(*91.504, 93.1)
	$. \sim (yR_{po} \, x)$	
T 93.601	$xB\breve{R} . \equiv . x \, \epsilon \, (\overline{D'R} \cap \alpha' R)$	(*33.2, .21, 93.1)
T 93.602	$\overrightarrow{B'\breve{R}} = \overline{D'R} \cap \alpha' R$	(*32.18, T93.601)
T 93.605	$R^2 = \wedge . \supset . D'R = \overrightarrow{B'R} .$	(*93.101, T93.602)
	$. \alpha' R = \overrightarrow{B'\breve{R}}$	
T 93.61	$E!B'R . \supset . \iota' B'R = \overrightarrow{B'R}$	(*53.31)
T 93.611	$E!B'R . \supset . B'R \, \epsilon \, \overrightarrow{B'R}$	(*51.16, 53.31)
T 93.6111	$E!B'R . \supset . B'R \, \epsilon \, D'R$	(*22.46, 93.101, T93.611)
T 93.612	$E!B'R . \supset . R \neq \wedge$	(*30.21, 33.24, 93.1)
T 93.613	$E!B'R . \supset : B'R \, \epsilon \, \alpha .$	(*51.31, 53.31)
	$. \equiv . \overrightarrow{B'R} \cap \alpha \neq \wedge$	
T 93.615	$E!B'R . \supset . (\overleftarrow{I \lceil C'R})'B'R =$	(*32.181, 33.16, 51.01,
	$= \iota' B'R$	T93.6111)
T 93.6201	$\alpha' S \cap \alpha = \wedge . \supset . D'S \cap \alpha =$	(*93.101)
	$= \overrightarrow{B'S} \cap \alpha$	
T 93.6202	$D'S \cap \alpha = \wedge . \supset . \alpha' S \cap \alpha =$	(T93.602)
	$= \overrightarrow{B'\breve{S}} \cap \alpha$	
T 93.621	$R \subseteq S . \alpha' S \cap \alpha = \wedge .$	(*33.264, 93.101)
	$. \supset . D'R \cap \alpha = \overrightarrow{B'R} \cap \alpha$	
T 93.622	$R \subseteq S . D'S \cap \alpha = \wedge .$	(*33.263, T93.602)
	$. \supset . \alpha' R \cap \alpha = \overrightarrow{B'\breve{R}} \cap \alpha$	
T 93.623	$R \subseteq S . D'S = \overrightarrow{B'S} . \supset .$	(*33.263, 93.101)
	$. D'R = \overrightarrow{B'R}$	

T 93.624 $R \subseteq S \cdot \overrightarrow{\alpha'S} = B'\overrightarrow{\check{S}} \cdot \supset \cdot$ (*33.264, T93.602)

$\cdot \overrightarrow{\alpha'R} = B'\check{R}$

T 93.63 $(R): R \epsilon \kappa \cdot \supset \cdot B'R \epsilon \alpha : (P):$ (*22.46, 24.311)

$: P \epsilon \lambda \cdot \supset \cdot B'P \epsilon \beta \therefore \supset : \alpha \cap \beta =$

$= \wedge \cdot \supset \cdot \kappa \cap \lambda = \wedge$

T 93.65 $\overrightarrow{B'R} \cap \alpha \neq \wedge \cdot \equiv \cdot R \epsilon \check{B}''\alpha$ (*31.11, 32.18, 37.1, 93.1

T 93.7 $xLk\,R \cdot \equiv \cdot x \epsilon (D'R \cap \alpha'R)$ (Definition)

T 93.701 $\overrightarrow{Lk'R} = D'R \cap \alpha'R$ (*32.18, T93.7)

T 93.702 $\overrightarrow{Lk'R} = \wedge \cdot \equiv \cdot R^2 = \wedge$ (*34.531, T93.701)

T 93.71 $R \subseteq S \cdot \supset \cdot \overrightarrow{Lk'R} \subseteq \overrightarrow{Lk'S}$ (*22.49, 93.101,

T93.701)

T 94.211 $(S|R)_{po} = S|R \cup S|(R|S)_{po}|R$ (*34.25, .26, 41.171,

43.11, .112, .43, 50.73,

53.04, 91.05, 94.21)

T 94.212 $S|(R|S)_{po} = (S|R)_{po}|S$ (*34.25, .26, 91.57,

T94.211)

T 94.221 $D'S \subseteq \alpha'R \cdot \vee \cdot \alpha'R \subseteq D'S :$ (*43.112, .43, 91.05, .55,

$: \supset \cdot (S|R)_{po} = S|(R|S)_*|R$ 94.22)

T 94.23 $(R \cup Q)_{po} = R_{po} \cup Q_{po} \cup$ (*34.25, .26, .34, 91.502,

$(R_{po}|Q_{po})_{po} \cup (Q_{po}|R_{po})_{po} \cup$ .56, .575, T91.591, .592,

$R_{po}|(Q_{po}|R_{po})_{po} \cup (Q_{po}|R_{po})_{po}|Q_{po}$ .597, .63, 94.212)

T 94.24 $D'Q \cap D'T = \wedge \cdot \alpha'Q \cap \alpha'T = \wedge$ (*24.402, 33.32, 34.36,

$\cdot \supset : (S \cup T)_{po} \cap Q_{po} =$ 91.504, T91.507, 94.23)

$= [S_{po} \cup S_{po}|(T_{po}|S_{po})_{po}] \cap Q_{po}$

T 96.161 $\alpha'(R \restriction \overleftarrow{R_*}'x) = \overleftarrow{R_{po}}'x$ (*96.15, .16)

T 96.164 $x \epsilon D'R \cdot \supset \cdot C'(R \restriction \overleftarrow{R_*}'x) =$ (*96.151, .16)

$= \overleftarrow{R_*}'x$

T 96.171 $xRy \cdot \equiv \cdot x(R \restriction \overleftarrow{R_*}'x)y \cdot$ (*36.13, 90.12)

$\cdot \equiv \cdot x(R \restriction \overrightarrow{R_*}'y)y$

T 96.1711 $x \epsilon D'R \cdot \equiv \cdot x \epsilon D'(R \restriction \overleftarrow{R_*}'x)$ (T96.171)

T 96.1712 $y \epsilon \alpha'R \cdot \equiv \cdot y \epsilon \alpha'(R \restriction \overrightarrow{R_*}'y)$ (T96.171)

T 96.172 $xR_{po}y . \equiv . x(R \mathbin{\char"5B} \overleftarrow{R_*}'x)_{po}\, y .$ (*90.12, 91.504, 96.15,

        $\equiv . x(R \mathbin{\char"5B} \overrightarrow{R_*}'y)_{po}\, y$ .16)

T 96.1721 $\overleftarrow{R_{po}'x} = \overleftarrow{(R \mathbin{\char"5B} \overleftarrow{R_*}'x)_{po}}'x$ (T96.172)

T 96.1722 $\overrightarrow{R_{po}'y} = \overrightarrow{(R \mathbin{\char"5B} \overrightarrow{R_*}'y)_{po}}'y$ (T96.172)

T 96.173 $x \in D'R . \supset : xR_*y .$ (*90.13, 96.131, .16)

        $\equiv . x(R \mathbin{\char"5B} \overleftarrow{R_*}'x)_*y$

T 96.1732 $x \in D'R . \supset . \overleftarrow{R_*}'x = \overleftarrow{(R \mathbin{\char"5B} \overleftarrow{R_*}'x)_*}'x$ (T96.173)

T 96.181 $xB(R \mathbin{\char"5B} \overleftarrow{R_*}'x) . \equiv . x \in D'R .$ (*93.101, T96.161, .171 1)

        $. \sim (xR_{po}x)$

T 96.182 $yB(R \mathbin{\char"5B} \overleftarrow{R_*}'x) . \equiv . y = x .$ (*32.35, 33.161, 91.54,

        $. y \in D'R . \sim (yR_{po}y)$ 96.15, .16)

T 96.183 $xB(R \mathbin{\char"5B} \overleftarrow{R_*}'x) . \equiv .$ (*30.31, T96.182)

        $. x = B'(R \mathbin{\char"5B} R_*'x)$

T 96.261 $\mathcal{C}'R \subseteq \overleftarrow{R_{po}}'B'R . \supset . D'R$ (*33.16, 32.33, 90.14,

        $\subseteq \overleftarrow{R_*}'B'R . C'R \subseteq \overleftarrow{R_*}'B'R$ 53.31, 91.54, 93.1,

                                          T30.38, 93.615)

T 96.262 $R \in 1 \to Cls . \supset : \mathcal{C}'R =$ (*14.18, .21, 35.66,

        $= \overleftarrow{R_{po}}'B'R . \equiv . C'R = \overleftarrow{R_*}'B'R$ 96.16, .2, T96.161, .1721)

T 96.2301 $R \in 1 \to Cls . xBR .$ (*96.13, .16, .23)

        $\supset . (R \mathbin{\char"5B} \overleftarrow{R_*}'x)_{po} \subseteq J$

C 99.101 $R \in sym . \equiv . R \subseteq \breve{R}$ (D31-1a, SLA)

C 99.102 $R \in as . \equiv . R \subseteq \breve{R}$ (D31-1b, SLA)

C 99.103 $R \in trans . \equiv . R^2 \subseteq R$ (D31-2a, SLA), (*201.1)

C 99.104 $R \in intr . \equiv . R^2 \subseteq \bar{R}$ (D31-2b, SLA)

C 99.105 $R \in refl . \equiv . I \mathbin{\char"5B} C'R \subseteq R$ (D31-3a, SLA)

C 99.106 $R \in irr . \equiv . R \subseteq J$ (D31-3b, SLA)

C 99.114 $R \in as . \equiv . R^2 \in irr$ (T31-1e, SLA)

C 99.115 $as \subseteq irr$ (T31-1f, SLA)

C 99.116 $trans \cap as = trans \cap irr$ (T31-1g, SLA)

C 99.2 $R_* \in trans$ (T31-1j1, SLA)

C 99.201 $R_{po} \in trans$ (T36-1j2, SLA)

T 99.21	$R_{po} \ \epsilon \ irr \ . \supset . \ R \ \epsilon \ as$	(*91.503, C99.114)
T 99.211	$R_{po} \ \epsilon \ irr \ . \equiv . \ R_{po} \ \epsilon \ as$	(C99.116, .201)
T 99.23	$R \ \epsilon \ 1 \to Cls \ . \ R_{po} \ \epsilon \ irr \ . \supset . \ R \ \epsilon \ intr$	(*71.17, 91.502, C99.104 .106)
T 99.24	$R^2 = \wedge \ . \equiv . \ R \ \epsilon \ (trans \ \cap \ intr)$	(*25.12, .21, C99.103, .104)

I. POSTULATES ON MITOSIS, MEIOSIS AND GAMETIC FUSION

Before we can decide on a particular set of postulates concerning mitosis, meiosis and gametic fusion, we first must state what kind of logical constants they represent. Following Woodger's axiom system (AMB), we shall consider these concepts as two-termed relations. Thus, mitosis, for instance, is the relation that holds between a cell that undergoes mitosis and another that results from it, or between a nucleus that undergoes mitosis and another that results from it. Unfortunately, there are no verbs corresponding to 'mitosis' and 'meiosis' in English; such verbs would make it easier to express our formal statements in common terms.

In keeping with accepted biological usage, the terms 'mitosis', 'meiosis', and 'gametic fusion' are taken here as applicable to either cells or nuclei. No attempt will be made in this paper to distinguish between these two possibilities, we will always refer to 'one individual standing in mitosis to another' rather than to 'one cell or one nucleus standing in mitosis to another cell or nucleus'. In this way the scope of this system becomes wider, but without losing any accuracy, because it is applicable not only to organisms that consist entirely of uninucleate cells, but also to organisms with coenocytic or dikaryotic phases in their life cycles. One must keep in mind, however, that any one statement should not be interpreted as referring to both kinds of individuals, cells and nuclei, i.e., if we are speaking of the fusion of two cells, then the individual formed in the process must also be a cell and not a nucleus.

The individuals of our system should not be interpreted as merely spatial units, cells or nuclei which may be seen on a fixed microscope slide, but as time-extended entities. We must, therefore, be concerned not only with their spatial but also with their temporal boundaries. Woodger, in his axiom system (AMB), used the convention that an individual comes into being either by the splitting of another individual or by the union of

432

some other individuals, and the individual ceases to exist when it, in turn, splits into or unites with some other individuals. Thus cells or nuclei are to be regarded, at least in our system, to have their temporal beginning with a division or a fusion, and their end when they divide or fuse or die. Adopting this convention would fill a gap in biological usage, since there do not seem to be any uniform rules in this respect. For example, a zygote is usually considered as a cell different from either of the gametes from which it arose when the gametes are of similar size, but a fertilized egg is frequently regarded as the same cell as the unfertilized egg when the sperm is much smaller than the egg. Similarly, in the case of cell division the mother cell is sometimes considered to be distinct from its daughter cells, while at other times the mother cell is identified with one of the daughters (as in the budding of yeast).

The primitive notions of the axiom system to be presented are 'mitosis', 'meiosis' and 'gametic fusion', all other biological terms mentioned will be defined in terms of the primitives. The universe of individuals of the lowest type is constituted by the fields of these relations, in other words, no individuals of this kind will be mentioned except those which stand in one of these relations to something or to which something stands in one of these relations. Beside the biological terms, the logical terms to be used are those introduced in the previous section.

Some theorems derivable from the present axiom system are given in formal terms at the end of each section. In the formal treatment 'N' stands for 'mitosis', 'M' stands for 'meiosis' and 'Fu' stands for 'gametic fusion'. The advantage of writing our statements in logical notation is brevity and conciseness, as well as accuracy that is due to the clearly stated axioms and definitions by which this notation is governed.

Postulates 1.01, 1.02, 1.03. The first three postulates state that mitosis is a one-two relation, while meiosis is a one-four relation and gametic fusion is a two-one relation. The following diagrams may make it easier to visualize these kinds of relations.

Accordingly, the term 'mitosis' cannot be applied, at least within this system, to division processes which result in more than two, or less than two cells or nuclei. This excludes from our consideration incomplete division and endomitosis. Similarly, 'meiosis' is applicable only when exactly four cells or nuclei arise by meiotic division (this does not rule out, however, cases where three out the four nuclei disintegrate afterwards).

433

'Gametic fusion' applies only to cases where exactly two cells or nuclei fuse. Multiple fusions, such as those leading to endosperm formation in angiosperms, could still be admitted by considering them as sequential fusions of pairs of nuclei, but Postulate 1.10 rules them out even in that case.

Postulate 1.04. This postulate states that if any two individuals stand in any sequential combination of any finite number of mitotic, meiotic or fusion steps to each other, then these two individuals cannot be the same. In the formula, 'J' stands for the usual sign of non-identity '$\neq$' (*50.11). This postulate assures that after a sequence of division or fusion steps one can never get back to the same cell or nucleus that one started from. Thus all powers of mitosis, meiosis or fusion (including any combinations of them) are irreflexive relations (cf. C99.106). In Woodger's axiom system (AMB) the irreflexivity of his cell-division and cell-fusion relations and their powers, was guaranteed by the fact that they were included in the relation "before in time", which, in turn, was postulated to be irreflexive. In the present system it was possible to dispense with the introduction of a time relation by adopting Postulate 1.04.

Postulates 1.05, 1.06 and 1.07 state that no individual can be a member of both the domain of mitosis and the domain of meiosis, and no individual can be a member of both the converse domain of mitosis and the converse domain of meiosis; and similarly for meiosis and fusion, and for fusion and mitosis. This condition, that neither the domains nor the converse domains of two relations R and P have any members in common, is a stronger requirement than the condition that the Boolean product of R and P be empty. These postulates assure that any cell or nucleus can arise by only one of the three processes considered, namely, mitosis, meiosis or fusion, and also that any individual can undergo only one of the three processes. They help, therefore, to establish the temporal boundaries of the individuals of this system, in accordance with that which has already been said above.

Postulates 1.08, 1.09 and *1.10.* The first of these states that if an individual is a member of the domain of mitosis then it must also be a member of the converse domain of either mitosis or meiosis or fusion. The second of these postulates states that if an individual is member of the domain of meiosis then it must also be a member of the converse domain either of fusion directly, or of fusion followed by a finite number of mitotic steps. And the last postulate says that if an individual is a member of the domain of fusion, then it must also be a member of the converse domain of either meiosis, or of meiosis followed by a finite number of mitotic steps. The two latter postulates incorporate into our system the rule of the sequential alternation of meiosis and gametic fusion in life cycles.

According to the last three postulates every individual in this system must be a member of the converse domain of one of our three relations. These postulates imply, therefore, that an infinite number of individuals must precede each individual.* In other words, the possibility of cells or nuclei originating *de novo* is excluded. Whether this is a desirable consequence of our postulates, or not, is left to the reader to decide.

Another consequence of the last three postulates must also be pointed out. This is that whenever an "irregular" division or fusion occurs in an organism, i.e., a division or fusion that cannot be considered in this system because it does not satisfy some of our postulates, then none of the divisions or fusions subsequent to it can be considered either. For instance, if a polyploid cell arises in a plant by endomitosis, then none of the divisions or fusions that this cell or its descendants may undergo would satisfy the last three postulates, even if they are perfectly regular mitotic or meiotic divisions, or gametic fusions.

Various conditions have been mentioned in this section regarding the cells or cellular processes, or nuclei and nuclear processes, to which this axiom system can be applied. These statements concern the interpretation of our axiom system. Our position is that we use the common cytological interpretation of the terms 'mitosis', 'meiosis' and 'gametic fusion', except for the restrictions posed by the postulates and elaborated above. It may be asked whether these restrictions are too severe and thus seriously impair the usefulness of our axiom system. In my view, these axioms have the great advantage of being extremely simple, while they and the theorems

* This has been shown to be derivable from our postulates, but the proof cannot be included in this article.

derived from them are still applicable to a wide range of observations. If thought necessary, this system could always be extended by introducing additional primitive notions and postulates to account for those aspects of life cycles which are disregarded in this treatment.

We can now turn to the theorems which are the immediate consequences of our postulates. Accompanying each theorem are the numbers of the statements listed which are necessary for its derivation, except for those of the basic sentential and functional calculus of PM. The proofs of the theorems are, however, not given, they can be easily reconstructed from the supporting theorems.

We have theorems first of all about the irreflexivity of the powers of mitosis, meiosis and fusion (1.111, .114, .117), which follow from Postulate 1.04 and from T91.591. (These theorems are the equivalents of Theorems 1.11, .113, .116, by the definition of irreflexive relations, C99.106). We also see that the powers of mitosis, meiosis and fusion are all asymmetric relations (Theorems 1.112, .115, .118). This follows by T99.211 from the irreflexivity of these relations.

The impossibility of having one meiosis followed by another is indicated by Theorem 1.12, and similarly the impossibility of fusion followed by another fusion by Theorem 1.122. These theorems follows from Postulates 1.05, 1.06, 1.07, 1.09, and 1.10. A direct consequence of these two theorems is that there is only one power of meiosis or of fusion, namely the first power (Theorems 1.121, .123).

Theorems 1.13, 1.131, 1.132 and 1.133 enumerate the characteristics of the relation of mitosis itself (in contrast to 1.11, .111 and .112 which dealt with the powers of mitosis). According to these theorems, mitosis is a one-many, asymmetric, intransitive and irreflexive relation. The one-many-ness of mitosis follows from it being a one-two relation (Postulate 1.01). The intransitivity of mitosis follows from it being one-many, and from the fact that all its powers are irreflexive, by T99.23. Since all powers of mitosis are asymmetric and irreflexive, its first power, which is mitosis itself, must also be asymmetric and irreflexive.

Theorems 1.14, 1.141, 1.142 and 1.143 give the characteristics of meiosis as a relation, according to which meiosis has the same properties as mitosis (being one-many, asymmetric, intransitive and irreflexive), but in addition it is also a transitive relation. As mentioned above, a relation may be both intransitive and transitive, since these two classes are not mutually exclusive. Meiosis is both transitive and intransitive because by 1.12 the case xMy and yMz can never be realized, thus meiosis fulfils both definitions (cf. T99.24).

Theorems 1.15, 1.151, 1.152 and 1.153 state that gametic fusion is a many-one, asymmetric, transitive, intransitive and irreflexive relation. Since gametic fusion is a two-one relation (Postulate 1.03), it also must be a many-one relation. The reason for it being both transitive and intransitive is the same as in the case of meiosis, i.e., it being impossible to have two fusion steps following each other (Theorem 1.122). Fusion is asymmetric and irreflexive because it has been shown that all its powers are asymmetric and irreflexive (Theorems 1.117, .118).

Theorems 1.16, 1.161, and 1.162 express the fact that no individual can stand in both mitosis and meiosis to another individual, or in both meiosis and fusion, or in both fusion and mitosis. These statements are simply consequences of Postulates 1.05, 1.06 and 1.07.

Theorems 1.20 and 1.21 state that two meiotic steps or two fusion steps cannot follow each other even when there are mitotic steps in between, as a consequence of Postulates 1.07 and 1.09, and of Postulates 1.05 and 1.10, respectively.

Theorem 1.23 is an important one because of its further consequences. It gives all the possible combinations of any finite number of consecutive mitotic or meiotic steps, which are: (1) all mitotic steps, (2) a single meiotic step, (3) mitotic steps followed by meiosis, (4) meiosis followed by mitotic steps, and (5) mitotic steps followed by meiosis followed by mitotic steps. These five combinations represent, furthermore, mutually exclusive components of a Boolean sum (a partition). Most our derivations concerning life cycles depend on this theorem, and it was not one easy to prove. Its proof depends mainly on T94.23 as well as on Theorems 1.121 and 1.20. Theorem 1.231 is identical to the previous one but for substituting 'fusion' wherever 'meiosis' occurs. In the present system such substitutions between 'meiosis' and 'fusion' are often possible, and it can be traced back to a certain symmetry between these terms in the postulates. Theorem 1.232 gives the possible combinations of consecutive meiotic and fusion steps, and these, of course, do not correspond at all to the two previous ones.

According to Theorem 1.24, the class of beginners of mitosis is identical to the class of the members of the domain of mitosis which are also members of the converse domain of either meiosis or of fusion (*93.101 gives the definition of the class of beginners of a relation). Furthermore, the class of the beginners of meiosis is the same as the domain of meiosis, and the class of the beginners of fusion is the same as the domain of fusion (Theorems 1.25, 1.252). On the other hand, the class of the terminals of meiosis is the same as the converse domain of meiosis, and similarly for fusion (Theorems 1.251 and 1.253). The reason for these characteristics of the meiotic and fusion relations is again the fact that two meiotic steps or two fusion steps cannot follow each other (Theorems 1.12 and 1.122), and thus every individual that stands in the domain of meiosis is a beginner of meiosis, and every individual that stands in the converse domain of meiosis is a terminal of meiosis, the same being true for fusion.

The rest of the theorems in this section need no further comment, many of them are lemmas of subsequent theorems.

Postulates and Theorems on N, M and Fu

1.01	$N \in 1 \to 2$	(Postulate)
1.02	$M \in 1 \to 4$	(Postulate)
1.03	$Fu \in 2 \to 1$	(Postulate)
1.04	$(N \cup M \cup Fu)_{po} \subseteq J$	(Postulate)
1.05	$D'N \cap D'M = \wedge . \mathbf{\mathit{a}}'N \cap \mathbf{\mathit{a}}'M = \wedge$	(Postulate)
1.06	$D'M \cap D'Fu = \wedge . \mathbf{\mathit{a}}'M \cap \mathbf{\mathit{a}}'Fu = \wedge$	(Postulate)
1.07	$D'Fu \cap D'N = \wedge . \mathbf{\mathit{a}}'Fu \cap \mathbf{\mathit{a}}'N = \wedge$	(Postulate)
1.08	$D'N \subseteq (\mathbf{\mathit{a}}'N \cup \mathbf{\mathit{a}}'M \cup \mathbf{\mathit{a}}'Fu)$	(Postulate)
1.09	$D'M \subseteq (\mathbf{\mathit{a}}'Fu \cup \mathbf{\mathit{a}}'Fu|N_{po})$	(Postulate)
1.10	$D'Fu \subseteq (\mathbf{\mathit{a}}'M \cup \mathbf{\mathit{a}}'M|N_{po})$	(Postulate)
1.11	$N_{po} \subseteq J$	(1.04, T91.591)
1.111	$N_{po} \in irr$	(1.11, C99.106)
1.112	$N_{po} \in as$	(1.111, T99.211)
1.113	$M_{po} \subseteq J$	(1.04, T91.591)
1.114	$M_{po} \in irr$	(1.113, C99.106)
1.115	$M_{po} \in as$	(1.114, T99.211)
1.116	$Fu_{po} \subseteq J$	(1.04, T91.591)
1.117	$Fu_{po} \in irr$	(1.116, C99.106)
1.118	$Fu_{po} \in as$	(1.117, T99.211)

1.12	$M^2 = \wedge$	(1.05, .06, .09, *34.36, .531, 91.504)
1.121	$M_{po} = M$	(1.12, T91.61)
1.122	$Fu^2 = \wedge$	(1.06, .07, .10, *34.36, .531, 91.504)
1.123	$Fu_{po} = Fu$	(1.122, T91.61)
1.13	$N \,\epsilon\, 1 \to Cls$	(1.01, *70.11, T71.104)
1.131	$N \,\epsilon\, as$	(1.111, T99.21)
1.132	$N \,\epsilon\, intr$	(1.111, .13, T99.23)
1.133	$N \,\epsilon\, irr$	(1.131, C99.115)
1.14	$M \,\epsilon\, 1 \to Cls$	(1.02, *70.11, T71.104)
1.141	$M \,\epsilon\, as$	(1.114, T99.21)
1.142	$M \,\epsilon\, (trans \cap intr)$	(1.12, T99.24)
1.143	$M \,\epsilon\, irr$	(1.141, C99.115)
1.15	$Fu \,\epsilon\, Cls \to 1$	(1.03, *70.11, T71.105)
1.151	$Fu \,\epsilon\, as$	(1.117, T99.21)
1.152	$Fu \,\epsilon\, (trans \cap intr)$	(1.122, T99.24)
1.153	$Fu \,\epsilon\, irr$	(1.151, C99.115)
1.16	$N \cap M = \wedge$	(1.05, *33.32)
1.161	$M \cap Fu = \wedge$	(1.06, *33.32)
1.162	$Fu \cap N = \wedge$	(1.07, *33.32)
1.19	$\Box'M\|N_{po} \cap \Box'Fu\|N_{po} = \wedge$	(1.05, .06, .07, .13, T92.313)
1.191	$D'N_{po}\|M \cap D'N_{po}\|Fu = \wedge$	(1.05, .06, .07, .09, .10, .13, .19, T92.313)
1.20	$M\|N_{po}\|M = \wedge$	(1.07, .09, .19, *34.301, .36, 91.504)
1.201	$(M\|N_{po})^2 = \wedge$	(1.20, *34.02, .32)
1.202	$(N_{po}\|M)^2 = \wedge$	(1.20, *34.02, .32)
1.203	$(N_{po}\|M\|N_{po})^2 = \wedge$	(1.20, *34.02, .32)
1.21	$Fu\|N_{po}\|Fu = \wedge$	(1.05, .10, .19, *34.301, .36, 91.504)
1.211	$(Fu\|N_{po})^2 = \wedge$	(1.21, *34.02, .32)
1.212	$(N_{po}\|Fu)^2 = \wedge$	(1.21, *34.02, .32)
1.213	$(N_{po}\|Fu\|N_{po})^2 = \wedge$	(1.21, *34.02, .32)
1.22	$(M\|N_{po}) \,\epsilon\, 1 \to Cls$	(1.05, .13, .14, T92.411)
1.221	$(\Box'Fu \uparrow N_{po}) \,\epsilon\, 1 \to Cls$	(1.07, .13, T92.41)

1.23 $(N \cup M)_{po} = N_{po} \cup M \cup$ (1.121, .20, .201, .202,
         $\cup\ N_{po}|M \cup M|N_{po} \cup N_{po}|M|N_{po}$ T94.23)

1.231 $(N \cup Fu)_{po} = N_{po} \cup Fu \cup$ (1.123, .21, .211, .212,
         $\cup\ N_{po}|Fu \cup Fu|N_{po} \cup N_{po}|Fu|N_{po}$ T94.23)

1.232 $(M \cup Fu)_{po} = M \cup Fu \cup$ (1.121, .123, T94.23)
         $\cup\ (M|Fu)_{po} \cup (Fu|M)_{po} \cup$
         $\cup\ M|(Fu|M)_{po} \cup (Fu|M)_{po}|Fu$

1.24 $\vec{B}'N = (\mathcal{C}'M \cup \mathcal{C}'Fu) \cap \mathcal{C}'N$ (1.08, *93.101)

1.25 $\vec{B}'M = D'M$ (1.12, T93.605)

1.251 $\vec{B}'\breve{M} = \mathcal{C}'M$ (1.12, T93.605)

1.252 $\vec{B}'Fu = D'Fu$ (1.122, T93.605)

1.253 $\vec{B}'\breve{F}u = \mathcal{C}'Fu$ (1.122, T93.605)

1.26 $\mathcal{C}'N \cap D'M = \mathcal{C}'Fu|N_{po} \cap D'M$ (1.07, .09, T91.506)

1.261 $\mathcal{C}'N \cap D'Fu = \mathcal{C}'M|N_{po} \cap D'Fu$ (1.05, .10, T91.506)

1.28 $D'Fu|N \cap D'Fu|M = \wedge$ (1.05, .15, T72.494)

1.29 $\mathcal{C}'N|M \cap \mathcal{C}'Fu|M = \wedge$ (1.07, .14, T72.493)

1.291 $\mathcal{C}'N^2 \cap \mathcal{C}'M|N = \wedge$ (1.05, .13, *34.02,
         T72.493)

1.292 $\mathcal{C}'N^2 \cap \mathcal{C}'Fu|N = \wedge$ (1.07, .13, *34.02,
         T72.493)

1.30 $D'Fu \subseteq \mathcal{C}'Fu|(M \cup N_{po}|M$ (1.09, .10, *33.261,
         $\cup\ M|N_{po} \cup N_{po}|M|N_{po})$ 34.25, .36, 37.2, .22, .25,
         .32)

1.301 $\mathcal{C}'Fu = \mathcal{C}'Fu|(M \cup N_{po}|\ M$ (1.30, *34.36, 37.2, .25,
         $\cup\ M|N_{po} \cup N_{po}|M|N_{po})|Fu$ .32)

1.31 $D'N_{po}|M \subseteq \mathcal{C}'Fu \cup \mathcal{C}'Fu|N_{po}$ (1.05, .09, *33.14, 37.2,
         .32, 91.504)

1.311 $D'N_{po}|Fu \subseteq \mathcal{C}'M \cup \mathcal{C}'M|N_{po}$ (1.07, .10, *33.14, 37.2,
         .32, 91.504)

1.33 $\mathcal{C}'M = \mathcal{C}'Fu|M \cup \mathcal{C}'Fu|N_{po}|M$ (1.09, *34.36, 37.22, .25,
         .32)

1.331 $\mathcal{C}'Fu = \mathcal{C}'M|Fu \cup \mathcal{C}'M|N_{po}|Fu$ (1.10, *34.36, 37.22, .25,
         .32)

II. THE DIVISION RELATION

The concept of 'division' (Di) is introduced now by our definition 2.01, which states that the relation of division is identical to the Boolean sum of mitosis and meiosis. The introduction of this new concept, even if it seems to be trivial, is necessary for the development of further theorems in a concise form. It is also useful for pointing out similarities between the present system and that of Woodger's (AMB) where the relations cell division (D) and cell fusion (F) exhibit the same characteristics as division (Di) and fusion (Fu) do in our system.

Theorems 2.111 and 2.112 state that all powers of this division relation are irreflexive and asymmetric, which follows from postulate 1.04. This is also true for Woodger's D relation (derivable from his definitions 1.7.1 and 2.1.1, and his Theorem 1.6.13).

Theorems 2.12, 2.121, 2.122, and 2.123 indicate that division is a one-many, asymmetric, intransitive and irreflexive relation in our system (D has the same properties in Woodger's system, according to his Theorem 2.1.19). The one-many-ness of division follows from the one-many-ness of mitosis and meiosis (Theorems 1.13 and 1.14) plus from the fact that their converse domains have no members in common (Postulate 1.05 according to *71.24). The intransitivity of the division relation, in turn, follows from its one-many-ness, and from the irreflexivity of all its powers (Theorem 2.111 by T99.23).

Theorem 2.13 states that the domains and the converse domains of division and fusion have no members in common (which is also true for Woodger's D and F relations by his Theorems 2.1.4 and 2.1.5). This is a consequence of Postulates 1.06, 1.07 and 1.08.

Among the rest of the theorems in this section, 2.16, 2.17 and 2.19 are the most important ones. Theorem 2.16 is merely a rephrased version of Theorem 1.23, and presents the various possible division sequences. Theorem 2.17, on the other hand, gives the division sequences for those cases when fusion steps precede and follow the division steps, i.e., the division sequences that one may find in a sexual life cycle. According to this theorem just four such division sequences are possible, (1) a meiotic step, (2) mitotic steps followed by a meiotic one, (3) a meiotic step followed by mitotic ones, and (4) mitotic steps followed by a meiotic one and then followed again by mitotic ones. Theorem 2.19, finally, gives a breakdown of the possible division and fusion sequences, encompassing not only one life cycle from fusion to fusion, but any number of life cycles.

Definition and Theorems on Di

2.01	$Di = N \cup M$	(Definition)
2.11	$Di_{po} \subseteq J$	(1.04, T94.23)
2.111	$Di_{po} \, \epsilon \, irr$	(2.11, C99.106)
2.112	$Di_{po} \, \epsilon \, as$	(2.111, T99.211)
2.12	$Di \, \epsilon \, 1 \rightarrow Cls$	(1.05, .13, .14, 2.01, *71.24)
2.121	$Di \, \epsilon \, as$	(2.111, T99.21)
2.122	$Di \, \epsilon \, intr$	(2.111, .12, T99.23)
2.123	$Di \, \epsilon \, irr$	(2.121, C99.115)
2.13	$D'Di \cap D'Fu = \wedge.$ $.\,\mathfrak{a}'Di \cap \mathfrak{a}'Fu = \wedge$	(1.06, .07, *33.26, .261)
2.131	$Di \cap Fu = \wedge$	(2.13, *33.32)
2.14	$D'Di \subseteq \mathfrak{a}'Di \cup \mathfrak{a}'Fu$	(1.08, .09, *33.26, .261, 34.36, 91.504)
2.15	$D'Fu \subseteq \mathfrak{a}'Di$	(1.10, *33.261, 34.36, 91.504)
2.151	$D'Fu \subseteq \mathfrak{a}'(Fu \vert Di_{po})$	(1.23, .30, *33.264, 34.34)
2.16	$Di_{po} = N_{po} \cup M \cup N_{po} \vert M$ $\cup \, M \vert N_{po} \cup N_{po} \vert M \vert N_{po}$	(1.23, 2.01)
2.17	$Fu \vert Di_{po} \vert Fu = Fu \vert (M \cup N_{po} \vert \, M$ $\cup \, M \vert N_{po} \cup N_{po} \vert M \vert N_{po}) \vert Fu$	(1.21, 2.16)
2.18	$(Fu \vert Di_{po})_{po} = Fu \vert (Di_{po} \vert Fu)_* \vert Di_{po}$	(2.15, *91.504, T94.221)
2.19	$(Di \cup Fu)_{po} = Di_{po} \cup Fu \cup$ $\cup \, (Di_{po} \vert Fu)_{po} \cup (Fu \vert Di_{po})_{po} \cup$ $\cup \, Di_{po} \vert (Fu \vert Di_{po})_{po} \cup (Fu \vert Di_{po})_{po} \vert Fu$	(1.123, T94.23)
2.191	$(Di \cup Fu)_{po} \subseteq J$	(1.04, 2.01)
2.20	$(\mathfrak{a}'Fu \upharpoonright Di_{po}) \, \epsilon \, 1 \rightarrow Cls$	(2.12, .13, T92.41)
2.21	$\vec{B}'Di = \mathfrak{a}'Fu \cap D'Di$	(2.14, *93.101)
2.22	$M \vert Di_{po} = M \vert N_{po}$	(1.12, .20, 2.16, *34.32)
2.221	$M \vert N_{po} \vert Di_{po} = M \vert N_{po}{}^2$	(1.20, 2.16, *34.32)

2.222	$Di_{po}|M = N_{po}|M$	(1.12, .20, 2.16, *34.32)
2.223	$Di_{po}|N_{po}|M = N_{po}^2|M$	(1.20, 2.16, *34.32)
2.24	$D'Fu \subseteq \vec{B}'\check{D}i$	(2.13, .15, T93.6201)

III. THE CLASSES HAPLOID, DIPLOID AND APLOID

Four more definitions are introduced in this section. The first one (3.01) furnishes merely a short notation for our universal set of individuals (c in the formal language). As it has been pointed out before, individuals may be either cells or nuclei. This definition corresponds to Woodger's Theorem 2.1.3 (AMB). The next expression (3.02) defines the set of haploid individuals (h) as the Boolean sum of the converse domain of meiosis or of mitotic divisions that were preceded by meiosis. Thus, within this system, all those individuals that originate by meiosis or by mitoses following meiosis are by definition haploid. Similarly, the set of diploid individuals (d) is defined (3.03) as the Boolean sum of the converse domain of fusion or of mitoses that were preceded by fusion. These definitions are, I believe, in complete agreement with biological convention, and their formulation was possible without any reference to chromosome numbers, or chromosome behavior.

As soon as one tries to work out theorems with these concepts, it becomes obvious that there are individuals in our universe which are not members of either the haploid or the diploid sets. These would be the cells or nuclei of asexually or vegetatively reproducing organisms, where the only observable divisions would be mitoses (these divisions would nevertheless have to be true equation divisions in order to qualify for our consideration here). Definition 3.04 defines, therefore, a class of "aploid" individuals (a) – to coin a new term – as the set of those individuals which are members of the converse domain of mitosis but are not members of the converse domains of mitotic divisions which were preceded by meiosis or by fusion.

With the help of these definitions the last three Postulates (1.08, .09, .10) can be put in shorter form, as shown by Theorems 3.10, .101, .102, which themselves could have served as our postulates.

According to Theorems 3.11 and 3.111 the domains of mitosis, meiosis and fusion are all included in the set of individuals c, and the Boolean sum of the fields of mitosis, meiosis and fusion is identical to c.

Theorem 3.12 states that the set c is identical to the set of those individuals which are derived by mitosis, meiosis or fusion from members of c. This theorem expresses in a more specific form the familiar principle of *omnis cellula e cellula.**

Theorems 1.14, 1.141 and 1.142 establish that the classes of haploid, diploid and aploid individuals are mutually exclusive, as one would expect them to be.

Theorem 3.15, on the other hand, shows that these three sets among themselves exhaust our entire set of individuals (set c), i.e., there are no individuals in our system that are not members of the haploid, diploid or aploid sets.

Beginning with Theorem 3.20 a number of theorems follow exhibiting the haploid, diploid and aploid classes of individuals as hereditary classes with respect to our relations.

Theorems 3.20 and 3.201 state that the diploid class is an hereditary class with respect to both mitosis and the converse of mitosis. Thus all the ancestors and descendants by mitosis of a diploid cell or nucleus must be diploid.

Theorems 3.21 and 3.211 state similarly that the haploid class is an hereditary class with respect to mitosis and the converse of mitosis.

Next, four groups of theorems are listed following the numbers 3.22, 3.23, 3.24 and 3.25 concerning the ancestors and descendants by meiosis or fusion of haploid and diploid individuals. As we would expect, the ancestors by meiosis of haploid individuals must be diploid (3.221), while the descendants by meiosis of diploid individuals must be haploid (3.241). Similar statements can be made with regard to fusion (3.231 and 3.251). It is also indicated that haploid individuals cannot undergo meiosis (3.222), and that nothing can stand in meiosis to diploid individuals (3.242), similar statements being again true for fusion (3.232 and 3.252).

Theorems 3.28 and 3.281 show that the haploid class is an hereditary class with respect to the division relation, while the diploid class is an hereditary class with respect to the converse of division. In other words, the descendants of haploid individuals by division are all haploid, and the ancestors of diploid individuals by division are all diploid. It obviously would not be true that all the descendants of diploid individuals by

---

* Attributed to R. Virchow (*Arch. Path. Anat. Physiol.*, 1855).

division are diploid, there may be haploid descendants as well since meiosis is included in the division relation. Similarly, it could not be the case that all the division ancestors of haploid individuals are haploid.

Theorems 3.29 and 3.291 state that the aploid class of individuals is also an hereditary class with respect to both the relation of mitosis and the converse of mitosis.

Finally it should be pointed out that, according to Theorems 3.35 and 3.351, the sum of the haploid and diploid classes is an hereditary class with respect to the relation of division and also with respect to the converse of division.

Many of the statements mentioned here on hereditary classes are needed below in the theorems dealing with various division hierarchies.

Definitions and Theorems on c, h, d and a

3.01	$c = \mathcal{C}'N \cup \mathcal{C}'M \cup \mathcal{C}'Fu$	(Definition)
3.02	$h = \mathcal{C}'M \cup \mathcal{C}'M \vert N_{po}$	(Definition)
3.03	$d = \mathcal{C}'Fu \cup \mathcal{C}'Fu \vert N_{po}$	(Definition)
3.04	$a = \mathcal{C}'N \cap \overline{\mathcal{C}'M \vert N_{po}} \cap \overline{\mathcal{C}'Fu \vert N_{po}}$	(Definition)
3.10	$D'N \subseteq c$	(1.08, 3.01)
3.101	$D'M \subseteq d$	(1.09, 3.03)
3.102	$D'Fu \subseteq h$	(1.10, 3.02)
3.11	$D'N \cup D'M \cup D'Fu \subseteq c$	(1.08, .09, .10, 3.01, *34.36, 91.504)
3.111	$C'N \cup C'M \cup C'Fu = c$	(3.01, .11, *33.16)
3.12	$c = [Cnv'(N \cup M \cup Fu)]''c$	(3.01, .11, *33.26, .261, 37.16, .2, .25)
3.121	$c = [Cnv'(N \cup M \cup Fu)_{po}]''c$	(3.12, *91.53, .711)
3.14	$h \cap d = \wedge$	(1.05, .06, .07, .19, 3.02, .03, *24.402, T91.506)
3.141	$h \cap a = \wedge$	(1.05, 3.02, .04)
3.142	$d \cap a = \wedge$	(1.07, 3.03, .04)
3.15	$c = h \cup d \cup a$	(1.19, 3.01, .02, .03, .04, *24.22, .26, T91.506)
3.16	$D'N_{po} \vert M \subseteq d$	(1.31, 3.03)

3.161	$D'N_{po}\vert Fu \subseteq h$	(1.311, 3.02)
3.17	$a \cap (D'M \cup D'N_{po}\vert\ M$	(3.141, .142, .16, .161,
	$\cup\ D'Fu \cup D'N_{po}\vert Fu) = \wedge$	*24.402)
3.18	$h \cup d = \mathcal{C}'Fu \cup \mathcal{C}'Fu\vert Di_{po}$	(1.33, .34, 2.16, 3.02, .03)
3.20	$\check{N}''d \subseteq d$	(3.03, *33.264, 34.34,
		37.22, .32, 91.502, .511)
3.201	$N''d \subseteq d$	(1.07, .13, 3.03, *33.14,
		37.22, 91.502, T92.313)
3.21	$\check{N}''h \subseteq h$	(3.02, *33.264, 34.34,
		37.22, .32, 91.502, .511)
3.211	$N''h \subseteq h$	(1.05, .13, 3.02, *33.14,
		37.22, 91.502, T92.313)
3.22	$M''h = D'M$	(1.05, 3.02, *37.22, .25)
3.221	$M''h \subseteq d$	(3.101, .22)
3.222	$\check{M}''h = \wedge$	(3.14, .101, *24.402,
		37.261, .29)
3.23	$Fu''d = D'F$	(1.07, 3.03, *37.22, .25)
3.231	$Fu''d \subseteq h$	(3.102, .23)
3.232	$\check{F}u''d = \wedge$	(3.14, .102, *24.402,
		37.261, .29)
3.24	$\check{M}''d = \mathcal{C}'M$	(1.33, 3.03, *37.22, .32)
3.241	$\check{M}''d \subseteq h$	(3.02, .24)
3.242	$M''d = \wedge$	(3.03, .14, *24.402,
		37.26, .29)
3.25	$\check{F}u''h = \mathcal{C}'Fu$	(1.34, 3.02, *37.22, .32)
3.251	$\check{F}u''h \subseteq d$	(3.03, .25)
3.252	$Fu''h = \wedge$	(3.02, .14, *24.402,
		37.26, .29)
3.26	$d = \check{F}u''h \cup \check{N}_{po}''\check{F}u''h$	(3.03, .25, *37.32)
3.261	$h = \check{M}''d \cup \check{N}_{po}''\check{M}''d$	(3.02, .24, *37.32)
3.28	$\check{D}i''h \subseteq h$	(2.01, 3.21, .222, *31.15,
		37.221)
3.281	$Di''d \subseteq d$	(2.01, 3.201, .242, *37.
		*37.221)
3.29	$\check{N}''a \subseteq a$	(1.291, .292, 3.04,
		*22.48, 34.36, 37.32,
		71.38, 91.57, T22.75)

446

3.291	$N''a \subseteq a$	(1.08, 3.04, *24.21, 33.14, 35.672, 91.502, .57)
3.31	$\breve{M}''a = M''a = \wedge$	(1.05, 3.04, .17, *24.402, 37.26, .261, .29)
3.311	$\breve{Fu}''a = Fu''a = \wedge$	(1.07, 3.04, .17, *24.402, 37.26, .261, .29)
3.32	$\breve{Di}''a = \breve{N}''a$	(2.01, 3.31, *31.15, 37.221)
3.321	$Di''a = N''a$	(2.01, 3.31, *37.221)
3.33	$\breve{Di}''h = \breve{N}''h$	(2.01, 3.222, *31.15, 37.221)
3.331	$Di''d = N''d$	(2.01, 3.242, *37.221)
3.34	$\breve{Di}''d \subseteq (h \cup d)$	(2.01, 3.02, .20, .24, *37.221)
3.341	$Di''h \subseteq (h \cup d)$	(2.01, 3.16, .211, .22, *37.221)
3.35	$\breve{Di}''(h \cup d) \subseteq (h \cup d)$	(3.28, .34, *37.22)
3.351	$Di''(h \cup d) \subseteq (h \cup d)$	(3.281, .341, *37.22)
3.37	$R \subseteq Di . \supset : C'R \cap d \neq \wedge . \\ . \equiv . D'R \cap d \neq \wedge$	(3.281, *24.56, 33.16, 37.1)
3.371	$R \subseteq Di . \supset : C'R \cap h \neq \wedge . \\ . \equiv . \mathcal{a}'R \cap h \neq \wedge$	(3.28, *24.56, 33.16, 37.1)

IV. ZYGOTE, GAMETE AND RELATED CONCEPTS

The domains and converse domains of the meiosis and fusion relations can be further employed to define other important classes of cells or nuclei.

It seems obvious, for instance, that the meaning of 'converse domain of fusion' coincides with that of the 'class of zygotes or zygotic nuclei'. We would not gain conciseness, however, by merely substituting a term like 'zygote' for '$\mathcal{a}'Fu$' in our formal system. We forego, therefore, a definition of zygotes in general and define instead two particularly useful subclasses of zygotes. The first of these is composed of those zygotes (i.e. members of the converse domain of fusion) which undergo mitosis, and is designated by 'zn' (Definition 4.02). The other class is made up of

447

zygotes that undergo meiosis and is named 'zm' (Definition 4.03). Cells of this latter class occur mainly in the fungi and algae and are called variously 'zygospores', 'oospores', 'chlamydospores' or 'resting spores'. I do not attempt to propose English names for the classes zn and zm, although such names would be very useful to have.

'The domain of fusion' seems, at first sight, to correspond to 'the class of gametes'. It must be realized, however, that 'the domain of fusion' designates the class of individuals which actually fuse. Thus cells which would be called gametes because of their shape or position in an organism are not members of the domain of fusion unless they fuse. We could not have an expression for gametes in our language without introducing morphological concepts. But we can proceed to define two subclasses of gametes, one made up of the members of the converse domain of meiosis and of the domain of fusion ('gm'), and the other of the members of the converse domain of mitosis and of the domain of fusion ('gn'), as shown in Definitions 4.06 and 4.07. The members of gm would ordinarily be called 'sperm' or 'eggs' in animals, while there is no general term for the members of gn, which occur mostly in plants.

The converse domain of meiosis represents an important class of cells or nuclei. Sharp proposed to call them 'gones'*, but this name has not found wide acceptance. We already dealt with one subclass (gm) belonging to this class. Another subclass to be defined is constituted of those individuals which are members of the converse domain of meiosis and of the domain of mitosis, designated as 'ms' (Definition 4.05). Cells of this kind are called 'meiospores' in the case of certain plant groups, especially in fungi.

For the members of the domain of meiosis the term 'meiocyte' has been proposed by Sharp.* The same comment applies here as in the case of the domain of fusion, i.e., only individuals that actually undergo meiosis can be members of the domain of meiosis, while 'meiocyte' should probably refer to any individual capable of meiosis. We have already discussed one subclass of the domain of meiosis, namely zm. Another subclass can be considered, which is composed of individuals standing both in the converse domain of mitosis and in the domain of meiosis, named here 'mc' (Definition 4.04). Such a cell is usually called a 'mother cell', as in the

* L. W. Sharp, *Introduction to Cytology*, 3rd edition, McGraw-Hill, New York, 1934, p. 251.

case of 'sperm mother cell' and 'microspore mother cell', or a 'cyte' as in 'spermatocyte' and 'microsporocyte'.

The domain and converse domain of mitosis have already been involved in some of the previous definitions, the only combination that has not been mentioned yet is that of the converse domain of mitosis with the domain of mitosis. This class is to be named 'lc' (from 'link cell') by Definition 4.01 (cf. AMB 2.1.8).

Finally, there remains the class of those individuals which neither divide any further nor fuse, but eventually die. These individuals must be members of the converse domains of mitosis, meiosis or fusion. This class as a whole is designated as 'tc' (from 'terminal cell') in Definition 4.08 (cf. AMB 2.1.9).

It is easy to show that these eight classes exhaust our entire class of individuals (Theorem 4.11), and that they are mutually exclusive (Theorem 4.12).

Life cycles may be characterized by the kinds of individuals that occur in them. There are certain rules one may deduce from the definitions, which determine which kinds of individuals can follow each other in a life cycle. Two of these rules are given in Theorems 4.22 and 4.221, which state that mc individuals always arise by mitotic divisions from zn individuals, and that gn individuals arise by mitotic divisions from ms individuals.

Definitions and Theorems on Some Classes of Individuals

4.01	$lc = \mathcal{C}'N \cap D'N$	(Definition)
4.02	$zn = \mathcal{C}'Fu \cap D'N$	(Definition)
4.03	$zm = \mathcal{C}'Fu \cap D'M$	(Definition)
4.04	$mc = \mathcal{C}'N \cap D'M$	(Definition)
4.05	$ms = \mathcal{C}'M \cap D'N$	(Definition)
4.06	$gm = \mathcal{C}'M \cap D'Fu$	(Definition)
4.07	$gn = \mathcal{C}'N \cap D'Fu$	(Definition)
4.08	$tc = c \cap \overline{D'N} \cap \overline{D'M} \cap \overline{D'Fu}$	(Definition)
4.11	$c = lc \cup zn \cup zm \cup mc \cup$	(1.08, .09, .10, 3.01,
	$\cup ms \cup gm \cup gn \cup tc$	4.01–.08)
4.12	$(\iota'lc \cup \iota'zn \cup \iota'zm \cup \iota'mc \cup \iota'ms \cup$	(1.05, .06, .07, 4.01–.08,
	$\cup \iota'gm \cup \iota'gn \cup \iota'tc) \in Cls^2 excl$	*84.1)

4.14	$zn \cup zm \cup mc \subseteq d$	(3.03, .101, 4.02, .03, .04)
4.141	$gm \cup gn \cup ms \subseteq h$	(3.02, .102, 4.05, .06, .07)
4.16	$D'N = lc \cup zn \cup ms$	(1.08, 4.01, .02, .05)
4.161	$D'M = zm \cup mc$	(1.09, .26, 4.03, .04)
4.162	$D'Fu = gm \cup gn$	(1.10, .261, 4.06, .07)
4.21	$\vec{B}'Di = zn \cup zm$	(2.01, .21, 4.02, .03)
4.22	$mc \subseteq \check{N}_{po}''zn$	(1.26, 4.02, .04, *37.32)
4.221	$gn \subseteq \check{N}_{po}''ms$	(1.261, 4.05, .07, *37.32)

Va. HIERARCHICAL RELATIONS

At this point I would like to introduce the concept of hierarchies into our discussion. This concept was defined in formal logical terms by Woodger (AMB) as follows. A relation R is a member of the class of hierarchical relations if and only if R is one-many, and the beginner of R stands in some power of R to every member of the converse domain of R.

The following diagram may help visualizing this kind of a relation.*

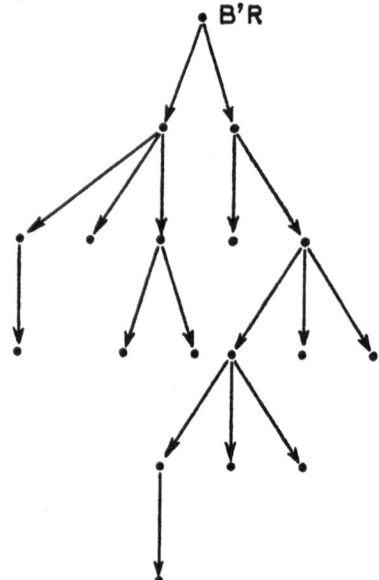

* For further elaboration of this concept, cf. J. H. Woodger, 'Problems arising from the application of mathematical logic to biology', In: *Actes de 2ieme Colloque Internationale de Logique Mathématique*, Gauthier-Villars, Paris, 1954.

450

The name chosen by Woodger for these relations already indicates where examples may be found for them, namely in the organization of a feudal state or of an army (originally 'hierarchy' referred to the ranks of angels). The head of such an organization is the beginner of the hierarchy (there must be only one beginner), and this individual stands in some relation of command to one or several individuals under him, who in turn stand in the same relation to other individuals under them, and so on. Now, according to the definition, every individual that stands in the converse domain of this command relation can be reached by a finite number of command steps from the beginner. We would not consider somebody to be a member of an army unless he is under the actual command of the commanding officer of that army, while the chain of command may consist of any finite number of steps. The difficulty in translating the logical concept of hierarchies into everyday language is caused by the fact that we usually consider an organization like an army a class of ndividuals, while it could be designated more accurately as a class of ordered pairs of individuals. In the same way, as we shall see later, a particular multicellular organism can be designated as a class of ordered pairs of mother and daughter cells, which is what a division hierarchy is.

Any relation which is one-many and the powers of which are irre- flexive includes hierarchical relations as its subrelations. To find a suitable subrelation R of such an S, one needs only to choose an individual x in the domain of S, and then designate R as identical to S with its converse domain limited to the class of individuals to which x stands in some power of S. This R is an hierarchy, and x is its beginner.

Hierarchies generated in this way from a relation S will be called 'hierarchies fully included in S' (in short, 'S-hierarchies'). The field of such an hierarchy is comprised of the beginner of the hierarchy and of those of the individuals to which the beginner stands in some power of the relation in which the hierarchy is fully included (cf. SH.01, .37, .371, .372). Such hierarchies were first introduced by Woodger (cf. AMB 2.2.1, which is the definition of his term 'cellhier'). The formal development of this concept in this paper is, however, not based on AMB.

Definitions and Theorems on Hierarchies

H.01 $R \, \epsilon \, hier \, . \equiv . \, R \, \epsilon \, 1 \rightarrow Cls \, . \, \mho'R =$ (Definition, cf. 0.68

        $= \overleftarrow{R}_{po} \, 'B \, 'R$ AMB)

SH.11 $R \, \epsilon \, hier \, . \, \supset \, . \, E!B'R$ (H.01, T30.38)

.14 $R \, \epsilon \, hier \, . \, \supset \, . \, R \neq \wedge$ (H.11, T93.612)

.16 $R \, \epsilon \, hier \, . \, \supset \, . \, R_{po} \subseteq J$ (H.01, T91.763, 96.262, .2301)

.18 $hier \subseteq as \cap intr \cap irr$ (H.16, C99.106, T99.21.13)

Definitions and Theorems on Hierarchies Fully Included in a Relation

SH.01 $R \, \epsilon \, S\text{-}hier \, . \equiv . \, R \, \epsilon \, 1 \rightarrow Cls \, .$ (Definition)
$. \, R = S \, [\overleftarrow{S}_*'B'R$

.02 $R \, \epsilon \, S\text{-}hier\text{-}max \, . \equiv . \, R \, \epsilon \, S\text{-}hier \, .$ (Definition)
$. \sim (\exists P) \, . \, P \, \epsilon \, S\text{-}hier \, .$
$. \, R \subseteq P \, . \, R \neq P$

.11 $R = S \, [\overleftarrow{S}_*'x \, . \, \supset \, .$ (*13.191, T96.161,
$. \, \mathcal{C}'R = \overleftarrow{R}_{po}'x = \overleftarrow{S}_{po}'x$ .1721)

.111 $R = S \, [\overleftarrow{S}_*'x \, . \, x \, \epsilon \, D'S \, . \, \supset \, .$ (*13.191, T96.164,
$. \, C'R = \overleftarrow{R}_*'x = \overleftarrow{S}_*'x$ .1732)

.13 $R = S \, [\overleftarrow{S}_*'x \, . \, \supset \, : x \, \epsilon \, D'S \, .$ (*13.191, T96.1711)
$. \equiv . \, x \, \epsilon \, D'R$

.21 $R = S \, [\overleftarrow{S}_*'B'R \, . \, \supset \, . \, B'R \, \epsilon \, D'S$ (*33.263, T30.38, 36.44, 93.6111)

.222 $R = S \, [\overleftarrow{S}_*'B'R \, . \equiv . \, (\exists x) \, . \, R =$ (*13.191, .195, T96.181)
$= S \, [\overleftarrow{S}_*'x \, . \, x \, \epsilon \, D'S \, . \sim (x S_{po} x)$

.23 $R = S \, [\overleftarrow{S}_*'B'R \, . \, \supset \, .$ (SH.11, .111, .21)
$. \, \mathcal{C}'R = \overleftarrow{R}_{po}'B'R = \overleftarrow{S}_{po}'B'R \, .$
$. \, C'R = \overleftarrow{R}_*'B'R = \overleftarrow{S}_*'B'R$

.25 $R = S \, [\overleftarrow{S}_*'B'R \, . \, \supset \, : R \, \epsilon \, hier \, .$ (H.01, SH.23)
$. \equiv . \, R \, \epsilon \, 1 \rightarrow Cls$

.301 $C'R = \overleftarrow{R}_*'x \, . \, \supset \, : R \subseteq S \, .$ (*32.19, 36.24, .241, .25,
$. \equiv . \, R \subseteq S \, [\overleftarrow{S}_*'x$ T36.44)

SH.31 $x \in D'S . \supset : R = S \, \overleftarrow{\lceil S_*}'x .$ (*22.43, .58, .59, 32.31,

         $. \equiv . R = (R \cap S) \, \overleftarrow{\lceil (R_* \cap S_*)}'x.$ 36.2, .25, T36.44)

         $. S \, \overleftarrow{\lceil S'}_*x \subseteq R$

.32 $R = S \, \overleftarrow{\lceil S_*}'B'R . \equiv . R \subseteq S .$ (SH.23, .301)

         $. S \, \overleftarrow{\lceil S_*}'B'R \subseteq R .$

         $. C'R = \overleftarrow{R_*}'B'R$

.321 $R \in S\text{-}hier . \equiv . R \in hier .$ (H.01, SH.01, .32,

         $. R \subseteq S . S \, \overleftarrow{\lceil S_*}'B'R \subseteq R$ T96.262)

.33 $S\text{-}hier \subseteq hier$ (SH.321)

.331 $R \in S\text{-}hier . \supset . R_{po} \subseteq J$ (H.16, SH.33)

.333 $R \in S\text{-}hier . \supset . R \subseteq S$ (SH.321)

.334 $R \in S\text{-}hier . \supset . E!B'R$ (H.11, SH.33)

.335 $R \in S\text{-}hier . \supset . R \neq \wedge$ (H.14, SH.33)

.37 $S \in 1 \rightarrow Cls . \supset : R \in S\text{-}hier .$ (*71.22, SH.01, .333)

         $. \equiv . R = S \, \overleftarrow{\lceil S_*}'B'R$

.371 $S \in 1 \rightarrow Cls . S_{po} \subseteq J . \supset :$ (*50.24, SH.222, .37)

         $: R \in S\text{-}hier . \equiv .$

         $. (\exists x) . x \in D'S . R = S \, \overleftarrow{\lceil S_*}'x$

.372 $R \in S\text{-}hier . \equiv . R \in 1 \rightarrow Cls .$ (SH.01, .222)

         $. (\exists x) . x \in D'S .$

         $. R = S \, \overleftarrow{\lceil S_*}'x . \sim (x S_{\rho} x)$

.38 $R \in S\text{-}hier . \supset : x R y .$ (SH.01, *96.16)

         $. \equiv . x S y . (B'R) S_* x$

.39 $R, P \in S\text{-}hier . \supset :$ (*22.49, 30.37, 33.263,

         $: B'R = B'P . \equiv . R = P$ .264, 35.43, .431, 51.23,

         53.31, 93.1, SH.01)

.391 $R, P \in S\text{-}hier . S \in 1 \rightarrow Cls .$ (*71.36, T91.61, SH.01,

         $S^2 = \wedge . \supset : \sigma'R \cap \sigma'P \neq \wedge .$ .21, .39)

         $. \equiv . R = P$

.41 $S \in 1 \rightarrow Cls . S_{po} \subseteq J . \supset :$ (*21.54, T96.1711,

         $x \in D'S . \equiv . (\exists R) . R \in S\text{-}hier .$ SH.13, .22, .37)

         $. x = B'R$

SH.411 $\quad S \in 1 \to Cls . S_{po} \subseteq J . \supset :$ $\qquad$ (*21.54, 37.25, 91.504,

$\quad y \in \mathbb{C}'S . \equiv . (\exists R) . R \in S\text{-}hier .$ $\qquad$ T96.1712, SH.11, .13,

$\quad . y \in \mathbb{C}'R$ $\qquad$ .22, .37)

.412 $\quad S \in 1 \to Cls . S_{po} \subseteq J . \supset : x Sy.$ $\quad$ (T96.171, SH.22, .37)

$\quad . \equiv . (\exists R) . R \in S\text{-}hier . xRy .$

$\quad . x = B'R$

.51 $\quad R = S \restriction \overleftarrow{S_*}'B'R . \mathbb{C}'S \cap \mathbb{C}'Q =$ $\quad$ (*24.402, 37.1, .32, 51.31,

$\quad = \wedge . \supset : Q|R_{po}|P \neq \wedge .$ $\qquad$ 53.31, 91.504, T30.38,

$\quad . \equiv . Q|R \neq \wedge . R|P \neq \wedge$ $\qquad$ 34.362, SH.23)

.53 $\quad R = S \restriction \overleftarrow{S_*}'x . \supset : \mathbb{C}'R \cap \alpha \neq$ $\quad$ (*37.1, 96.15, .16)

$\quad \neq \wedge . \equiv . x \in S_{po}''\alpha$

.531 $\quad R = S \restriction \overleftarrow{S_*}'x . x \in D'S . \supset :$ $\quad$ (*37.1, 96.151, .16)

$\quad : C'R \cap \alpha \neq \wedge . \equiv . x \in S_*''\alpha$

.55 $\quad R = S \restriction \overleftarrow{S_*}'x . x \in D'S . \check{S}''\alpha \subseteq$ $\quad$ (*33.161, T90.241,

$\quad \subseteq \alpha . \supset : x \in \alpha . \equiv . C'R \subseteq \alpha$ $\qquad$ SH.111, .13)

.551 $\quad R = S \restriction \overleftarrow{S_*}'x . x \in D'S . S''\alpha \subseteq$ $\quad$ (*91.71, SH.531)

$\quad \alpha . \supset : x \in \alpha . \equiv . C'R \cap \alpha \neq \wedge$

.552 $\quad R = S \restriction \overleftarrow{S_*}'x . x \in D'S . \check{S}''\alpha \subseteq$ $\quad$ (SH.55, .551)

$\quad \alpha . S''\alpha \subseteq \alpha . \supset : C'R \cap \alpha \neq \wedge .$

$\quad . \equiv . C'R \subseteq \alpha$

.553 $\quad R = S \restriction \overleftarrow{S_*}'x . x \in D'S . \check{S}''\alpha \subseteq$ $\quad$ (*22.48, 24.58, 33.161,

$\quad \alpha . S''\alpha \subseteq \alpha . \supset : \overrightarrow{Lk}'R \cap \alpha \neq \wedge .$ $\quad$ 91.71, T24.581, 93.701,

$\quad . \equiv . \overrightarrow{Lk}'R \subseteq \alpha . \overrightarrow{Lk}'R \neq \wedge$ $\qquad$ SH.23, .53, .55)

.61 $\quad R = S \restriction \overleftarrow{S_*}'x . x \in D'S . D'R \cap$ $\quad$ (*34.36, 91.504, T94.23,

$\quad \cap D'T = \wedge . R|T = \wedge . \supset .$ $\qquad$ 96.171, SH.11, .13)

$\quad . \overleftarrow{(S \cup T)_{po}}'x = \overleftarrow{S_{po}}'x$

.62 $\quad R = (S \cup T) \restriction \overleftarrow{(S \cup T)_*}'x .$ $\quad$ (*23.58, T36.44, 72.39)

$\quad . (S \cup T) \in 1 \to Cls . \supset .$

$\quad . \mathbb{C}'R \cap \mathbb{C}'T = \mathbb{C}'(R \cap T)$

.621 $\quad R = (S \cup T) \restriction \overleftarrow{(S \cup T)_*}'x$ $\quad$ (*33.25, 96.16, SH.111)

$$. x \in D'(S \cup T) . \supset .$$
$$. D'R \cap D'T = D'(R \cap T)$$

SH.622 $R = (S \cup T) \, [\overleftarrow{(S \cup T)_*}'x \,.$ (*33.241, SH.62, .621)
$$. x \in D'(S \cup T) . (S \cup T) \in$$
$$\epsilon \, 1 \to Cls . \supset : R \cap T = \wedge .$$
$$. \equiv . D'R \cap D'T = \wedge . \equiv .$$
$$. \mathit{Q}'R \cap \mathit{Q}'T = \wedge$$

.623 $R = (S \cup T) \, [\overleftarrow{(S \cup T)_*}'x . \supset :$ (*91.54, .56, .575, 96.15,
$$: R|T \neq \wedge . \supset . D'R \cap D'T \neq \wedge \quad \text{.16})$$
$$. \mathit{Q}'R \cap \mathit{Q}'T \neq \wedge$$

.624 $R = (S \cup T) \, [\overleftarrow{(S \cup T)_*}'x .$ (SH.622, .623)
$$. x \in D'(S \cup T) . (S \cup T) \in$$
$$\epsilon \, 1 \to Cls . \supset : R|T \neq \wedge . \supset .$$
$$. R \cap T \neq \wedge$$

.65 $(S \cup T) \in 1 \to Cls . D'S \cap D'T$ (*22.58, 32.19, .31,
$$= \wedge . \mathit{Q}'S \cap \mathit{Q}'T = \wedge . \qquad \text{34.36, 35.14, 36.2, .24,}$$
$$. \supset : R \in (S \cup T)\text{-}hier . R \cap T = \qquad \text{.241, 71.22, 90.18,}$$
$$= \wedge . \equiv . R \in S\text{-}hier . R|T = \wedge \qquad \text{*91.504, 96.16, .2,}$$

                                        T33.361, .362, 94.212,
                                        .23, .24, 96.172, SH.01,
                                        .21, .31, .37, .61, .622,
                                        .624)

.71 $P = S \, [\overleftarrow{S_*}'x . R = S \, [\overleftarrow{S_*}'y .$ (*30.37, 32.121, 33.161,
$$. x, y \in D'S . \sim (y S_{po} y) . \qquad \text{36.24, 91.504, .54, .574,}$$
$$\supset : R \subseteq P . R \neq P . \equiv . x S_{po} y . \qquad \text{.59, 96.16, T23.812,}$$
$$. \equiv . y \in \mathit{Q}'P \qquad \text{36.44, SH.11, .111, .13})$$

.7101 $P = S \, [\overleftarrow{S_*}'x . R = S \, [\overleftarrow{S_*}'y .$ (*33.161, 90.163, .17,
$$. x, y \in D'S . \supset : R \subseteq P . \qquad \text{96.16, T36.44, SH.111,}$$
$$. \equiv . C'R \subseteq C'P . \equiv . x S_* y . \qquad \text{.13, .55})$$
$$. \equiv . y \in C'P$$

.711 $R, P \in S\text{-}hier . \supset : R \subseteq P .$ (SH.01, .22, .71)
$$. R \neq P . \equiv . B'P \, S_{po} \, B'R .$$
$$. \equiv . B'R \in \mathit{Q}'P$$

.713 $P = S \, [\overleftarrow{S_*}'x . R = S \, [\overleftarrow{S_*}'y .$ (*2.11, 92.31, SH.111,

$$. x, y \in D'S . S \in 1 \to Cls . \qquad .7101)$$
$$. \supset : R \subseteq P . \lor . P \subseteq R . \lor .$$
$$. C'R \cap C'P = \land$$

.72 $\quad S \in 1 \to Cls . S_{po} \subseteq J . R \in S\text{-}hier .$ (SH.01, .21, .411, .711)
$$. \supset : (B'R)BS . \equiv . \sim (\exists P) .$$
$$. P \in S\text{-}hier . R \subseteq P . R \neq P$$

SH.721 $\quad S \in 1 \to Cls . S_{po} \subseteq J . \supset :$ (SH.02, .72)
$$. R \in S\text{-}hier\text{-}max . \equiv . R \in S\text{-}hier .$$
$$. (B'R)BS$$

vb. MITOTIC MEIOTIC AND DIVISION HIERARCHIES

Since in our system mitosis, meiosis and division are all relations which are one-many and the powers of which are irreflexive, we can obtain classes of mitotic hierarchies, meiotic hierarchies and division hierarchies in the same manner we obtain S-hierarchies in general. A relation R is thus a mitotic hierarchy if and only if there is an individual x which is a member of the domain of mitosis and R is identical to the relation of mitosis with its converse domain limited to the individuals to which x stands in some power of mitosis (Theorem 5.11). Similar statements can be made with reference to meiotic hierarchies and division hierarchies (Theorems 5.111 and 5.112) (cf. Woodger's Definition 2.2.1 in AMB). For example, every cell in the gametophyte of a fern can be considered to be a member of the field of a mitotic hierarchy, the beginner of which may be the spore from which the gametophyte develops. But one does not need to start necessarily with the spore, one could start with any cell of the gametophyte and consider the mitotic hierarchy for which that cell is the beginner. Similarly, one can speak of the division hierarchy of a fern that begins with the zygote, and the converse domain of which is composed of all the cells which are derived by division (mitotic or meiotic) from that zygote, which would include the cells of both the sporophyte and the gametophyte. While one can choose any cell or nucleus which is in the domain of mitosis or division to begin a mitotic or division hierarchy, one cannot terminate those hierarchies at any arbitrary point, because as a consequence of definition SH.01 the field of an S-hierarchy that begins with an individual x must include all the individuals to which x stands in any power of S.

The expressions given in Theorems 5.12, 5.121 and 5.122 for mitotic, meiotic, and division hierarchies are equivalent to those mentioned above. Either one of these two sets of theorems could be used as definitions, but I preferred to define first S-hierarchies in general, and to consider these statements as theorems.

Theorem 5.13 states that an individual x is a member of the domain of mitosis if and only if there is a single relation R which is a mitotic hierarchy and the beginner of which is x. Similar statements can be made for members of the domain of meiosis and of division (Theorems 5.131, .132; cf. also Woodger's Theorem 2.2.4 in AMB). Therefore every individual in the domain of mitosis defines one and only one mitotic hierarchy, of which it is the beginner. But such an individual may be member of the domains of many different mitotic hierarchies without being their beginner. By the same token, a member of the converse domain of mitosis may be member of the converse domains of many mitotic hierarchies, and it must be a member of the converse domain of at least one (Theorem 5.14). The same holds for division hierarchies (Theorem 5.142). A member of the converse domain of meiosis can, however, be a member of the converse domain only of a single meiotic hierarchy (Theorem 5.141).

Theorems 5.19, .191, .192 show what kinds of individuals may begin the various hierarchies, on the basis of the classification in Section 4. Many other theorems of this nature could be listed (cf. 3.23, .231).

The concept of maximal mitotic, or meiotic, or division hierarchies is introduced next. In general, maximal hierarchies fully included in a relation S may be defined as those S-hierarchies which are not included in other S-hierarchies (SH.02). If S is a one-many relation and its powers are irreflexive, then those S-hierarchies the beginners of which are beginners of S are also maximal S-hierarchies (SH.721). This is one of Woodger's concepts, unpublished. To come back to mitotic hierarchies, any mitotic hierarchy the beginner of which is a member of the converse domain of fusion or meiosis will be a maximal mitotic hierarchy (Theorem 5.22). Thus the mitotic hierarchy which begins with the spore in a fern gametophyte is a maximal mitotic hierarchy, since the spore would be a member of the converse domain of meiosis. On the other hand, those mitotic hierarchies which begin with some other cell in the gametophyte are not maximal ones, because the cell which is the beginner of such a hierarchy would have to be a member of the converse domain of mitosis. A similar theorem (5.222) states that maximal division hierarchies are those division hierarchies which begin with an individual in the converse domain of fusion.* Every division hierarchy beginning with a zygote is thus a maximal one (5.231). Finally, every meiotic hierarchy is also a maximal meiotic hierarchy (Theorem 5.221), because no meiotic hierarchy can be included in another one, without being identical to it.

It seems that the concept of maximal mitotic and division hierarchies is useful whenever whole sections of life cycles must be designated, i.e., a section beginning with fusion or meiosis and comprised of mitotic or division steps. These are the portions of life cycles usually called "generations" or "phases", the alternation of which is one of the main points of interest in life cycles of plants.

A number of theorems follow which are not discussed here in order, rather the main types of hierarchies will be listed and their chief characteristics mentioned.

If one considers all possible kinds of mitotic, meiotic and division hierarchies, one finds that the first obvious way of classifying them is by forming the various Boolean products of these three classes. Thus we obtain the following possibilities: (1) mitotic hierarchies which are also division hierarchies, (2) mitotic hierarchies which are not division hierarchies, (3) meiotic hierarchies which are division hierarchies, (4) meiotic hierarchies which are not division hierarchies, and (5) division hierarchies which are neither mitotic nor meiotic hierarchies (Theorem 5.36).

The difference between a mitotic hierarchy which is a division hierarchy and one which is not is that the latter has a member of its converse domain standing in the domain of meiosis, while the former has not (Theorems 5.361 and 5.363). The reason for this is that while a mitotic hierarchy is terminated by an individual which undergoes meiosis, a division hierarchy is not terminated by such an individual, on the contrary, Theorem 5.122 requires it to proceed as long as there are any individuals to which the beginner stands in some power of division, which includes meiosis. Now, there are three kinds of mitotic hierarchies, according to the kind of individuals that make up their fields, which may be all haploid, diploid or aploid (Theorem 5.273). It is impossible for a

---

* Thus '*Di-hier-max*' is analogous to Woodger's '*D-hier*', according to AMB 2.2.12.

mitotic hierarchy to have more than one of these three types of individuals in its field (Theorems 5.27, 5.271 and 5.272). Mitotic hierarchies which are also division hierarchies may, therefore, have their fields included in any one of the three classes of haploid, diploid or aploid individuals. On the other hand, mitotic hierarchies which are not division hierarchies can only have diploid individuals in their fields (Theorem 5.382). One can also say that if the field of a relation has no diploid individuals in it then this relation is a mitotic hierarchy whenever it is a division hierarchy, and *vice versa* (Theorem 5.38).

The difference between a meiotic hierarchy which is a division hierarchy and one which is not is, similarly, that the latter has a member of its converse domain standing in the domain of mitosis, while the former has not (Theorems 5.362 and 5.364).

Theorems 5.35 and 5.351 express rather concisely the convertibility conditions among mitotic, meiotic and division hierarchies. Accordingly, every mitotic hierarchy that is not followed by meiosis is also a division hierarchy which does not contain meiosis; and every meiotic hierarchy that is not followed by mitosis is also a division hierarchy which does not contain mitosis. No relation can be both a mitotic and a meiotic hierarchy (Theorem 5.16).

Perhaps the most interesting of the five classes mentioned above is that of the division hierarchies which are neither mitotic nor meiotic hierarchies. The relations which are members of this class are division hierarchies which contain both mitotic and meiotic steps (Theorems 5.365, .366). It is possible, therefore, to show that some members of the domains of these hierarchies must be diploid, and that some members of their converse domains must be haploid (Theorem 5.39). Three kinds of these hierarchies are possible: (1) the links of which are all haploid, (2) the links of which are all diploid, and (3) those which have both haploid and diploid links (Theorems 5.402, .403). The links of a relation, as we recall, are those individuals which are members of both the domain and converse domain of the relation (T93.701). It is necessary to consider the links of these hierarchies rather than just the members of their fields because according to the previously mentioned theorem there are always haploid and diploid members of the fields of all these hierarchies but there are not both kinds of links in every one of them. Theorem 5.40 states, moreover, that there are always links in these hierarchies, so that the statement that all the links are diploid or haploid is not tautological when applied to these hierarchies (if a relation has no links, then its links can be said to be included in any class, because the empty class is included in every class).

Theorems on N-hier, M-hier and Di-hier

5.11	$R \in N\text{-}hier . \equiv . (\exists x) . x \in D'N .$	(1.11, .13, SH.371)
	$. R = N \restriction \overleftarrow{N_*}'x$	
5.111	$R \in M\text{-}hier . \equiv . (\exists x) . x \in D'M .$	(1.113, .14, SH.371)
	$. R = M \restriction \overleftarrow{M_*}'x$	
5.112	$R \in Di\text{-}hier . \equiv . (\exists x) . x \in D'Di .$	(2.11, .12, SH.371)
	$. R = Di \restriction \overleftarrow{Di_*}'x$	
5.12	$R \in N\text{-}hier . \equiv . R = N \restriction \overleftarrow{N_*}'B'R$	(1.13, SH.37)
5.121	$R \in M\text{-}hier . \equiv . R = M \restriction \overleftarrow{M_*}'B'R$	(1.14, SH.37)
5.122	$R \in Di\text{-}hier . \equiv . R = Di \restriction Di_*'B'R$	(2.12, SH.37)
5.13	$x \in D'N . \equiv . \therefore (\exists R) . R \in N\text{-}hier .$	(1.11, .13, SH.39, .41)
	$. x = B'R : (P,Q): P,Q \in N\text{-}hier .$	
	$. x = B'P . x = B'Q . \supset . P = Q$	
5.131	$x \in D'M . \equiv . \therefore (\exists R) . R \in M\text{-}hier .$	(1.113, .14, SH.39, .41)
	$. x = B'R : (P, Q) : P, Q \in M\text{-}hier .$	
	$. x = B'P . x = B'Q . \supset . P = Q$	
5.132	$x \in D'Di . \equiv . \therefore (\exists R) . R \in Di\text{-}hier .$	(2.11, .12, SH.39, .41)
	$. x = B'R : (P, Q): P, Q \in Di\text{-}hier .$	
	$. x = B'P . x = B'Q . \supset . P = Q$	
5.14	$y \in \mathcal{C}'N . \equiv . (\exists R) . R \in N\text{-}hier . y \in \mathcal{C}'R$	(1.11, .13, SH.411)
5.141	$y \in \mathcal{C}'M . \equiv . \therefore (\exists R) . R \in M\text{-}hier .$	(1.113, .12, .14, SH.391,
	$. y \in \mathcal{C}'R: (P, Q): P, Q \in M\text{-}hier .$	.411)
	$. y \in \mathcal{C}'P . y \in \mathcal{C}'Q . \supset . P = Q$	
5.142	$y \in \mathcal{C}'Di . \equiv . (\exists R) . R \in Di\text{-}hier .$	(2.11, .12, SH.411)
	$. y \in \mathcal{C}'R$	
5.16	$N\text{-}hier \cap M\text{-}hier = \wedge$	(1.16, T24.581, SH.335, .333)
5.17	$R \in Di\text{-}hier . \supset : R \cap N = \wedge . \equiv .$	2.01, .12, 5.112, SH. 622)
	$. D'R \cap D'N = \wedge . \equiv . \mathcal{C}'R \cap$	
	$\cap \mathcal{C}'N = \wedge$	

5.171 $R \in Di\text{-}hier . \supset : R \cap M = \wedge .$ (2.01, .12, 5.112, SH.622)
          $. \equiv . D'R \cap D'M = \wedge .$
          $. \equiv . \mathcal{C}'R \cap \mathcal{C}'M = \wedge$

5.172 $R \in Di\text{-}hier . \supset : R|N \neq \wedge . \supset .$ (2.01, .12, 5.112, SH.624)
          $. R \cap N \neq \wedge$

5.173 $R \in Di\text{-}hier . \supset : R|M \neq \wedge . \supset .$ (2.01, .12, 5.112, SH.624)
          $. R \cap M \neq \wedge$

5.19 $R \in N\text{-}hier . \supset . B'R \in (lc \cup ms \cup zn)$ (1.08, 4.01, .02, .05,
                                                      SH.21)

5.191 $R \in M\text{-}hier . \supset . B'R \in (zm \cup mc)$ (1.09, 4.03, .04, SH.21)

5.192 $R \in Di\text{-}hier . \supset . B'R \in$ (1.08, .09, 4.01, .02, .03,
          $(lc \cup zn \cup zm \cup mc \cup ms)$ .04, .05, SH.21)

5.21 $R \in N\text{-}hier . \supset : B'R \in$ (1.24, SH.21)
          $(\mathcal{C}'M \cup \mathcal{C}'Fu) . \equiv . (B'R) BN$

5.211 $R \in M\text{-}hier . \supset . (B'R) B M$ (1.25, SH.21)

5.212 $R \in Di\text{-}hier . \supset : B'R \in \mathcal{C}'Fu . \equiv .$ (2.21, SH.21)
          $. (B'R) B Di$

5.22 $R \in N\text{-}hier\text{-}max . \equiv . R \in N\text{-}hier .$ (1.11, .13, 5.21, SH.721)
          $. B'R \in (\mathcal{C}'M \cup \mathcal{C}'Fu)$

5.221 $M\text{-}hier = M\text{-}hier\text{-}max$ (1.113, .14, 5.211,
                                                    SH.721)

5.222 $R \in Di\text{-}hier\text{-}max . \equiv . R \in Di\text{-}hier .$ (2.11, .12, 5.212, SH.721)
          $. B'R \in \mathcal{C}'Fu$

5.23 $R \in N\text{-}hier\text{-}max . \equiv . R \in N\text{-}hier .$ (4.02, .05, 5.12, .22,
          $. B'R \in (ms \cup zn)$ SH.21)

5.231 $R \in Di\text{-}hier\text{-}max . \equiv . R \in Di\text{-}hier .$ (2.01, 4.02, .03, 5.122,
          $. B'R \in (zn \cup zm)$ .222, SH.21)

5.24 $R \in (N\text{-}hier \cup Di\text{-}hier) . \supset . \mathcal{C}'Fu \cap$ (1.07, 2.13, SH.33, .333,
          $\cap D'R = \mathcal{C}'Fu \cap \iota'B'R$ .334, T93.61, 621).

5.241 $R \in N\text{-}hier . \supset . \mathcal{C}'M \cap D'R =$ (1.05, SH.33, .333, .334,
          $= \mathcal{C}'M \cap \iota'B'R$ T93.61, .621)

5.242 $R \in M\text{-}hier . \supset . D'R = \iota'B'R$ (1.25, SH.334, T93.61,
                                                    .621)

5.25 $R \in (N\text{-}hier \cup Di\text{-}hier) . \supset .$ (1.07, 2.13, SH.333,

          $. \mathcal{C}'R \cap D'Fu = \vec{B'}\check{R} \cap D'Fu$ T93.622)

5.251 $R \,\epsilon\, N\text{-}hier \,.\, \supset \,.\, \alpha'R \,\cap\, D'M =$ (1.05, SH.333, T93.622)

       $= \vec{B}'\breve{R} \,\cap\, D'M$

5.252 $R \,\epsilon\, M\text{-}hier \,.\, \supset \,.\, \alpha'R = \vec{B}'\breve{R}$ (1.251, SH.333, T93.624)

5.26 $R \,\epsilon\, (N\text{-}hier \,\cup\, Di\text{-}hier) \,.\, \supset \,:$ (2.222, 5.12, .122, SH.53,

       $: \alpha'R \,\cap\, D'M \neq \wedge \,.\, \equiv \,.$ *37.32)

       $.\, B'R \,\epsilon\, D'N_{po}|M$

5.261 $R \,\epsilon\, N\text{-}hier \,.\, \supset \,: \alpha'R \,\cap\, D'Fu \neq \wedge \,.$ (5.12, SH.53, *37.32)

       $.\, \equiv \,.\, B'R \,\epsilon\, D'N_{po}|Fu$

5.262 $R \,\epsilon\, Di\text{-}hier \,.\, \supset \,: D'R \,\cap\, D'Fu \neq \wedge \,.$ (5.122, SH.53, *37.32)

       $.\, \equiv \,.\, B'R \,\epsilon\, D'Di_{po}|Fu$

5.27 $R \,\epsilon\, N\text{-}hier \,.\, \supset \,: C'R \,\cap\, h \neq \wedge \,.$ (3.21, .211, 5.12, SH.552)

       $.\, \equiv \,.\, C'R \subseteq h$

5.271 $R \,\epsilon\, N\text{-}hier \,.\, \supset \,: C'R \,\cap\, d \neq \wedge \,.$ (3.20, .201, 5.12, SH.552)

       $.\, \equiv \,.\, C'R \subseteq d$

5.272 $R \,\epsilon\, N\text{-}hier \,.\, \supset \,: C'R \,\cap\, a \neq \wedge \,.$ (3.29, .291, 5.12, SH.552)

       $.\, \equiv \,.\, C'R \subseteq a$

5.273 $R \,\epsilon\, N\text{-}hier \,.\, \supset \,: C'R \subseteq h \,.\, \vee \,.$ (3.111, .15, 5.27, .271,

       $.\, C'R \subseteq d \,.\, \vee \,.\, C'R \subseteq a$ .272, SH.335, .333,

       T24.581)

5.28 $R \,\epsilon\, Di\text{-}hier \,.\, \supset \,: C'R \,\cap$ (3.35, .351, 5.122,

       $(h \cup d) \neq \wedge \,.\, \equiv \,.\, C'R \subseteq (h \cup d)$ SH.552)

5.281 $R \,\epsilon\, Di\text{-}hier \,.\, \supset \,: C'R \,\cap\, a \neq \wedge \,.$ (3.29, .291, .32, .321,

       $.\, \equiv \,.\, C'R \subseteq a$ 5.122, SH.552)

5.30 $R \,\epsilon\, Di\text{-}hier \,.\, \supset \,: B'R \,\epsilon\, h \,.\, \equiv \,.\, C'R \subseteq h$ (3.28, 5.122, SH.55)

5.301 $R \,\epsilon\, Di\text{-}hier \,.\, \supset \,:$ (3.281, 5.122, SH.551)

       $: B'R \,\epsilon\, d \,.\, \equiv \,.\, C'R \,\cap\, d \neq \wedge$

5.31 $R \,\epsilon\, Di\text{-}hier \,.\, \supset \,: \vec{Lk}'R \,\cap\, (h \cup d) \neq$ (3.35, .351, 5.122,

       $\neq \wedge \,.\, \equiv \,.\, \vec{Lk}'R \subseteq (h \cup d) \,.\, \vec{Lk}'R \neq \wedge$ SH.553)

5.311 $R \,\epsilon\, Di\text{-}hier \,.\, \supset \,: \vec{Lk}'R \,\cap\, a \neq \wedge \,.\, \equiv$ (3.29, .291, .32, .321,

       $\equiv \,.\, \vec{Lk}'R \subseteq a \,.\, \vec{Lk}'R \neq \wedge$ 5.122, SH.553)

5.32 $R \,\epsilon\, Di\text{-}hier \,.\, \supset \,: \vec{Lk}'R \,\cap\, h \neq \wedge \,.$ (5.31, *22.58, 24.49, .561

       $.\, \vec{Lk}'R \,\cap\, d = \wedge \,.\, \equiv \,.\, \vec{Lk}'R \subseteq h \,.$ T24.581)

       $.\, \vec{Lk}'R \neq \wedge$

5.321	$R \, \epsilon \, Di\text{-}hier \,.\, \supset :$	(5.31, *22.58, 24.49,
	$.\,\overrightarrow{Lk'}R \cap d \neq \wedge \,.\, \overrightarrow{Lk'}R \cap h = \wedge \,.\,$	.561, T24.581)
	$.\equiv .\, \overrightarrow{Lk'}R \subseteq d \,.\, \overrightarrow{Lk'}R \neq \wedge$	
5.33	$R \, \epsilon \, M\text{-}hier \,.\, \supset .\, \overrightarrow{Lk'}R = \wedge$	(1.12, SH.333, *34.531,
		T93.701, .71)
5.331	$R \, \epsilon \, M\text{-}hier \,.\, \supset .\, D'R \subseteq d \,.\, \overline{d}'R \subseteq h$	(3.02, .16, SH.335, .333,
		*24.58, T24.581)
5.35	$R \, \epsilon \, N\text{-}hier \,.\, R|M = \wedge \,.\,$	(1.05, 2.01, .12, SH.65)
	$.\equiv .\, R \, \epsilon \, Di\text{-}hier \,.\, R \cap M = \wedge$	
5.351	$R \, \epsilon \, M\text{-}hier \,.\, R|N = \wedge \,.\,$	(1.05, 2.01, .12, SH.65)
	$.\equiv .\, R \, \epsilon \, Di\text{-}hier \,.\, R \cap N = \wedge$	
5.36	$N\text{-}hier \cup M\text{-}hier \cup Di\text{-}hier =$	(5.16)
	$= (N\text{-}hier \cap Di\text{-}hier) \cup$	
	$(N\text{-}hier \cap \overline{Di\text{-}hier}) \cup (M\text{-}hier \cap Di\text{-}hier) \cup$	
	$(M\text{-}hier \cap \overline{Di\text{-}hier}) \cup$	
	$(Di\text{-}hier \cap \overline{N\text{-}hier} \cap \overline{M\text{-}hier})$	
5.361	$R \, \epsilon \, (N\text{-}hier \cap Di\text{-}hier) \,.\,$	(1.16, 5.35, SH.333)
	$.\equiv .\, R \, \epsilon \, N\text{-}hier \,.\, R|M = \wedge \,.\,$	
	$.\equiv .\, R \, \epsilon \, Di\text{-}hier \,.\, R \cap M = \wedge$	
5.362	$R \, \epsilon \, (M\text{-}hier \cap Di\text{-}hier) \,.\,$	(1.16, 5.351, SH.333)
	$.\equiv .\, R \, \epsilon \, M\text{-}hier \,.\, R|N = \wedge \,.\,$	
	$.\equiv .\, R \, \epsilon \, Di\text{-}hier \,.\, R \cap N = \wedge$	
5.363	$R \, \epsilon \, (N\text{-}hier \cap \overline{Di\text{-}hier}) \,.\,$	(1.16, 5.35, SH.333)
	$.\equiv .\, R \, \epsilon \, N\text{-}hier \,.\, R|M \neq \wedge$	
5.364	$R \, \epsilon \, (M\text{-}hier \cap \overline{Di\text{-}hier}) \,.\,$	(1.16, 5.351, SH.333)
	$.\equiv .\, R \, \epsilon \, M\text{-}hier \,.\, R|N \neq \wedge$	
5.365	$R \, \epsilon \, (Di\text{-}hier \cap \overline{N\text{-}hier}) \,.\,$	(5.361)
	$.\equiv .\, R \, \epsilon \, Di\text{-}hier \,.\, R \cap M \neq \wedge$	
5.366	$R \, \epsilon \, (Di\text{-}hier \cap \overline{M\text{-}hier}) \,.\,$	(5.362)
	$.\equiv .\, R \, \epsilon \, Di\text{-}hier \,.\, R \cap N \neq \wedge$	
5.37	$R \, \epsilon \, Di\text{-}hier \,.\, \supset \,\therefore\, C'R \cap h = \wedge$	(2.01, 3.02, .101, 5.122,
	$.\,\vee.\, C'R \cap d = \wedge : \supset .\, R \cap M = \wedge$	.171, .361, *33.161,
	$.\, R|M = \wedge$	22.58, 24.13)
5.38	$C'R \cap d = \wedge \,.\, \supset : R \, \epsilon \, Di\text{-}hier \,.$	(3.101, 5.35, .37, *24.13,

$. \equiv . R \in N\text{-}hier$ 33.161)

5.381 $C'R \cap h = \wedge . \supset :$ (5.35, .37)

     $: R \in Di\text{-}hier . \supset . R \in N\text{-}hier$

5.382 $R \in (N\text{-}hier \cap \overline{Di\text{-}hier}) . \supset . C'R \subseteq d$ (5.271, .38)

5.39 $R \in (Di\text{-}hier \cap \overline{N\text{-}hier}) .$ (3.37, .371, 5.28, .365,

     $. \supset . D'R \cap d \neq \wedge . \mathcal{C}'R \cap h \neq \wedge$ .37, SH.333)

     $. C'R \subseteq (h \cup d)$

5.391 $R \in (Di\text{-}hier \cap \overline{N\text{-}hier}) . \supset . B'R \in d$ (3.37, 5.301, .39, SH.333)

5.40 $R \in (Di\text{-}hier \cap \overline{N\text{-}hier} \cap M\text{-}hier) .$ (1.05, 5.122, .365, .366,

     $\supset . \overrightarrow{Lk'}R \neq \wedge$ SH.21, .23)

5.401 $R \in (Di\text{-}hier \cap \overline{N\text{-}hier}) . \supset :$ (5.33, .40)

     $: R \in M\text{-}hier . \equiv . \overrightarrow{Lk'}R = \wedge$

5.402 $R \in (Di\text{-}hier \cap \overline{N\text{-}hier} \cap \overline{M\text{-}hier}) .$ (5.32, .321, .39, .401,

     $\supset \therefore \overrightarrow{Lk'}R \subseteq d : \vee : \overrightarrow{Lk'}R \subseteq h : \vee :$ *24.56, T24.581)

     $: \overrightarrow{Lk'}R \cap d \neq \wedge . \overrightarrow{Lk'}R \cap h \neq \wedge$

5.403 $R \in (Di\text{-}hier \cap \overline{N\text{-}hier} \cap \overline{M\text{-}hier}) .$ (5.39, .401, .402, *24.56,

     $\supset \therefore \overrightarrow{Lk'}R \cap h = \wedge : \vee : \overrightarrow{Lk'}R \cap d =$ T24.581)

     $= \wedge : \vee : \overrightarrow{Lk'}R \cap d \neq \wedge .$

     $. \overrightarrow{Lk'}R \cap h \neq \wedge$

5.42 $R \in Di\text{-}hier . \supset : Fu|R_{po}|Fu \neq \wedge .$ (2.13, 5.122, .24, .25,

     $. \equiv . Fu|R \neq \wedge . R|Fu \neq \wedge .$ .262, SH.51, *33.21,

     $. \equiv . B'R \in (\mathcal{C}'Fu \cap D'Di_{po}|Fu) .$ T93.613, .65)

     $. \equiv . R \in \breve{B}''\mathcal{C}'Fu . \breve{R} \in \breve{B}''\mathcal{C}\breve{F}u$

5.421 $R \in N\text{-}hier . \supset :$ (1.05, .07, 5.12, .24, .241,

     $(M \cup Fu)|R_{po}|(M \cup Fu) \neq \wedge .$ .25, .251, .26, .261,

     $. \equiv . (M \cup Fu)|R \neq \wedge .$ SH.51, *31.15, 33.21,

     $. R|(M \cup Fu) \neq \wedge . \equiv . B'R \in$ .26, .261, 34.25, .3,

     $[\mathcal{C}'(M \cup Fu) \cap D'N_{po}|(M \cup Fu)] .$ T93.613, .65)

     $. \equiv . R \in \breve{B}''\mathcal{C}'(M \cup Fu) .$

     $. \breve{R} \in \breve{B}''\mathcal{C}'Cnv'(M \cup Fu)$

5.43 $R \in Di\text{-}hier . Fu|R_{po}|Fu \neq \wedge .$ (5.222, .24, .42, *34.3,

     $. \equiv . R \in Di\text{-}hier\text{-}max . R|Fu \neq \wedge$

5.431 $R \epsilon$ *N-hier* . $(M \cup Fu)|R_{po}|(M \cup Fu)$ (5.22, .24, .241, .421,

        $\neq \wedge$. $\equiv$. $R \epsilon$ *N-hier-max* .   *33.261, 34.26, .3,

        . $R|(M \cup Fu) \neq \wedge$ T93.613)

5.44 $R \epsilon$ *N-hier* . $\supset$. $Fu|R_{po}|Fu = \wedge$.   (1.20, .21, SH.333,

        . $M|R_{po}|M = \wedge$ *24.13, 34.34, 91.59)

5.441 $R \epsilon$ *Di-hier* . $\supset$: $Fu|R_{po}|Fu \neq \wedge$.   (5.361, .44)

        . $\supset$. $R \cap M \neq \wedge$

5.49 $R, P \epsilon$ (*N-hier* $\cup$ *M-hier* $\cup$ *Di-hier*).   (1.05, .111, .114, .12,

        . $C'R = C'P$. $\supset$. $R = P$ .121, .20, .23, 2.01, .111,

                                         .222, 3.14, 5.16, .351,

                                         .382, .39, SH.01, .21, .23,

                                         .333, .7101, *22.45,

                                         24.58, 33.241, .251,

                                         37.32, 91.574, T93.6111)

VI. LIFE CYCLES AS HIERARCHICAL RELATIONS

Life cycles are usually depicted by diagrams in which arrows are arranged in a circular pattern, connecting subsequent "stages". Such "stages" may be haploid and diploid "generations", or certain developmental stages, or host-parasite relationships, or some other aspects that one may want to emphasize in the description of an organism. In these life cycle diagrams the positions of meiosis and fusion are usually marked, these being the "cardinal events" of sexual life cycles. In the present paper we have concentrated on these "cardinal events", and we want to see now how the concepts developed so far will help in elucidating the concept of life cycles in general, as well as the related concepts of sporophyte and gametophyte.

Statements like "Mammals have diplontic life cycles" occur frequently in biological literature. But what does it mean for a class of organisms to *have* a certain kind of life cycle? One could try to explain this by saying that all those members of this class which attain maturity exhibit a developmental pattern which is called a 'diplontic life cycle'. Aside from the vagueness of the phrase 'developmental pattern', this explanation would suffice, provided that we can formalize the concept of diplontic life cycles. Before doing this, however, we need a further concept, a primitive notion that has not appeared in our theory before. This is the relation of "a cell

465

or a nucleus being a part of a whole organism", to be designated by 'Cen'. Woodger in AMB 4.1.1 formally defined a similar notion (Ce), the relation of "a cell being a part of a whole organism". We are going to use the concept Cen in complete analogy to Ce, but for the inclusion of nuclei as well as cells in the domain of Cen; no postulates are therefore given for it. It should be pointed out that in Woodger's system whole organisms are time-extended entities and the "part of" relation is used in a temporal as well as in a spatial sense. Thus, e.g., the zygote from which an animal develops and the gametes to which it gives rise, as well as the cells in between, are all parts of the same whole organism. We shall use the variable 'X' to stand for unspecified whole organisms, the class of which is distinct from our previous class of individuals (cells or nuclei) for which variables 'x', 'y', 'z', etc., have been employed.

We proceed now to define a relation that will assign a division or mitotic or meiotic hierarchy to whole organisms in such a way that every cell (or nucleus) which is part of the organism is a member of the field of the hierarchy, and every individual in the field of the hierarchy is a part of the organism. This relation is designated by 'Hu', and is defined formally as follows:

6.01 $\qquad R \, Hu \, X \, . \equiv . \, R \, \epsilon \, (\textit{Di-hier} \, \cup \, \textit{N-hier} \, \cup \, \textit{M-hier}) \, . \, \overrightarrow{\textit{Cen}'} \, X = C'R$

It is found, by theorem 5.49, that Hu is a one-many relation:

6.11 $\qquad Hu \, \epsilon \, 1 \rightarrow Cls$

Thus, every organism that has a hierarchy standing in Hu to it can have only one such hierarchy, by *71.163, and theorem 6.11.

6.12 $\qquad (\exists R) \, . \, R \, Hu \, X \colon \supset . \, E \, ! \, Hu' \, X$

It is clear, however, that there are organisms to which no division or mitotic or meiotic hierarchy stands in Hu. These may be organisms which developed from a graft union or arose by cell aggregation, like slime molds, and their cells (or nuclei) are derived from several beginners.

Formal expressions for life cycles are introduced next, based entirely on our formal N, M, Fu system. First of all, we define sexual life cycles as division hierarchies that stretch from fusion to fusion:

5.51 $\qquad R \, \epsilon$ sexual life cycle $. \equiv . \, R \, \epsilon \, \textit{Di-hier} \, . \, Fu|R_{po}|Fu \neq \wedge$

Furthermore, haplontic, diplontic and haplo-diplontic life cycles are defined as follows:

5.52 $\qquad R \, \epsilon$ haplontic life cycle $. \equiv . \, R \, \epsilon \, \textit{Di-hier} \, .$
$\qquad . \, B' \, R \, \epsilon \, [\overrightarrow{a}' \, Fu \, \cap \, D'(M|N_{po}|Fu) \, \cap \, \overline{D'(M|Fu)}]$

466

5.53 $R \,\epsilon$ diplontic life cycle $. \equiv . \, R \,\epsilon$ *Di-hier* .

   $. \, B' \, R \,\epsilon \left[\mathcal{Q}'Fu \, \cap \, D'(N_{po}|M|Fu) \, \cap \, \overline{D'(N_{po}|M|N_{po}|Fu)} \right]$

5.54 $R \,\epsilon$ haplo-diplontic life cycle $. \equiv . \, R \,\epsilon$ *Di-hier* .

   $. \, B' \, R \,\epsilon \left[\mathcal{Q}' \, Fu \, \cap \, D'(N_{po}|M|N_{po}|Fu) \, \cap \, \overline{D'(N_{po}|M|Fu)} \right]$

It is seen that while in a diplontic life cycle meiosis must always be followed by fusion, in a haplo-diplontic life cycle this is never the case. On the other hand, in a haplontic life cycle, meiosis follows fusion directly, but it is never followed by fusion.

We are in a position now to present a formal statement equivalent to "Mammals have diplontic life cycles", namely:

   (X): $X \,\epsilon$ mammal $. \, Hu' \, X \,\epsilon$ sexual life cycle .

   $\supset . \, Hu' \, X \,\epsilon$ diplontic life cycle

This statement refers only to those mammals to which a sexual life cycle can be assigned (standing to it in *Hu*), and asserts that to all of these organisms division hierarchies can be assigned which are diplontic life cycles. As we can see, we have disposed of the necessity of referring to "developmental patterns" in this statement, without impairing its explicitness.

An additional problem presents itself when one is considering lower plants or animals. This difficulty may be discussed by considering this sentence: "*Allomyces arbuscula* has a haplo-diplontic life cycle". This fungus species has distinct and independent sporophytes and gametophytes and the question is: Are we to consider sporophytes and gametophytes as themselves members of the class *Allomyces arbuscula*, or are we to assume this class to be made up of some sort of ideal "complete organisms", each consisting of a sporophyte and of the gametophytes to which the sporophyte gave rise. Under the latter assumption, the approach outlined above is directly applicable, and we can write the corresponding formal sentence:

   (X): $X \,\epsilon$ *Allomyces arbuscula* $. \, Hu' \, X \,\epsilon$ sexual life cycle .

   $\supset . \, Hu'X \,\epsilon$ haplo-diplontic life cycle

On the other hand, if this solution should be found distasteful, on account of the necessity to invoke "complete organisms", which are really not organisms at all in the case of *Allomyces arbuscula*, then the following alternative proposal may be adopted. Let us take a certain sporophyte and the gametophytes it gives rise to, consider them to constitute the

467

class α, and consider them to be individually members of *Allomyces arbuscula*. Then the term '*Cen''α*' designates the class of all the nuclei of the organisms in α (it is preferable to talk about their nuclei since they are coenocytic organisms). We state then that if there is a division hierarchy R the field of which is identical to *Cen''α*, and if R is a sexual life cycle, then R is a haplo-diplontic life cycle.

(α, R): $\alpha \subseteq$ *Allomyces arbuscula* . *Cen''α* $= C'R$. $R \in$ sexual
life cycle . $\supset$. $R \in$ haplo-diplontic life cycle.

When dealing with higher animals or plants, the first kind of expression may be preferred (but then in case of higher plants all the gametophytes – pollen and embryo sacs – produced on one plant must be thought of as parts of that plant). For thallophytes and unicellular organisms the second approach is definitely better suited, since in this case one does not need to specify the organisms in order to speak of their life cycles, e.g., entire clones of cells may make up the field of a single division hierarchy.

Definitions can also be obtained for "complete sporophytes" and "complete gametophytes" in the following manner:

6.02 $X \in$ complete sporophyte .
        . $\equiv$. $(\exists R)$. $R \in$ *N-hier* . $(Fu|R_{po}|M) \neq \wedge$. R *Hu* X
6.03 $X \in$ complete gametophyte .
        . $\equiv$. $(\exists R)$. $R \in$ *N-hier* . $(M|R_{po}|Fu) \neq \wedge$. R *Hu* X

In these definitions, 'complete' refers to the fact that the mitotic hierarchies associated with these organisms stretch all the way from fusion to meiosis or from meiosis to fusion. (They are also maximal mitotic hierarchies by theorem 5.431).

Further inspection of definitions 5.51 to 5.54 shows that the classes of haplontic, diplontic and haplo-diplontic life cycles do not account for the whole class of sexual life cycles. In fact, three other classes of life cycles may be readily obtained simply by interchanging or omitting the complementation signs (bars) over the symbols for the various domains in these expressions. No commonly known names are available for these classes.

5.55 $R \in A$ life cycle . $\equiv$. $R \in$ *Di-hier* .
        . $B' R \in [\alpha' Fu \cap D'(M|Fu) \cap D'(M|N_{po}|Fu)]$
5.56 $R \in B$ life cycle . $\equiv$. $R \in$ *Di-hier* .
        . $B' R \in [\alpha' Fu \cap D'(M|Fu) \cap D'(M|N_{po}|Fu)]$

5.57 $R \in C$ life cycle $. \equiv . \ R \in Di\text{-}hier$.

$. \ B' \ R \in \left[\mathbb{C}' \ Fu \cap D'(N_{po}|M|Fu) \cap D'(N_{po}|M|N_{po}|Fu) \right]$

The six classes of life cycles so defined (in 5.52, .53, .54, .55, .56 and .57) are mutually exclusive and they jointly exhaust the class of sexual life cycles; this being, of course, not the only such classification. Descriptions could also be obtained corresponding to asexually reproducing organisms, or to organisms that have repeating asexual phases in their life cycles, or to organisms that have as beginners cells other than zygotes (like identical twins).

As for A and B life cycles, they occur rarely and only among some unicellular organisms, like certain yeasts and Chlamydomonas. Interestingly enough, the class of C life cycles is applicable to Hymenoptera species with diploid females and haploid males. It may be noted in this connection that these haploid males, if they mate, are "complete gametophytes", and the diploid females (without their eggs as parts) are "complete sporophytes" according to our definitions, unorthodox as this terminology may seem to a zoologist. Sterile females (workers) on the other hand, do not fall into any of the categories we have discussed.

It should be pointed out again, in conclusion, that all the formal expressions of this theory were obtained without any recourse to morphological, physiological or genetical concepts, just with the help of the three primitive notions: mitosis, meiosis and gametic fusion. These three simple concepts were used strictly in accordance with the ten postulates given for them, and all the formal statements were obtained by rigorous deductive reasoning. While the meaning of the concepts contained herein is completely explicit and unambiguous, their usefulness for biologists remains to be seen. Much more powerful theoretical concepts might be needed to cope with biological complexities, and if this exercise helps to prepare the way towards them, it serves its purpose.

Queens College of the City University of New York, New York, U.S.A.

BIBLIOGRAPHY

The abbreviations in parentheses serve as reference in the text.

Carnap, R., *Introduction to Symbolic Logic and Its Applications,* Dover Publications, New York, 1958. (SLA)

Whitehead, A. N. and Russell, B., *Principia Mathematica,* Vol. 1, Second edition, Cambridge University Press, 1957. (PM)

Woodger, J. H., *The Axiomatic Method in Biology,* Cambridge University Press, 1937. (AMB)

BIBLIOGRAPHICAL APPENDIX

PUBLICATIONS BY JOSEPH HENRY WOODGER

I. BOOKS

A. Original

Elementary Morphology and Physiology for Medical Students, Oxford, 1924.

Biological Principles: A Critical Study, London, 1929.

The Axiomatic Method in Biology, Cambridge, 1937.

'The Technique of Theory Construction', *International Encyclopedia of Unified Science,* Vol. II, No. 5, Chicago, 1939.

Biology and Language: An Introduction to the Methodology of the Biological Sciences Including Medicine, The Tarner Lectures for 1949–1950, Cambridge, 1952.

Physics, Psychology and Medicine: A Methodological Essay, Cambridge, 1956.

Abstraction in Natural Science. Logic, Methodology and Philosophy of Science: Proceedings of the 1960 International Congress, Stanford University Press, Stanford, 1962.

B. Translations

L. von Bertalanffy, *Modern Theories of Development,* Oxford, 1933.

Felix Mainx, 'Foundations of Biology', *International Encyclopedia of Unified Science,* Vol. I, No. 9, Chicago, 1955.

A. Tarski, *Logic, Semantics, Metamathematics.* Papers published between 1923 and 1938, Oxford, 1956.

II. ARTICLES

'On the Relationship between the Formation of Yolk and the Mitochondria and Golgi Apparatus during Oogenesis', *J. Roy. Microscopical Soc.* 1920, 129–156. (With J. B. Gatenby).

'Notes on a Cestode Occurring in the Haemocoele of Houseflies in Mesopotamia', *Ann. Appl. Biol.* 7 (1921) 345–351.

'On the Origin of the Golgi Apparatus on the Middle-piece of the Ripe Sperm of Cavia', *Quart. J. Microscopical Science* **65** (1921) 265–291. (With J. B. Gatenby)

'Observations on the Origin of the Germ-cells of the Fowl, Studied by means of their Golgi Bodies', *Quart. J. Microscopical Science* **69** (1925) 445–462.

'Some Problems of Biological Methodology', *Proc. Aristotelian Soc.* **29** (1929) 331–358.

'The 'Concept of Organism' and the Relation between Embryology and Genetics', *Quart. Rev. Biol.* **5** (1931) 1–22 and 438–463, and **6** (1932) 178–207.

'The Early Development of the Skull of the Rabbit', *Phil. Trans. Roy. Soc.,* Ser. B. **218** (1930) 373–414. (With G. R. de Beer)

'Mr. Russell's Theory of Perception', *The Monist* **40** (1930) 621–636.

'The Relation between Descriptive and Experimental Embryology', *Science Progress* **26** (1931) 306–324.

'Some Apparently Unavoidable Characteristics of Natural Scientific Theory', *Proc. Aristotelian Soc.* **32** (1932) 95–120.

'A Simple Method of Testing Truth Functions', *Analysis* **3** (1936) 92–96. (With W. F. Floyd)

'The Formalization of a Psychological Theory', *Erkenntnis* **7** (1938) 195–198.

'The Origin of the Endoderm in the Sparrow', *Biomorphosis* 1938. (With J. P. Hill)

'Remarks on Method and Technique in Theoretical Biology', *Growth* Supplement 1940, 97–99.

'Notes on the First Symposium on Development and Growth', *Growth* Supplement 1940, 101–111.

'On Biological Transformations', *Essays on Growth and Form, presented to D'Arcy Wentworth Thompson,* Oxford University Press, 1945, pp. 95–120.

'Observations on the Present State of Embryology', Society for Experimental Embryology Symposium II, *Growth* 1948.

'Science without Properties', *Brit. J. Philos. Science* **2** (1952) No. 7.

'Problems arising from the Application of Mathematical Logic to Biology', *Applications scientifiques de la logique mathématique* (Actes du 2e Colloque Internationale de Logique Mathématique, Paris, 1952) pp.

133–139. Gauthier-Villars, Paris; E. Nauwelaerts, Louvain, 1954.
'What do we mean by 'Inborn'?', *Brit. J. Philos. Science* **3** (1953) No. 12.
'From Biology to Mathematics', *Brit. J. Philos. Science* **3** (1953) No. 9.
"Proper Objects", *Mind* **65** (1956) N. S. No. 260.
'Formalization in Biology', *Logique et Analyse,* Nouvelle Série, **1** (1958) 3–4.
'Studies in the Foundations of Genetics', *The Axiomatic Method, with Special Reference to Geometry and Physics,* Proceedings of an International Symposium held at the University of California, Berkeley, Dec. 26, 1957–Jan. 4, 1958, Amsterdam, 1959.
'Biology and Physics', *Brit. J. Philos. Science* **11** (1960) 42, 89–100.
'Taxonomy and Evolution', *La Nuova Critica,* Ser. 3 (1961) No. 12, 67–78.
'Biology and the Axiomatic Method', *Ann. N.Y. Acad. Sci.* **96** (1962) 1093–1104.

III. LETTERS

'Biological Principles', Letter to the Editor *Mind* 1930, N.S. 39, 403–405.
'Mental Health and The Basic Sciences', *Brit. Med. J.* 1955, 419–420.
'A Reply to Professor Haldane' (*Brit. J. Philos. Science,* Notes and Comments VI, 1955, 245–248), *Brit. J. Philos. Science,* Notes and Comments VII, 1956, 149–155.
'Time and the Nervous System', *Lancet* 1958, 2, 44.

IV. REVIEWS

A. D. Ritchie, *The Natural History of Mind,* Longmans Green, 1936, in: *Mind* 1936, N.S. 45, 399.
Philipp Frank, *Modern Science and Its Philosophy,* Harvard University Press, 1949, in: *Brit. J. Philos. Science* **2** (1951) 6, 168.
Herbert Butterfield, *The Origins of Modern Science,* G. Bell and Sons, London, 1950, in: *Brit. J. Philos. Science* **1** (1951) 4, 332.
Oscar Becker (editor), *Einführung in die Logistik, vorzüglich in den Modalkalkül,* Westkulturverlag Anton Hain, Meisenheim, 1951, in: *Brit. J. Philos. Science* **2** (1951) 8, 337.
Anthony Flew (editor), *Essays on Logic and Language,* Blackwell, Oxford, 1951, in *Brit. J. Philos. Science* **2** (1952) 8, 338–342.
Haskell B. Curry, *Leçons de logique algébrique,* Collection de Logique

Mathématique, Série A, No. 2, Gauthier-Villars, Paris, 1952, in: *Brit. J. Philos. Science* **3** (1952) 11, 293.

Jan Łukasiewicz, *Aristotle's Syllogistic from the Stand Point of Modern Formal Logic,* Clarendon Press, Oxford, 1951, in: *Brit. J. Philos. Science* **4** (1953) 15, 251.

Nathaniel Lawrence, *Whitehead's Philosophical Development,* University of California Press, 1956, in: *Brit. J. Philos. Science* **8** (1958) 32, 348.

Rudolf Carnap, *Einführung in die symbolische Logik,* Springer-Verlag, Vienna, in: *Brit. J. Philos. Science* **9** (1958) 33, 70–72.

Karl Menger, *The Basic Concepts of Mathematics,* The Bookstore Illinois Institute of Technology, 1957, in: *Brit. J. Philos. Science* **9** (1958) 34, 172.

Michael Polanyi, *Science and Persons, Personal Knowledge: Towards a Postcritical Philosophy,* Routledge and Kegan Paul, London, 1958, in: *Brit. J. Philos. Science* **11** (1960) 41, 65–71.

R. M. Martin, *Truth and Denotation,* Routledge and Kegan Paul, London, 1958, in *Brit. J. Philos. Science* **10** (1959) 157–159.